A Century of Interspecific Hybridization and Introgression in Tobacco

Apoloniusz Berbeć

A Century of Interspecific Hybridization and Introgression in Tobacco

Springer

Apoloniusz Berbeć
Plant Breeding and Biotechnology
Institute of Soil Science and Plant
Cultivation-National Research Institute
Puławy, Poland

ISBN 978-3-031-54963-2 ISBN 978-3-031-54964-9 (eBook)
https://doi.org/10.1007/978-3-031-54964-9

This Springer imprint is published by the registered company Springer Nature Switzerland AG
The registered company address is: Gewerbestrasse 11, 6330 Cham, Switzerland

Paper in this product is recyclable.

To the memory of my father, Professor Jan Berbeć

Preface

This book is probably the first authored monograph solely dedicated to the use of *Nicotiana* species in the genetic improvement of tobacco. Needless to say, the genus *Nicotiana* and hybridization in the genus have been taken as the subject of many excellent reviews and monographic essays. The most notable early comprehensive review of the field was *The Genetics of the Genus Nicotiana* written by E. M. East in 1928. In 1943, following 15 years of the stormy development of *Nicotiana* science, the monumental bilingual *Cytogenetics of the Genus* Nicotiana was published by the Bulgarian geneticist Doncho Kostoff. It was soon followed by *Cytotaxonomy of the Genus Nicotiana* written by Thomas H. Goodspeed. Several years later, *The Genus Nicotiana* was published, probably the most significant and well-known book on *Nicotiana* ever written. In 1968, the flourishing days of *Nicotiana* cytogenetics and breeding were summarized by Harold Smith in his *Recent Cytogenetic Studies in the Genus Nicotiana.* A collective review of the uses of *Nicotiana* species in tobacco genetics and breeding edited by RD Durbin and titled *Nicotiana. Procedures for Experimental Use* was published a decade later, in 1979. That date also marks the end of authored monographs on *Nicotiana*, which were replaced by chapters included in contributed books that covered wider topics. Although highly informative, those contributions were necessarily subject to the limitations of size and scope required of constituent parts of a larger edited book. The last monographs on the genetics of *Nicotiana* that provided complete lists of interspecific hybrids with references were Kostoff's *Cytogenetics of the Genus Nicotiana* and Goodspeed's *Cytotaxonomy of the Genus Nicotiana*. Goodspeed's last book *The Genus Nicotiana* also contained a list of *Nicotiana* hybrids, but it was not a separate inventory, the hybrids having been scattered among other entries in the book's subject index. As an additional bonus offered to the reader, the last chapter of this volume contains a comprehensive update of those earlier lists.

A Century of Interspecific Hybridization and Introgression in Tobacco is different from its predecessors on several accounts. As pointed out before, it is free of limitations that are faced by the authors of book chapters who, by necessity, treat some topics in a cursory way and skip others. What makes it stand out among other works on the subject is that, especially in its tabular part, it tries to take a species-oriented approach whereby particular topics and issues under discussion are reviewed on a species-by-species basis. It is a marked shift from the usual approach where topics and issues receive a preferred treatment while the role of species themselves is limited to a demonstrative and auxiliary function. In this book, the arrangement of material under discussion allows the reader to assess the impact of individual species in the studies on *Nicotiana* genetics, especially in those related to tobacco improvement.

In other respects, the author remained faithful to the long-established order by which topics have been organized in monographs of this type. It is an expanded, largely revised, and updated follow-up to the monographic chapter titled "The Use of Nicotiana Species in Tobacco Improvement" and included in the book *Tobacco Genome* published by Springer Nature in 2020. Like its immediate predecessor, this book almost exclusively concentrates its attention on tobacco and interspecific hybridization within *Nicotiana*. The author has deliberately avoided any references to the work done with other crops and in other genera as it would have greatly complicated the task and enlarged the size of this volume without adding much to its objective as it was signaled in the title.

Introgressive breeding in tobacco seems to be somewhat over the hill now, so there is an apt moment to summarize the immense wealth of *Nicotiana* research that has accumulated over the period of more than a century. Much emphasis in this review was placed on broadening and enlarging the geographical and historical framework of interspecific breeding in tobacco. It was done by citing reports from as many regions and dates as was technically possible. First and foremost, the book is a guide to worldwide literature on the subject. Nearly a thousand different items of literature are cited across the book and the lists of references occupy nearly one third of its content.

As his ultimate goal, the author wanted to pay a modest tribute to the long succession of eminent scholars and rank-and-file scientists who worked in different parts of the world and at different times. It is their ground-breaking achievements and small contributions that have made *Nicotiana* science what it is today.

Puławy, Poland Apoloniusz Berbeć

Contents

Chapter 1
Introductory Notes

1.1 The Importance of Tobacco (*Nicotiana tabacum L.*) and of Other Species in the Genus *Nicotiana*

The genus *Nicotiana* was only recently considered the fifth largest in Solanaceae (Knapp, 2020). Now, owing to a series of unprecedented new discoveries in Australia, it may have acquired the status of the fourth largest. *Nicotiana* constitutes a large and highly diversified assemblage of species ranging from the diminutive root-sprouting *N. acaulis* through annuals and short-lived perennials of different sizes and habits to a small-sized tree (*N. glauca*). The current number of *Nicotiana* species may vary in different publications. At present, the author of this review counted 108 taxa which have been given the specific status or at least treated provisionally as separate species and/or included in some research. Of these, 96 occur (or occurred) in natural habitats, two have been domesticated as commercial crops (*N. tabacum* and *N. rustica*) and one has been synthetized as highly heterogeneous ornamental (*N. sanderae*).

Nicotianae growing in the wild are presently indigenous to many parts of North and South America, Australia and to some offshore islands of those continents. One of those species, or its progenitor, made its way to as far as south-western Africa and became part of the Namibian flora. Most of the *Nicotiana* species are known from their diverse natural habitats such as roadsides, derelict lands, gravelly or rocky riverbeds and ravines, some of them prefer shaded places such as forest margins, shady slopes, rock crevices, cave entrances etc. (Tatemichi, 1990). *Nicotiana glauca* has spread naturally in many parts of the world as a persistent and toxic weed (Bogdanović et al., 2006). Some of the *Nicotiana* species, e. g. *N. obtusifolia*, *N. attenuata*, *N. quadrivalvis*, *N. benthamiana*, *N. gossei*, *N. excelsior*, *N. ingulba* have been transiently domesticated or their leaves were collected from the wild by natives chiefly for ceremonial drugs (Horton, 1981; Tatemichi, 1990). Some others, such as *N. alata* and its hybrids with other members of the section Alatae (*N. forgetiana*, *N. langsdorffii*) and also *N. sylvestris,* have become popular as

A. Berbeć, *A Century of Interspecific Hybridization and Introgression in Tobacco*,
https://doi.org/10.1007/978-3-031-54964-9_1

ornamental plants owing to their fragrant or sometimes large and showy flowers. The attraction as garden plants of some short-day *Nicotianae* from the section Tomentosae lies in their decorative purple flowers such as these of *N. setchellii*, but they are also valued because of the sheer height to which they can grow.

For a long time, *N. tabacum* was regarded as almost exclusively a cultivated species whose occurrence was, save for occasional escapes from cultivation, invariably associated with human habitation (Goodspeed, 1954). Since that time an increasing amount of evidence has been presented for tobacco becoming established in many parts of the world and over a range of different habitats as a weed infesting cultivated fields and gardens but also as a naturalized part of the local flora (Randall, 2012; CABI, 2019).

Tobacco (*Nicotiana tabacum* L.) takes a singular place among cultivated plants. It continues to be, along with cotton, one of the two major non-food cash crops worldwide. Another feature of *N. tabacum* is that it is one of two *Nicotianae* whose history of cultivation has reached far beyond its natural range of occurrence. The other is *N. rustica*, known as Aztec tobacco or makhorka. Both *N. rustica* and *N. ta*bacum are the sources of nicotine, but over time *N. tabacum* has outweighed *N. rustica* in importance as a cultivated crop finally to become the sole provider of tobacco leaves as an item of commerce. Cured leaves of tobacco are used in the manufacture of nicotine-containing products mainly in the form of cigarettes but also available in other presentations to be used for smoking, chewing and snuffing. More recently, some of the tobacco crop has been grown for nicotine as the end product to be used as highly diluted aqueous solutions called e-liquids. In a smoking-simulating device called e-cigarette the liquid is atomized into an aerosol that is inhaled by the user. The liquid, apart from nicotine, also contains other additives, some of them potentially carcinogenic, but e-cigarettes are now regarded as a less harmful option to the traditional smoking materials.

As an industrial crop and the source of widely used nicotine-based stimulants tobacco still provides livelihood for rural people in many parts of the world although its importance in many developed countries has been on the decline due to public health-motivated governmental pressures and economic constraints. Tobacco is also unique in that it has become a model plant in both fundamental and applied genetic studies. As a crop plant of substantial economic importance, *N. tabacum* has been subject to regular, conscious and research-based breeding effort that has lasted for more than a century. The latter aspect benefited to a considerable extent from the fact that numerous wild species of the genus *Nicotiana* provide a vast reservoir of potentially usable germplasm for their cultivated relative.

1.2 Hybridization within the Genus *Nicotiana*

The interest in hybrids of *Nicotiana* started in the eighteenth century, long before the breeding of tobacco advanced from lore to science. The major driving motive that made early *Nicotiana* investigators produce and explore the products of different interspecific crosses was concerned with *Nicotiana* phylesis and systematics. At that time, the study of affinities between various *Nicotiana* species drew to a large extent on the behaviour of meiotic chromosomes and on other aspects of cytogenetics in interspecific hybrids. Up to the early 70s of the last century the vast majority of hybrids in *Nicotiana* were probably created with that sole purpose in mind. The interest of geneticists in interspecific hybrids waned considerably with the advent of more sophisticated molecular methods.

Although scaled down, the research on interspecific hybridization in *Nicotiana* has not been abandoned altogether and the wild relatives of tobacco have continued to feature on the breeding agenda of many tobacco laboratories. A large number of *Nicotiana* species have been found to be resistant to common and destructive diseases of the tobacco crop, a valuable asset especially if no resistance could be available within the cultivated species. For the tobacco breeder, cytoplasmic male sterility (CMS) is yet another important benefit that could be accessed through interspecific hybridization. CMS is a prerequisite for technically feasible and economically viable seed production of hybrid cultivars. It looks like the wild *Nicotianae* are practically the sole providers of that important trait for tobacco.

Over more than a century, numerous attempts to hybridize one species of *Nicotiana* with another have resulted in an increasing number of different hybrid combinations and in the refinement of the methods by which those hybrids are obtained. The development of tissue and cell culture techniques and, more recently, also genetic engineering technologies allowed many of the hybridization barriers to be removed or circumvented thereby greatly enlarging the number of species the genomes of which could be combined in interspecific hybrids thus creating opportunities for increased biodiversity from which both nature and man could benefit.

1.3 Types of Interspecific Hybrids in *Nicotiana*

In this review, three processes are discussed by which the genome of one species can become united with the genome of another and thus form an interspecific hybrid. Two of these processes are well documented and generally accepted, the third was and, to some extent, still is the subject of major controversy:

(a) **sexual hybrids** that arise from union of two gametes, female and male, which is done by fertilizing the egg cells of the female parent by the pollen of the male parent. The process occurs in nature and is also imitated in experimental work. In the latter case, various modifications, unknown in nature, have been introduced to facilitate fertilization and, in many cases, to make it feasible.
(b) **somatic hybrids** that arise through the union of naked somatic protoplasts from different species that are isolated and induced to fuse into one cell (protoplast fusion). By dividing and organ differentiation, a process analogous to that occurring in sexual reproduction, the fused cell gives rise to a hybrid plant. Protoplast fusion bypasses the sexual reproductive path and generally requires a

sophisticated laboratory and refined methods to be accomplished. Until very recently, the basic mechanism of somatic hybridization was thought to be entirely alien to nature. Recent discoveries, however, seem to have thrown new light on the significance of what is called horizontal transfer in natural evolutionary processes and possibly created a new method by which interspecific hybrids can be produced.

(c) **graft hybrids** that arise through the fusion of two plants or their parts called stock and scion and which were purported to be a source of novel hereditary variation. Historically, graft hybrids were advocated in some communist-controlled countries as a breakthrough in plant genetics and breeding that had invalidated the principles of 'mendelism-morganism'. The tenets of the "revolutionary" biology and genetics, saturated with ideological apriorisms and ill-supported by scientific evidence, were generally treated with extreme skepticism by the majority of conventional geneticists of the time. The concept of graft hybrids, as it was understood in the 1940s and 1950s, was ultimately abandoned altogether but some of its aspects have been given a new life by recent discoveries that involved *Nicotiana*

1.4 Terminology and Usage in Relation to Interspecific Hybrids and Introgression in This Book

(a) hybrid e.g. *N. tabacum* × *N. glutinosa*, unless otherwise specified, usually means an amphihaploid hybrid from mating female parent *Nicotiana tabacum* with male (pollen) parent *Nicotiana glutinosa*; such an example hybrid contains a single (haploid) genome of *N. tabacum* (T $n = 24$) and a single (haploid) genome of *N. glutinosa* (G $n = 12$) and is an amphihaploid (TG). If the amphihaploid status of the hybrid is to be emphasized the notation 2x (*N. tabacum* x *N. glutinosa*) is used; Alternatively, F_1 (*N. tabacum* x *N. glutinosa*) or, simply, *N. tabacum* x *N. glutinosa* denotes the first generation from mating the two species; in either case, name/designation of the maternal species/genome comes first, unless indicated otherwise
Note: In this review, while due recognition is given to the allopolyploid origin of *N. tabacum* and several other *Nicotiana* species, they are treated as functional diploids, thereby avoiding semantic confusion when experimentally created hybrids of different ploidy levels are discussed.

(b) reciprocal hybrid: hybrid with the reversed order of the maternal and paternal species e.g., *N. glutinosa* × *N. tabacum* is the reciprocal of *N. tabacum* × *N. glutinosa*.

(c) amphidiploid means a hybrid in which two genomes of each parental species are combined, e.g.: 4x (*N. tabacum* × *N. glutinosa*), (TTGG).

(d) consequently, sesquidiploid contains a doubled chromosome complement of one parental species and a single complement of the other species, even though the author is aware that that, e.g., the sesquidiploid 3x (*N. tabacum* ×

N. glutinosa) and many others discussed in this review are phylogenetic allopentaploids. Thus (SylSylTomTomGlu) is composed of two ancestral doubled, tomentosoid and sylvestroid, subgenomes (SylSylTomTom) contributed by *N. tabacum* and a single genome contributed by *N. glutinosa* (Glu).

(e) allopolyploid signifies any hybrid that contains multiplications of the basic chromosome complements of distinct species without specifying the number of constituent genomes and species.

(f) aneuploidy refers to a deficiency or an excess of one or more chromosomes *vis-à-vis* normal or expected chromosome complement.

(g) polysomy/nullisomy refers to deficiency or multiplication for a particular chromosome in diploid genomes where normally all chromosomes are present in duplicate. Thus, nullisomic means deficient for the whole pair of homologous chromosomes, monosomic refers to the presence of only one chromosome of a particular chromosome pair; disomic means the normal diploid condition for a particular chromosome pair; trisomic, tetrasomic etc. denote respective extra multiplications of a particular chromosome in the genome.

(h) chromosome pairing: (in gametogenesis) association of structurally similar or identical chromosomes in the first reductional division also referred to as conjunction or, less frequently, as conjugation.

(i) "Drosera scheme" chromosome pairing – chromosome pairing characteristic of hybrids in which the number of paired chromosomes is the same as the haploid number of chromosomes of the parental species with the lower chromosome number (Goodspeed, 1945, 1954). It is found in interspecific hybrids between amphidiploids and one of their diploid progenitor species. Similar pairing behavior may be also shown by hybrids between diploids and autotriploids of the same species.

(j) trivalent, bivalent, univalent – associations of three, two or of a single chromosome left without a pair during the first reductional division, respectively.

(k) homology, homeology – homology refers to structural identity of chromosomes within the genome of the same species; homeology implies structural similarity of chromosomes belonging to different species.

(l) homoploidy, heteroploidy – refers to evolutionary events that involve hybridization as part of the speciation process: homoploid origin refers to hybridization that is not followed by the change in chromosome number; heteroploidy refers to multiplication of chromosome number in the speciation process.

(m) alleles – different variants of a gene; diploid organisms may be homozygous (two identical alleles), heterozygous (two different alleles) or hemizygous (only a single allele) for a given locus.

(n) alloplasmic – having a nuclear genome combined with a cytoplasm of another species.

(o) CMS, cms – cytoplasmic male sterility or related to cytoplasmic male sterility.

(p) manifestations of cytoplasmic male sterility:

    (i) staminal male sterility – refers to either total absence of male organs (stamens) or their different degrees of degeneration: carpeloid stamens that resemble female organs or their parts (e.g. stigmatoid anthers are anthers transformed into stigma-like structures); petaloid stamens refers to stamens transformed into petal-like structures, and feathery anthers means feather-like degenerative changes of anthers; staminal male sterility is usually pre-meiotic i.e. involves total suppression of male gametogenesis; staminal male sterility is also referred to as structural male sterility

(ii) post-meiotic male sterility – usually is related to normal or nearly normal flower morphology including morphologically normal female organs and normal or nearly normally developed stamens. In post meiotic male sterility the microgametogenesis collapses at various stages and apparently normal anthers are either void of pollen or contain aborted or morphologically normal pollen grains with disabled functionality.

Staminal and post-meiotic male sterility are also referred to as structural and sporogeneous sterility, respectively (Kaul, 1988), although sporogeneous male sterility is a broader concept because it also includes premeiotic stages.

(q) introgression – gene flow from species to species. Natural introgression is an important factor in plant evolution, experimental or applied inrogression is one of the tools used in genetic improvement of plants.

References

Bogdanovic, S., Mitić, B., Ruscic, M., & Dolina, K. (2006). *Nicotiana glauca* Graham (Solanaceae), a new invasive plant in Croatia. *Acta Botanica Croatica, 65*, 203–209.

CABI. (2019). Nicotiana tabacum (tobacco). Invasive species compendium. https://www.cabi.org/isc/datasheet/36326

Goodspeed, T. H. (1945). Cytotaxonomy of *Nicotiana*. *The Botanical Review, 11*, 533–592.

Goodspeed, T. H. (1954). *The genus Nicotiana*. Chronica Botanica Co.

Horton, P. (1981). A taxonomic revision of *Nicotiana* (Solanaceae) in Australia. *Journal of Adelaide Botanic Gardens, 3*, 1–56.

Kaul, M. H. L. (1988). *Male sterility in higher plants* (Monogr Theor Appl Genet) (p. 10). Springer Verlag.

Knapp, S. (2020). Biodiversity of *Nicotiana* (Solanaceae). In N. V. Ivanov, N. Sierro, & M. C. Peitsch (Eds.), *The tobacco plant genome, compendium of plant genomes*. Springer. https://doi.org/10.1007/978-3-030-29493-9_2

Randall, R. P. (2012). *A global compendium of weeds*. Department of Agriculture and Food Western Australia, 1124 pp. http://www.cabi.org/isc/FullTextPDF/2013/20133109119.pdf

Tatemichi, Y. (Ed.). (1990). *Illustrated book of the genus Nicotiana. Plant breeding and genetics research laboratory*. Japan Tobacco.

Chapter 2
Classification of the Genus *Nicotiana* and Origin of *Nicotiana* Species

2.1 Classification of the Genus *Nicotiana*

The name *Nicotiana* was invented by Francis the duke of Guise, a French aristocrat and general, to honor his countryman Jean Nicot, the French ambassador to Portugal. In 1559, Nicot sent dried leaves of tobacco to the king Francis II and his mother Catherine de Medici as a presumed miracle cure against persistent headaches. In 1565, the German botanist and physician Adam Lonitzer gave the name *Nicotiana* to the whole genus of tobacco, the name recognized later by Linnaeus (Julio, 2005). In the meantime, the French botanist Jacques Dalechamps listed tobacco under the name *Herba nicotiana*. Another Frenchman, Paul Reneaulme, a physician of Blois, in his *Specimen Historiae Plantarum* issued in 1611 was the first to call tobacco by its present name of *Nicotiana tabacum*. In the eighteenth century, Linnaeus described, along with *N. tabacum*, three more species of *Nicotiana*: *N. rustica*, *N. paniculata* and *N. glutinosa* (Knapp et al., 2004). Some nineteenth century classifications included up to 41 species (East, 1928). The first attempt to systematize the growing number of *Nicotiana* species was made by George Don in 1838 (Knapp et al., 2004). Don divided the *Nicotiana* species known to him into four sections: Tabacum, Rustica, Petunioides and Polydiclia. In 1912, William Setchell modified Don's division by dropping the section Polydiclia (Knapp et al., 2004). Edward M. East (East, 1928) adopted Setchell's classification scheme but gave his recognition to only 27 species, "*upon which some genetic work has been accomplished*", as he explained his criterion. The foundations for the present-day classification of *Nicotiana* were laid down by Doncho Kostoff (Kostoff, 1943) and Thomas H. Goodspeed (1954). Kostoff divided the genus into 8 sections and 47 species. Goodspeed's taxonomical division of the genus, while not differing much in its essentials from Kostoff's, was nevertheless much larger. It included 60 species divided into three subgenera and 14 sections, and that classification continued to provide the basic framework for subsequent additions and revisions. Based on

A. Berbeć, *A Century of Interspecific Hybridization and Introgression in Tobacco*,
https://doi.org/10.1007/978-3-031-54964-9_2

previous molecular studies (Aoki & Ito, 2000; Chase et al., 2003; Clarkson et al., 2004), Knapp and her coworkers (Knapp et al., 2004) proposed some significant modifications to Goodspeed's classification. In the revised systematics, the division into three subgenera was dropped, as it was not supported by DNA sequencing evidence (Aoki & Ito, 2000). Two of Goodspeed's 14 sections (Thyrsiflorae and Nudicaules) were removed, a new section was added (Sylvestres), and three sections were renamed (Genuinae to Nicotiana, Bigelovianae to Polydicliae, Acuminatae to Petunioides). Several species (*N. glauca*, *N. thyrsiflora*, *N. glutinosa*, *N. sylvestris*, and *N. nudicaulis*) were shifted from one section to another. The current inventory of *Nicotiana* species is considerably enlarged compared to the original Goodspeed's list: earlier additions included those by Burbidge (1960), Smith and Downs (1964), Merxmueller and Buttler (1975), D'Arcy (1976), Symon (1984), Clarkson and Symon (1991); more recent additions were contributed by Symon (1998, 2005), Stehmann et al. (2002), and Scarpa and Rosso (2011). In conformance with the rules laid down in the International Code of Botanical Nomenclature (Greuter cited after Knapp et al., 2004), the number of recognized *Nicotiana* species can be further enlarged by the inclusion of *N. monoschizocarpa*, a new taxon separated from *N. debneyi* (Horton, 1981; Symon & Lepschi, 2007). Another addition was proposed by Marks (2010), who produced evidence for separating *N. fatuhivensis* from the already recognized *N. fragrans,* although she had not supplied a regular description of the new species. Further additions included *N.* 'rastroensis' and *N.* 'Corunna'. *N.* 'rastroensis' lacks a valid description but was the object of several studies (Descorbeth, 2004; Descorbeth & McClure, 2005; Lee et al., 2008; Jimenez-Duran et al., 2013). 'N. Corunna' was first discovered by E. Symon and was renamed 'N. symonii' by Dodsworth (2015) and finally described as a new species under the name of *N. paulineana* by Bally et al. (2021). *N. leguiana* was separated from N. *tomentosa* by Knapp (2020).

Starting in 2018, there has been a virtual rash of new *Nicotianae*, which were discovered in the Australian outback by essentially the same team of scientists and classified in the section Suaveolentes. Their long succession was opened by *N. yandinga*, *N. faucicola*, *N. karijini*, *N. gascoynica*, *N. notha, N. truncata, N. hoskingii, N. walpa, and N. pila* (Chase & Christenhusz, 2018a, b, 2021a, b; Chase et al., 2018a, b, 2021a, b, c, d, e, f). In several cases, the newly announced species were separated from previously known taxa, *N. insecticida, N. murchisonica, and N. salina,* and were included by their discoverers in what the authors described as the *N. occidentalis* group (Chase & Christenhush, 2021a); *N. hoskingii* was separated from *N. debneyi* (now *N. forsteri*) (Chase et al., 2021f); and *N. gascoynica* and *N. walpa* were recognized as distinct from *N. simulans* Chase et al. (2021g). *N. pila* was found to be distinct from both *N. rosulata* and *N. ingulba,* although in some aspects, it resembled one or the other (Chase & Christenhush, 2021b). Likewise, *N. karijini* bore resemblance to both *N. umbratica* and *N. benthamiana* (Chase & Christenhusz, 2018a). *N. notha* is considered a product of hybridization between *N. suaveolens* and *N. velutina* (Chase et al., 2021c). The most recent additions include four species that have been separated from *N. benthamiana*: *N. candelabra*, *N. bilybara*, *N. rupestris*, and *N. scopulorum* (Chase et al., 2021h; Chase et al., 2022a). Chase et al. (2021g) were themselves surprised at the wealth of *Nicotiana* species they had been able to discover and wondered why such a great level of diversity had passed unnoticed by their predecessors in their studies of Australian *Nicotianae*. The authors named several reasons for that underestimation. Among them were the very size of the continent, few side-by-side comparative studies of sufficiently large numbers of collected accessions and peculiar germination requirements of some species that persist in the soil as seed banks most of the time and only occasionally germinate, come to flower, and can be collected for study. Chase et al. (2021g) estimated that currently recognized Suaveolentes species may account for approximately 60% of their actual total number. It remains to be seen how many of these newly recognized species will stand the test of time and to what extent the predictions regarding the extraordinary specific diversity of the Suaveolentes will materialize.

Another recent addition to the genus is *N. gandarela*, found in a single location in Brazil. Like the other species native to that country, it was classified in the section Alatae. According to the researchers who described *N. gandarela*, the species is under threat of extinction from extensive iron mining in the area (Augsten et al., 2022).

Essentially, the list in Table 2.1 is a compilation of those published recently by Knapp (2020) and Berbeć and Doroszewska (2020). The five species not listed by Knapp but included by Berbeć and Doroszewska are *N.* sp. 'Rastroensis', *N. sanderae*, *N. eastii*, *N.* sp. 'Corunna' (now *N. paulineana*) and *N. palmeri*. Table 2.1 also lists eleven Suaveolentes species most recently reported by Prof. Chase and his collaborators (Anon, 2021) and discussed in one of the previous paragraphs and the previously mentioned *N. gandarela*. The largest part of the *Nicotiana* species, approaching half of their total number, is classified within the section Suaveolentes. The classification presented in Table 2.1 is fairly liberal for it includes both those taxa which are known from collections and natural sites as well as those of which only herbarium specimens exist. The status of the latter is also liable to change, e.g., *N. ameghinoi* was rediscovered in the wild (Knapp 2013). Knapp's classification comprised eighty-two species, and the list in Table 2.1 of this volume elevates their number to 99. Not included in Table 2.1 are the most recent discoveries in the family Suaveolentes made by Prof. Chase and his associates (Chase et al., 2023). The newest nine species include: *N. bungonia* Chase & Tereski, *N. clarksonii* Chase & Christenhush ($2n = 36$), *N. erytheia* Chase & Christenhush, *N. gibosa* Chase, Andrew & Brull, *N. karakara* Chase & Christenhush, *N. latifolia* Chase & Christenhush ($2n = 36$), *N. latzii* Chase, Jobson & Christenhush, *N. olens* Chase & Christenhush ($2n = 30$), and *N. praecipitis* Chase & Durban.

The basic framework of the revised classifications, their differences in recognized taxa notwithstanding, has been accepted as standard by most researchers writing on *Nicotiana* (Lewis & Nicholson, 2007; Doroszewska et al., 2009; Lewis, 2011; Knapp, 2020), although some continued to use Goodspeed's original systematics (Khan & Narayan, 2007).

Table 2.1 Species of the genus *Nicotiana* classified by sections and in alphabetical order within a section

Species name	Described by	Origin	Year of first description	Diploid chromosome number	Genome size (pg) 2C DNA content after Hussain et al., 2023 (H2C); 1C and 4C DNA contents after Leitch et al., 2008 (L1C and 4C, respectively)
Alatae					
N. alata Link et Otto	Goodspeed (1954)	SAm	1828	18	4.53, 5.49 (H2C)
N. azambujae Sm. et Down	Smith and Downs (1964)[17]	SAm	1964	?	
N. bonariensis Lehm.	Goodspeed (1954)	SAm	1818	18	4.45 (H2C)
N. forgetiana Hemsl	Goodspeed (1954)	SAm	1905	18	
N. gandarela Augsten & Stehmann[1]	Augsten et al. (2022)	SAm	2022	?	
N. langsdorffii Weinm.	Goodspeed (1954)	SAm	1819	18	6.82 (H2C)
N. longiflora Cav	Goodspeed (1954)	SAm	1802	20	5.74 (H2C)
N. mutabilis Stehmann et Samir	Stehmann et al. (2002)[18]	SAm	2000	18[20]	
N. plumbaginifolia Viv. *	Goodspeed (1954)	SAm	1802	20	5.46 (H2C)
N. sp. 'Rastroensis'[2]	Murfett et al. (2005)	SAm	2005	?	
(*N.* x *sanderae*) Watson	Christoff (1928), Daly (1959)[19]	SAm	1904	18	
Nicotiana					
N. tabacum L.	Goodspeed (1954)	SAm	1753	48	5.2, 20.70 (L1C, 4C), 9.77 (H2C)
Noctiflorae					
N. acaulis Speg.	Goodspeed (1954)	SAm	1902	24	6.20 (H2C)
N. ameghinoi Speg.	Goodspeed (1954)	SAm	1902	12[20]	
N. glauca Graham	Goodspeed (1954)	SAm	1828	24	6.85 (H2C)

N. noctiflora Hock. *	Goodspeed (1954)	SAm	1827	24	9.53 (H2C)
N. paa MartCrov.	Martinez-Croveto (1978), Scarpa and Rosso (2011)	SAm	1978	?	
N. petunioides (Griseb.) Millán*	Goodspeed (1954)	SAm	1928	24	5.30 (H2C)
Paniculatae					
N. benavidesii Goodsp.	Goodspeed (1954)	SAm	1938	24	6.11 (H2C)
N. cordifolia Phil.	Goodspeed (1954)	SAm	1856	24	
N. cutleri D'Arcy	D'Arcy (1976)	Sam	1976	?	
N. knightiana Goodsp.	Goodspeed (1954)	SAm	1938	24	3.2, 12.64 (L1C, 4C) 6.57 (H2C)
N. paniculata Goodsp.	Goodspeed (1954)	SAm	1753	24	3.0, 11.78 (L1C, 4C) 6.40 (H2C)
N. raimondii Macbr.	Goodspeed (1954)	SAm	1930	24	
N. solanifolia Goodsp.	Goodspeed (1954)	SAm	1844	24	
Petunioides					
N. attenuata Wats.	Goodspeed (1954)	CAm, NAm	1871	24	2.5, 9.93 (L1C, 4C) 6.95 (H2C)
N. corymbosa Rémy	Goodspeed (1954)	SAm	1849	24	
N. longibracteata Phil.	Goodspeed (1954)	SAm	1891	?	
N. linearis Phil.	Goodspeed (1954)	SAm	1895	24	6.50 (H2C)
N. miersii Rémy	Goodspeed (1954)	SAm	1849	24	5.82 (H2C)
N. pauciflora Remy*	Goodspeed (1954)	SAm	1849	24	
N. spegazzinii Millán	Goodspeed (1954)	SAm	1926	24[20]	7.11 (H2C)
Polydicliae					
N. clevelandii Gray	Goodspeed (1954)	NAm	1878	48	4.1, 16.56 (L1C, 4C) 7.76 (H2C)
N. quadrivalvis Pursh*	Goodspeed (1954)	NAm	1871	48	4.3, 17.01 (L1C, 4C) 10.50 (H2C)

(continued)

Table 2.1 (continued)

Species name	Described by	Origin	Year of first description	Diploid chromosome number	Genome size (pg) 2C DNA content after Hussain et al., 2023 (H2C); 1C and 4C DNA contents after Leitch et al., 2008 (L1C and 4C, respectively)
Repandae					
N. nesophila Jonst.	Goodspeed (1954)	CAm (I)	1931	48	5.0, 20.13 (L1C, 4C) 10.33 (H2C)
N. nudicaulis Wats.	Goodspeed (1954)	CAm	1883	48	3.6, 14.22 (L1C, 4C) 7.05 (H2C)
N. repanda Wild.	Goodspeed (1954)	NAm	1818	48	5.4, 21.76 (L1C, 4C) 9.98 (H2C)
N. stocktonii Brandegee	Goodspeed (1954)	CAm (I)	1899	48	5.0, 19.99 (L1C, 4C) 10.00 (H2C)
Rusticae					
N. rustica L.	Goodspeed (1954)	CAm	1753	48	5.3, 21.19 (L1C, 4C) 10.82 (H2C)
Suaveolentes					
N. africana Merxm.	Merxmueller and Buttler (1975)	SAm	1975	46	9.66 (H2C)
N. amplexicaulis Burb.	Burbidge (1960)	Africa (Namibia)	1960	36	6.92 (H2C)
N. benthamiana Domin.	Goodspeed (1954)	Au	1929	38	6.92 (H2C)
N. bilybara M.W. Chase and Christenh[3]	Chase et al. (2022a)	Au	2022	36, 38[21]	
N. burbidgeae Symon	Symon (1984)	Au	1984	42	
N. cavicola Burb	Burbidge (1960)	Au	1960	40[22]	
N. candelabra M.W. Chase and Christenh[3]	Chase et al. (2022a)	Au	2022	36, 38[21]	

N. debneyi (*N. forsteri*) Roem et Schult.[4]	Goodspeed (1954), Marks (2010)	Su	1929	48	9.15 (H2C)
N. eastii Kostoff[5]	Kostoff (1939, 1943)	Au	1939	64	
N. excelsior (J.M. Black) J.M. Black	Goodspeed (1954)	Au	1926	38	6.65 (H2C)
N. exigua Wheeler	Goodspeed (1954)	Au	1936	32	6.95 (H2C)
N. fatuhivensis F. Br.	Marks (2010)	FPol	2010	?	
N. faucicola Chase and Christenh.[6]	Chase et al. (2018b)	Au	2018	30	
N. fragrans Hock.	Goodspeed (1954)	Au (I NC)	1855	48	
N. gascoynica M.W. Chase and Christenh.[7]	Chase and Christenhusz (2018b)	Au	2018	40, 44	
N. goodspeedii Wheeler	Goodspeed (1954)	Au	1935	40[23]	6.31 (H2C)
N. gossei Domin.	Goodspeed (1954)	Au	1929	36	6.89 (H2C)
N. hesperis Burbidge	Burbidge (1960)	Au	1960	42	
N. heterantha Kenneally and Symon	Symon and Keneally (1994)	Au	1994	48[24]	
N. hoskingii Chase, Palsson & Christenh [8]	Chase, Palsson, and Christenhush (2021f)		2021	?	
N. ingulba Black	Goodspeed (1954)	Au	1933	32	
N. insecticida Chase and Christenh.[9]	Chase and Christenhush (2021a)	Au	2021	42	
N. karijini Chjini,). Chase and Christenh. [10]	Chase and Christenhusz (2018a)	Au	2018	?	
N. maritima Wheeler	Goodspeed (1954)	Au	1932	32[25]	
N. megalosiphon Van Huerck and Müll.	Goodspeed (1954)	Au	1870		
N. monoschizocarpa (P. Horton) Symon and Lepschi	Horton (1981)	Au	1981	48[24]	

(continued)

Table 2.1 (continued)

Species name	Described by	Origin	Year of first description	Diploid chromosome number	Genome size (pg) 2C DNA content after Hussain et al., 2023 (H2C); 1C and 4C DNA contents after Leitch et al., 2008 (L1C and 4C, respectively)
N. murchisonica Chase and Christenh.[9]	Chase, Przeslawski, Falvey, and Christenhusz (2021d)	Au	2021	?	
N. notha Chase and Christenh.[11]	Chase et al. (2021c)	Au	2021	64	
N. occidentalis Wheeler	Burbidge (1960)	Au	1935	42	5.83 (H2C)
N. paulineana Newbegin & Waterh[12]	Bally et al. (2021)	Au	2021	32[24]	
N. pila Chase & Christenh[13]	Chase and Christenhush (2021b)	Au	2021	?	
N. rosulata Wheeler	Chase and Christenhush (2021c)	Au	1929	20	5.42 (H2C)
N. rotundifolia Lindl.	Goodspeed (1954)	Au	1838	44[26]	5.44 (H2C)
N. rupestris M.W. Chase and Christenh[3]	Chase et al. (2022)		2022	38[21]	
N. salina Chase and Christenh.[9]	Chase, Fay, and Christenhusz (2021b)	Au	2021	42	
N. scopulorum M.W. Chase and Christenh[3]	Chase et al. (2022a)	?	2022	?	
N. simulans Burbidge	Burbidge (1960)	Au	1960	40	3.28 (H2C)
N. stenocarpa Wheeler	Goodspeed (1954), Chase and Christenhush (2018c)	Au	1935	40	
N. suaveolens Goodsp.	Goodspeed (1954) He et al. (2019)	Au	1818	32 64[27]	11.88 (H2C)

N. truncata Symon	Symon (1998), Chase, Conran, and Christenhusz (2021a)	Au	1998	36	
N. umbratica Burbidge	Burbidge (1960)	Au	1960	46	
N. velutina Wheeler	Goodspeed (1954)	Au	1935	32	
N. walpa Chase, Dodsworth & Christenh.[14]	Chase, Dodsworth, and Christenhush (2021e)	Au	2021	20	
N. wuttkei Clarkson and Symon	Clarkson and Symon (1991)	Au	1991	28, 32 [28]	
N. yandinga[15] Chase and Christenh.	Chase et al. (2018a)	Au	2018	42	
Sylvestres					
N. sylvestris Speg.	Goodspeed (1954)	Sam	1899	24	2.7, 10.78 (L1C, 4C) 5.81 (H2C)
Tomentosae					
N. kawakamii Ohashi	Ohashi (1976)	Sam	1976	24	6.34 (H2C)
N. leguiana Macbride[16]	Macbride (1930)		1930		
N. otophora Griseb.	Goodspeed (1954)	Sam	1879	24	5.99 (H2C)
N. setchellii Goodsp.	Goodspeed (1954)	Sam	1941	24	
N. tomentosa Macbride	Goodspeed (1954)	Sam	1759	24	
*N. tomentosiformis** Goodsp.	Goodspeed (1954)	Sam	1933	24	2.7, 10.97 (L1C, 4C) 5.52 (H2C)
Trigonophyllae					
N. obtusifolia Martens and Galeotti*	Goodspeed (1954)	Cam, Nam	1852	24	1.5, 6.18 (L1C, 4C)
N. palmeri Gray	Goodspeed (1954)	NAm	1878	24	
Undulatae					
N. arentsii Goodsp.	Goodspeed (1954)	Sam	1944	48	5.1, 20.22 (L1C, 4C) 4.71 (H2C)

(continued)

Table 2.1 (continued)

Species name	Described by	Origin	Year of first description	Diploid chromosome number	Genome size (pg) 2C DNA content after Hussain et al., 2023 (H2C); 1C and 4C DNA contents after Leitch et al., 2008 (L1C and 4C, respectively)
N. glutinosa L.	Goodspeed (1954)	Sam	1753	24	2.2, 8.94 (L1C, 4C)
N. thyrsiflora Goodsp.	Goodspeed (1954)	Sam	1938	24	
N. undulata Ruiz and Pay	Goodspeed (1954)	Sam	1759	24	2.4, 9.66 (L1C, 4C) 10.30 (H2C)
N. wigandioides Koch and Fintelm.	Goodspeed (1954)	Sam	1858	24	2.9, 11.38 (L1C, 4C)

The data include systematic status, provenance, somatic chromosome number and estimated genome size. Based on the lists of *Nicotiana* species by Berbeć and Doroszewska (2020) and Knapp (2020), expanded and updated

In early literature on *Nicotiana* the following species are sometimes referred to under the following names:*N. alata* = *N. affinis*; *N. plumbaginifolia N.* = *cavanillesii*; *N. noctiflora* = *N. cavanillesii*; *N. petunioides* = *N. pampasana*, *N. pauciflora* = *N. caudigera*; *N. quadrivalvis* = *N. bigelovii*; *N. tomentosiformis* = *N. Rusbyi*; *N. obtusifolia* = *N. trigonophylla* (see Kostoff, 1943; Goodspeed, 1954; Knapp, 2020)

Abbreviations: SAm (South and Central America including offshore islands), NAm (North America), CAm (Central America), CAm (I) (Revilla Gigedo islands southwest of Baja California) Au (Australia), Au (NC) (New Caledonia archipelago, Au (M) (Marquesas archipelago)

[1]discovered in Brazil by Augsten et al. (2022)

[2]described by Murfett et al. (2005) as closely related to *N. bonariensis* and reported in two other studies on hybrids within the section *Alata*e (Descorbeth, 2004; Lee et al., 2008);

[3]separated from *N. benthamiana* (Chase et al. 2022a)

[4]according to Marks (2010) the name *N. forsteri* takes rightful precedence over *N. debneyi,* the name *N. debneyi*, as universally adopted, is also used throughout this review

[5] separate taxonomic status not recognized by Wheeler (1945) and Goodspeed (1954) who considered *N. eastii* a tetraploid race of *N. suaveolens*

[6]similar to and confused with *N. maritima*, apparently closely related to *N. suaveolens* (Chase et al., 2018b); raised from seeds supplied by C. Marks and studied by Dodsworth (2015); description provided by (Chase & Christenhusz, 2018c)

[7]confused earlier with *N. simulans* from which it differs for some morphological features (Chase & Christenhusz, 2018b)

[8] morphologically similar but not closely related to *N.* debneyi (Chase, Palsson, & Christenhush, 2021f)

[9]separated from 'the *N. occidentalis* group' as a distinct taxon

[10]confused earlier with *N. benthamiana* and *N. umbratica*, found distantly related to *N. umbratica* by restriction site-associated DNA sequencing (RAD-seq), but distinguishable from either by size and other morphological features (Chase & Christenhusz, 2018a)
[11]most similar to western forms of *N. suaveolens*, probably an allotetraploid of *N. suaveolens* and *N. velutina* (Chase, Fay, Nollet, & Christenhusz, 2021c)
[12]discovered by D. E. Symon in early 2010s (Bally et al., 2021) and named provisionally "Corunna". Studied by Dodsworth (2015) who proposed the name "*N. symonii*" for the new accession. Finally, the species was named *N. paulineana* and described as related to *N. maritima* and *N. velutina*
[13]related to *N. rosulata* and *N. ingulba* (Chase & Christenhush, 2021b)
[14]similar to *N. gossei* but related to *N. stenocarpa* Chase, Dodsworth, and Christenhush (2021e)
[15]closely related to *N. maritima* from which it differs for somatic chromosome number (Chase et al., 2018a)
[16]described by Macbride (1930) as a species closely related to *N. tomentosa* and recognized as a separate species by Knapp (2020); a botanical variety of *N. tomentosa* according to Goodspeed (1954)
[17]no valid description; reported by Smith and Downs (1964), included in the list of *Nicotiana* species by Knapp et al. (2004), known from a single herbarium specimen (Lim et al., 2006)
[18] no valid description; reported by Smith and Downs (1964), included in the list of *Nicotiana* species by Knapp et al. (2004), known from a single herbarium specimen (Lim et al., 2006)
[19]in this work referred to as '*N. sanderae*' throughout the text and in tables to avoid graphic ambiguity when giving hybrid formulas involving this form; description by Christoff cites an earlier work by WB Hemsley published in 1905; no recognized status; a polymorphic form and ephemeral form—in most publications treated as an artificial horticultural hybrid of *N. alata* with *N. forgetiana* (Daly, 1959); in some strains of *N. sanderae* involvement of *N. langsdorffii* and other 18-chromosome members of the section Alatae is also possible
[20] after Knapp (2020)
[21]Chase et al. (2022a) give what seems to be dual monoploid chromosome numbers (18 and 19) for *N. bilybara* and *N. candelabra*. They follow their statements on chromosome numbers by a reference to another paper (Chase et al. 2022b). However, in the latter paper no direct mention is made of either of these two new species or of the third one (*N. rupestris*)
[22]originally reported by Burbidge (1960) as $2n = 46$, corrected by Williams (1975) to $2n = 40$, correction confirmed by Tatemichi (1990)
[23] as reported by Goodspeed (1945) and confirmed by Tatemichi (1990), Marks et al. (2011) report the number as $2n = 32$
[24]reported by Marks et al. (2011)
[25]reported by Goodspeed (1954) and confirmed by Tatemichi (1990); Marks et al. (2011) report the number as $2n = 30$
[26]reported by Goodspeed (1954) and confirmed by Tatemichi (1990); Marks et al. (2011) report the number as $2n = 32$
[27]tetraploid lineages reported to occur (Wheeler, 1945; He et al. 2019)
[28] as reported by Clarkson and Symon (1991) and Laskowska and Berbeć (2003), respectively

2.2 Spontaneous Interspecific Hybridization and Origin of Allopolyploid *Nicotiana* Species

2.2.1 Introductory Notes

Spontaneous interspecific hybridization is considered the basic evolutionary mechanism that led to the formation of today's *Nicotiana* species having from 9 to 24 chromosome pairs (Table 2.1). According to Goodspeed (1954), the genus *Nicotiana* originated from some hypothetical ancestors of the genera *Cestrum* and *Petunia* that gave rise to the "pre-*Nicotiana*" complex. More precisely, Goodspeed's theory postulated a dual ancestry of *Nicotiana*—the existence of two centers of divergence: "precestroid" and "prepetunioid" from which "petunioid" and "cestroid" present-day *Nicotiana* species had evolved. Goodspeed theorized that the base chromosome number of pre-*Nicotiana* was $n = 6$. Indeed, early researchers (Kostoff, 1943) found haploids of certain 12-paired *Nicotiana* species to form up to 5 chromosome pairs in meiosis, which was interpreted as remnants of ancient homology between two ancestral six-chromosome sets.

Plastid DNA analyses and molecular trees constructed by Clarkson et al. (2004) demonstrated that the involvement of *Cestrum* and *Petunia* in the origin of *Nicotiana* is unlikely and that it is the Australian tribe Anthocercidae that shows the closest affinity to *Nicotiana*.

The genus *Nicotiana* is now thought to have emerged as a separate solanaceous group approximately 24–28 million years ago (Särkinen et al., 2013). The evolution and speciation within the genus *Nicotiana* is considered to be largely driven by hybridization. Interspecific hybridization is thought to have contributed to the speciation process in *Nicotiana* species along two different routes, heteroploid and homoploid. Hybridization followed by chromosome doubling (heteroploid route) ultimately results in an allopolyploid species. The homoploid speciation process involves hybridization between two species with equal chromosome numbers and—essentially through recombination—ultimately generates a new species that retains the chromosome number of its parents (Goodspeed, 1954; Kelly et al., 2010; Runemark et al., 2019). Homoploid evolution is much more difficult to trace than the allopolyploid origin, and tentative evidence for homoploid hybrid ancestry was suggested for a few species only (Kelly et al., 2010; Clarkson et al., 2010; McCarthy et al., 2015). Of the present-day *Nicotianae*, the homoploid pathway of evolution has been proposed for *N. glauca*, *N. linearis* and *N. glutinosa*. The first two evolved following matings between the ancestral forms of *N. noctiflora* and *N. petunioides* on one side and a group involving *N. acuminata*, *N. attenuata*, *N. miersii* and *N. pauciflora* on the other. The origin of *N. glutinosa* appears to be equally complex and ensuing from ancient hybridization events that involved three tomentosoid species (*N. otophora*, *N. setchellii* and *N. tomentosiformis*) and two species of the section Undulatae (*N. undulata* and *N. wigandioides*) (McCarthy et al., 2015).

The major tenets of Goodspeed's theory on the origin and evolution of the genus and the arrangement of its species into groups of lower order have thus retained their validity, but numerous revisions, corrections and clarifications based on biochemical and, first of all, molecular methods, unavailable at the time of Goodspeed and his contemporary *Nicotiana* researchers, have since been introduced.

Over the next decades, a strong body of evidence has accumulated supporting the allopolyploid origin of several 24-chromosome pair species, including the most important one, the cultivated tobacco *N. tabacum*.

2.2.2 Origin of Nicotiana Tabacum

Anastasia, Christoff and Hachaturov (Kostoff, 1943) published the first views on the origin of *N. tabacum* in the early twentieth century, but their speculations were subsequently disproved. The first scientifically sound theories concerning the origin of *N. tabacum* were essentially built around the hypothesis advanced earlier by Winge (Barker et al., 2016), who proposed that the species with polyploid chromosome number had evolved from spontaneous sterile interspecific hybrids to which chromosome pairing and fertility were restored by chromosome doubling. These early hypotheses, based on morphological and cytogenetic evidence, pointed to *Nicotiana sylvestris* and one of the *Tomentosa* group species as the likely progenitors of *N. tabacum*: *N. tomentosa* (Clausen, 1928; Goodspeed & Clausen, 1928) and *N. otophora* (Goodspeed & Bradley, 1942; Goodspeed, 1945). Clausen (1932) was the first to propose *N. tomentosiformis* as the putative tomentosoid parent of *N. tabacum,* and the proposition was supported by Kostoff (1936, 1938). Kostoff based his assertion on the striking morphological similarity of his amphidiploid 4x (*N. sylvestris* × N. *tomentosiformis*) to *N. tabacum* and on its self-fertility (Kostoff, 1936). Incidentally, the 'Kostoff's amphidiploid' or 'Kostoff's hybrid', as it came to be known in later times, was demonstrated to carry a considerable amount of introgression from *N. tabacum* (Sheen, 1972; Lim et al., 2006; Moon et al., 2008). A hypothesis was also advanced regarding the involvement of *N. tomentosa* rather than *N. tomentosiformis* in Kostoff's amphidiploid (Slana et al., 1977; Stavely, 1979; Stavely et al., 1977). Those controversies notwithstanding, Gerstel (1960) furnished strong cytogenetic evidence in favor of *N. tomentosiformis* based on comparative homologies, pairing rates and segregation ratios in the offspring of allopolyploids of 6x (*N. tabacum* × *N. otophora*) and 6x (*N. tabacum* × *N. tomentosiformis*) when the two alloploids were backcrossed to *N. tabacum*. The segregation ratios for recessive markers were consistently lower for the alloploids involving N. *tomentosiformis* than for those involving *N. otophora*. These data indicated that *N. tomentosiformis* showed a greater homology with the tomentosoid subgenome of *N. tabacum* than did *N. otophora.* Hence, *N. tomentosiformis* was more likely to be the tomentosoid parent of *N. tabacum* even though the amphidiploid 4x (*N. sylvestris* × *N. otophora*) was fully self-fertile and the areas of the natural occurrence of those two species overlapped, while *N. tomentosiformis* was known to occur further up north (Goodspeed, 1954).

Cameron (1965) demonstrated that the interactions of the *N. sylvestis* plasmon with the *N. tabacum* genome and *vice versa* did not produce flower modifications commonly associated with the effect of alien cytoplasmic factors (see Chap. 5), indicating that *N. tabacum* and *N. sylvestris* shared the same type of cytoplasm. This provided a clue that *N. sylvestris* was the likely maternal parent in the amphidiploid from which *N. tabacum* evolved. Sheen (1972) compared *N. tabacum*, amphihaploids 2x (*N. sylvestris* × *N. otophora*) and 2x (*N. sylvestris* × *N. tomentosiformis*), and Kostoff's hybrid (alleged amphidiploid 4x (*N. sylvestris* × *N. tomentosiformis*) for the similarity index of their eight isoenzyme systems. The index was higher for *N. sylvestris* × *N. tomentosiformis* than for *N. sylvestris* × *N. otophora* but was the highest for Kostoff's amphidiploid. Sheen concluded that *N. tomentosiformis* is the likely progenitor of *N. tabacum* and explained the very high similarity between *N. tabacum* and Kostoff's hybrid by inadvertent introgression from *N. tabacum* in the latter, a possibility indirectly admitted by Kostoff himself (Kostoff, 1938) and confirmed later (see the previous paragraph but also Sect. 4.6.4 where Kostoff's hybrid is discussed at some length).

In the 1970s, a series of studies lent more support to *N. sylvestris and N. tomentosiformis* as respective putative maternal and paternal parents of *N. tabacum*. Gray et al. (1974) compared *N. tabacum*, *N. sylvestris*, *N. otophora* and *N. tomentosiformis* for the polypeptide composition of ribulose bisphosphate carboxylase-oxygenase (RuBisCO), an enzyme performing a dual function of oxygenation and carboxylation and a vital component of both photosynthetic and photorespiratory systems (Kung, 1977). The enzyme, the most abundant protein in higher plants, previously commonly known as Fraction 1 protein, is also unique in that it is composed of two subunits, the larger being coded by chloroplast DNA and the smaller by nuclear DNA. Thus, it can be used as a marker for both cytoplasmic and nuclear genomes (Kung, 1977). Gray et al. (1974) found that the polypeptide compositions of the large subunits in *N. tabacum* and *N. sylvestris* were identical. The small subunits were identical in *N. tomentosiformis* and *N. tabacum* and different in *N. otophora*. The parentage of *N. tabacum* originating from ancestral forms of *N. sylvestris* and *N. tomentosiformis* was confirmed in similar RuBisCO composition studies by Iwai et al. (1976) and Kawashima et al. (1976) and by comparing the content of soluble b-proteins induced by TMV infection in *N. tabacum* and in the amphidiploid 4x (*N. sylvestris* x *N. tomentosiformis*) (Ahl et al., 1982).

The major argument against *N. tomentosiformis* as the paternal progenitor of *N. tabacum* was that the synthetic amphidiploids 4x (*N. sylvestris* × *N. tomentosiformis*) produced by Greenleaf (1941), Lilienfeld (1952, 1953), Gerstel (after Sheen, 1972), Burk (1973), Lim et al. (2006) including those produced somatically by protoplast fusion (Liao & Lai, 1994a, b, c) were consistently reported to be pollen fertile but sterile on the female side due to the collapse of embryo sac development during meiotic divisions (Greenleaf, 1942; Ar-Rushdi, 1955; Liao & Lai, 1994a, b). Obviously, the original amphidiploid must have retained some ability to perpetuate itself as a lineage from which a new species could have evolved. Burk (1973) found his original amphidiploid plant of *N. sylvestris-tomentosiformis* to show vestigial self-fertility that tended to increase slightly over subsequent generations. He speculated that the spontaneous ancestral amphidiploid may have also had a chance to produce some selfed offspring, given that its self-fertility levels and longevity were approximately equal to those of his synthetic hybrid. Over generations, the fertility of the amphidiploid would have progressively increased owing to reverse mutations of the recessive sterility system (Clausen, 1941; Ar-Rushdi, 1955) as well as because of other genomic rearrangements as it continued to evolve into a new species (Bindler et al., 2011). Indeed, after several generations of selfing, Burk's amphidiploid regained enough fertility for its seeds to be eventually pooled into a seed stock designated the 'Th line' (Skalicka et al., 2003, 2005). The accession of 4x (*N. sylvestris* x *N. tomentosiformis*) of undeclared provenance used in the study by Sievert (1972) was implied by the author to be "fertile".

Since the advent of molecular methods, a sizeable new body of evidence on the ancestry of *N. tabacum* has accumulated. On the maternal side, *N. sylvestris* remained an undisputed candidate, and DNA sequencing data reconfirmed its status (Bland et al., 1986; Yukawa et al., 2006). Most of the molecular evidence, starting with the measurements of DNA content in *N. tabacum* and in its putative progenitors (Narayan & Rees, 1974), supported *N. tomentosiformis* as the donor of the tomentosoid (T) subgenome of *N. tabacum* (Okamuro & Goldberg, 1985; Murad et al., 2002; Lim et al., 2006; Khan & Narayan, 2007; Sierro et al., 2014). Some studies, while confirming the major role of *N. tomentosiformis*, also yielded evidence for introgression from *N. otophora* (Kenton et al., 1993; Riechers & Timko, 1999; Kitamura et al., 2001; Ren & Timko, 2001; Raju et al., 2012). Based on the transferability of *N. tabacum* microsatellite (SSR) probes, Moon et al. (2008) found *N. kawakamii* and *N. tomentosiformis* to show the greatest similarity to *N. tabacum* but also did not exclude the possible involvement of *N. otophora* in the evolution of the cultivated species. The issue of whether *N. tomentosiformis* alone or with some participation from other tomentosoid species contributed to the T subgenome of *N. tabacum* may be difficult to resolve. From the evolutionary perspective, *N. tabacum* is a relatively young species, although the estimates regarding its age differed substantially: from less than 6 million years (Okamuro & Goldberg, 1985) to approximately 200 thousand years (Clarkson et al., 2005; McCarthy et al., 2015) or even much less than that (Petit et al., 2007). The most recent investigations put the age of *N. tabacum* ca. 600 thousand years (Clarkson et al., 2017). The actual time span notwithstanding, since its inception, *Nicotiana tabacum* has undergone substantial changes in its genomic structure and organization (Bombarely et al., 2012). Analysis of repetitive DNA sequences indicated that the T genome of *N. tabacum* has experienced greater sequence loss than the S genome (Macke et al., 2004; Renny Byfield et al., 2011). Skalicka et al. (2003, 2005), in their studies of the 'synthetic tobacco' (amphidiploid hybrid 4x (*Nicotiana sylvestris* × *N. tomentosiformis*)), found the changes in DNA sequences to be very rapid and preferentially targeted to the *N. tomentosiformis* genome. Since the polyploidization event, the *N. tabacum* genome has decreased by 3.7–8% of its original size, and the sylvestroid and tomentosoid subgenomes now account for 53 and 47%, respectively, of the total *N. tabacum* genome (Wang & Bennetzen,

2015). This notwithstanding, due to its young age, *N. tabacum* still retains a sizeable and complex genome, with repeat elements accounting for approximately 70% of 4.5 Gb of its total size (Bromley et al., 2017). On the other hand, the changes that the *N. tabacum* genome has undergone may have blurred to some extent the molecular fingerprints originally left by its ancestral parents, especially by the tomentosoid one, the more so as the genomes of the putative parents themselves must have undergone some significant alterations during that time (Bindler et al., 2011).

The evolution of *Nicotiana tabacum* into contemporary market types of cultivated tobacco was summarized by Lewis and Nicholson (2007).

2.2.3 *Polyploid Origin of some Other* Nicotianae

There are several other 24-pair chromosome *Nicotiana* species, the origin of which has been traced to particular 12-chromosome-pair progenitors.

Origin of N. rustica Another cultivated tobacco (*N. rustica*) is recognized as the product of hybridization between the direct ancestors of *N. paniculata* and *N. undulata* (Goodspeed, 1954; Gray, 1978; Lim et al., 2004, 2005; Kovarik et al., 2008; Leitch et al., 2008). With the exception of Gray (1978), who, based on serological evidence, pointed to the female parentage of *N. undulata* in the hybrid, all subsequent reports agree on *N. paniculata* as the closest extant relative of the maternal parent of *N. rustica* (Lim et al., 2004; Kovarik et al., 2008; Knapp, 2020). This notwithstanding, introgression from *N. knightiana* (Aoki & Ito, 2000; Chase et al., 2003) on the maternal side and from *N. glutinosa* on the paternal side (Chase et al., 2003) was also suggested. Indeed, according to a more recent study by Sierro et al. (2018), present-day *N. rustica* inherited ca. 60% of its genome from the common ancestor of both *N. paniculata* and *N. knightiana*, and of the two, *N. knightiana* seems to be more closely related to *N. rustica*. The maternal parentage of *N. rustica* is also indirectly suggested by the similarity of flower morphologies in alloplasmic lineages of *N. tabacum* that carry the cytoplasm of *N. rustica*, *N. paniculata* or *N. knightiana* (Hart, 1965; Kubo, 1985; Berbeć, 2001; Nikova & Vladova, 2002). The cytoplasm of *N. undulata* affects the floral development in *N. tabacum* in a markedly different way (see Chap. 5). *N. rustica* is a relatively young allopolyploid whose age is estimated to range from 200 thousand (Mac McCarthy et al., 2015) to 700 thousand years (Clarkson et al., 2017).

Origin of *N. arentsii* N. *arentsii* is considered to be the youngest natural allopolyploid of *Nicotiana* that arose approximately 400 thousand years ago (Clarkson et al., 2017) and was determined to have descended from the union of *N. undulata* and *N. wigandioides* (Goodspeed, 1954, Gray, 1978, Leitch et al., 2008, Mc McCarthy et al., 2015). Spontaneous amphidiploid hybrids between some ancestral 24-chromosome forms of *N. obtusifolia* and *N. attenuata* gave rise to two separate 48-chromosome lineages from which contemporary species of the section

Polydicliae, *N. clevelandii* and *N. quadrivalvis* evolved (Lim et al., 2004; Anssour et al., 2009) approximately 1.2 million years ago (Clarkson et al., 2017).

Origin of the Repandae Species The four extant species of the Repandae section (*N. nudicaulis, N. repanda, N. nesophila* and *N. stocktonii*) have differentiated from the ancestral hybrid of *N. sylvestris* with *N. obtusifolia*: (Leitch et al., 2008; Clarkson et al., 2017), and those events are estimated to have taken place approximately 4 million years ago.

Origin of the Suaveolentes Species The most ancient allopolyploids in *Nicotiana* belong to the present-day section of Suaveolentes (Leitch et al., 2008; Liu & Marubashi, 2014; Clarkson et al., 2017), and their age is estimated at 6–10 million years (Clarkson et al., 2017; McCarthy et al., 2015; D'Andrea et al., 2023). While the majority of *Nicotiana* allopolyploids retained their original chromosome number, most species in the section Suaveolentes underwent a reduction in chromosome number (Gottula et al., 2014). That most numerous and most highly diversified group in the genus probably originated from a sequence of hybridization events that involved the hybrid between two ancient and as yet undetermined species of the sections Noctiflorae and Suaveolentes on the maternal side and the progenitor of *N. sylvestris* on the paternal side (Clarkson et al., 2017; Schiavinato et al., 2020). The most recent data also suggest contributions from some ancestral forms of the sections Alatae and Petunioides. According to the study by d'Andrea et al. (2023), the single hybridization event that ultimately gave rise to the section Suaveolentes occurred before the split between the ancient clades of Alatae/Sylvestres and Noctiflorae/Petunioides. Thus, the ancestral progenitor of Suaveolentes was a hybrid between the common ancestor of the sections Noctiflorae and Petunioides as the female parent and the progenitor of the sections Alatae and Sylvestres as the male parent (D'Andrea et al., 2023).

References

Ahl, P., Cornu, A., & Gianinazzi, S. (1982). Soluble proteins as markers in studies of resistance and phylogeny in *Nicotiana*. *Phytopathology, 72*, 80–85.

Anon. (2021). Botanists discover seven new tobacco species in Australia. *Science News* https://www.sci.news/biology/nicotiana-species-australia-09955.html

Anssour, S., Krugel, T., Sharbel, T. F., Saluz, H. P., Bonaventure, G., & Baldwin, I. T. (2009). Phenotypic, genetic and genomic consequences of natural and synthetic polyploidization of *Nicotiana attenuata* and *Nicotiana obtusifolia*. *Annals of Botany (London), 103*, 1207–1217. https://doi.org/10.1093/aob/mcp058

Aoki, S., & Ito, M. (2000). Molecular phylogeny of *Nicotiana* (Solanaceae) based on the nucleotide sequence of the matK gene. *Plant Biology, 2*, 316–324.

Ar-Rushdi, A. H. (1955). Inheritance in *Nicotiana tabacum*. XXVI. Sterility genes from *tomentosa*e species. *Journal of Genetics, 54*, 9–22.

Augsten, M., Burkowski Meyer, P., Freitas, L. B., Batista, J. A. N., & Stehmann, J. R. (2022). *Nicotiana gandarela* (Solanaceae), a new species of 'tobacco' highly endangered from the Quadrilátero Ferrífero in Brazil. *PhytoKeys, 190*, 113–129. https://doi.org/10.3897/phytokeys.190.76111

Bally, J., Marks, C. E., Huangtaek, J., Fangzi, J., Roden, S., et al. (2021). *Nicotiana paulineana*, a new Australian species in Nicotiana section Suaveolentes. *Australian Systematic Botany, 34*, 477–484. https://doi.org/10.1071/SB20025

Barker, M. S., Husband, B. C., & Pires, J. C. (2016). Spreading Winge and flying high: The evolutionary importance of polyploidy after a century of study. *American Journal of Botany, 103*, 1139–1145.

Berbeć, A. (2001). Floral morphology and some other characteristics of iso-genomic alloplasmics of *Nicotiana tabacum* L. *Beitrage zur Tabakforschung International/ Contributions to Tobacco Research, 19*, 309–314.

Berbeć, A., & Doroszewska, T. (2020). The use of *Nicotiana* species in tobacco improvement. In N. Ivanov, N. Sierro, & M. Peitsch (Eds.), *The tobacco plant genome. Compendium of plant genomes*. Springer.

Bindler, G., Plieske, J., Bakaher, N., et al. (2011). A high density genetic map of tobacco (*Nicotiana tabacum* L.) obtained from large scale microsatellite marker development. *Theoretical and Applied Genetics, 123*, 219. https://doi.org/10.1007/s00122-011-1578-8

Bland, M. M., Matzinger, D. F., & Levings, C. S. (1986). Comparison of the mitochondrial genome of *Nicotiana tabacum* with its progenitor species. *Theoretical and Applied Genetics, 69*, 535–541.

Bombarely, A., Edwards, K. D., Sanchez-Tamburrino, J., & Mueller, L. A. (2012). Deciphering the complex leaf transcriptome of the allotetraploid species *Nicotiana tabacum*: A phylogenomic perspective. *Genomics, 13*, 406.

Bromley, J. R., Evans, A. D., Fernandez-Pozo, N., Humphry, M. E., Bombarely, A., Hosmani, P. S., Diamantakis, S. K., Mueller, L. A., & Edwards, K. D. (2017). *Tobacco genomic resources for gene discovery*. CORESTA Meeting, Agronomy/Phytopathology, 2017, Santa Cruz do Sul, AP 25.

Burbidge, N. T. (1960). The Australian species of *Nicotiana*. *Australian Journal of Botany, 8*, 342–380.

Burk, L. G. (1973). Partial self-fertility in a theoretical amphidiploid progenitor of *Nicotiana tabacum*. *The Journal of Heredity, 64*, 348–350.

Cameron, D. R. (1965). Cytoplasmic effects in *Nicotiana*. *Proceedings of the XI International Congress of Genetics, 1*, 202–204.

Chase, M. W., & Christenhush, M. (2021a). Nicotiana insecticida. Solanaceae. *Curtis's Botanical Magazine, 38*. https://doi.org/10.1111/curt.12402

Chase, M. W., & Christenhush, M. J. M. (2021b). 998. *Nicotiana pila*. Solanaceae. *Curtis's Botanical Magazine, 38*, 394–404.

Chase, M. W., & Christenhush, M. J. M. (2021c). 991. *Nicotiana rosulata*. *Curtis's Botanical Magazine, 38*, 319–328.

Chase, M. W., & Christenhusz, M. (2018a). 883. *Nicotiana karijini*. Solanaceae. *Curtis's Botanical Magazine, 35*, 228–236. https://doi.org/10.1111/curt.12242

Chase, M. W., & Christenhusz, M. (2018b). 885. *Nicotiana gascoynica*. Solanaceae. *Curtis's Botanical Magazine, 35*, 245–252. https://doi.org/10.1111/curt.12244

Chase, M. W., & Christenhusz, M. J. M. (2018c). 894. *Nicotiana stenocarpa*. Solanaceae. *Curtis's Botanical Magazine, 35*, 319–327.

Chase, M. W., Knapp, S., Cox, A. V., Clarkson, J. J., Butsko, Y., Joseph, J., Savolainen, V., & Parokonny, A. S. (2003). Molecular systematics, GISH and the origin of hybrid taxa in *Nicotiana* (Solanaceae). *Annals of Botany, 92*, 107–127.

Chase, M. W., Conran, J., & Christenhusz, M. (2018a). 884. *Nicotiana yandinga*. Solanaceae. *Curtis's Botanical Magazine, 35*, 237–244. https://doi.org/10.1111/curt.12243

Chase, M. W., Conran, J., & Christenhusz, M. (2018b). 886. *Nicotiana faucicola*. *Curtis's Botanical Magazine, 35*, 253–260. https://doi.org/10.1111/curt.12243

Chase, M. W., Conran, J., & Christenhusz, M. (2021a). 988. *Nicotiana truncata*. Solanaceae. *Curtis's Botanical Magazine, 38*. https://doi.org/10.1111/curt.12399

Chase, M. W., Fay, M. F., & Christenhusz, M. (2021b). *Nicotiana salina*. Solanaceae. *Curtis's Botanical Magazine, 38*, 415–423. https://doi.org/10.1111/curt.12408

Chase, M. W., Fay, M. F., Nollet, F., & Christenhusz, M. (2021c). Nicotiana notha. Solanaceae. *Curtis's Botanical Magazine, 38*. https://doi.org/10.1111/curt.12401

Chase, M. W., Przeslawski, R. A., Falvey, R., & Christenhusz, M. (2021d). *Nicotiana murchisonica*. Solanaceae. *Curtis's Botanical Magazine, 38*, 382–392. https://doi.org/10.1111/curt.12405

Chase, M. W., Dodsworth, S., & Christenhush, M. (2021e). 989. *Nicotiana walpa*. Solanaceae. *Curtis's Botanical Magazine, 38*, 298–308.

Chase, M. W., Palsson, R. L., & Christenhush, M. J. M. (2021f). *Nicotiana hoskingii*. Solanaceae. *Curtis's Botanical Magazine, 38*, 365–373.

Chase, M. W., Christenhusz, M. J. M., Palsson, R. L., Fay, M. F., Dodsworth, S., Conran, J. G., et al. (2021g). Species delimitation in *Nicotiana* sect. Suaveolentes (Solanaceae): Reciprocal illumination leads to recognition of many new species. *Curtis's Botanical Magazine, 38*, 266–286.

Chase, M. W., Knapp, S., Cauz-Santos, L. A., & Christenhusz, M. J. M. (2021h). (2845) proposal to conserve the name *Nicotiana benthamiana* (*N. Suaveolens* var. *cordifolia*) (Solanaceae) with a conserved type. *Taxon, 70*, 1146–1147. https://doi.org/10.1002/tax.12587

Chase, M. W., Cauz-Santos, L. A., Dodsworth, S., & Christenhusz, M. J. M. (2022a). Taxonomy of the Australian *Nicotiana benthamiana* complex (*Nicotiana* section Suaveolentes; Solanaceae): Five species, four newly described, with distinct ranges and morphologies. *Australian Systematic Botany, 35*, 345–363. https://doi.org/10.1071/SB22009.

Chase, M. W., Samuel, R., Leitch, A. R., Guignard, M. S., Conran, J. G., Nollet, F., Fletcher, P., Jakob, A., Cauz-Santos, L. A., Vignolle, G., Dodsworth, S., Christenhusz, M. J. M., Buril, M. T., & Paun, O. (2022b). Down, then up: Non-parallel genome size changes and a descending chromosome series in a recent radiation of the Australian allotetraploid plant species, *Nicotiana* section Suaveolentes (Solanaceae). *Annals of Botany, 20*, 1–20.

Chase, M. W., Christenhush, M. J. M., Cauz-Santos, L. A., Nollet, F., et al. (2023). *Australian Journal of Botany, 36*, 167–205.

Christoff, M. (1928). Cytological studies in the genus *Nicotiana*. *Genetics, 13*, 233–277.

Clarkson and Symon. (1991). *Nicotiana wuttkei* (Solanaceae), a new species from North-Eastern Queensland with an unusual chromosome number. *Austrobaileya, 3*(3), 389–392.

Clarkson, J. J., Knapp, S., Garcia, V. F., Olmstead, A. G., Leitch, A. R., & Chase, M. W. (2004). Phylogenetic relationships in *Nicotiana* (Solanaceae) inferred from multiple plastid DNA regions. *Molecular Phylogenetics and Evolution, 33*, 75–90.

Clarkson, J. J., Lim, K. Y., Kovarik, A., Chase, M. W., Knapp, S., & Leitch, A. R. (2005). Long-term genome diploidization in allopolyploid *Nicotiana* section *Repanda*e (Solanaceae). *The New Phytologist, 168*, 241–252.

Clarkson, J. J., Kelly, L. J., Leitch, A. R., Knapp, S., & Chase, M. W. (2010). Nuclear glutamine synthetase evolution in Nicotiana: Phylogenetics and the origins of allotetraploid and homoploid (diploid) hybrids. *Molecular Phylogenetics and Evolution, 55*, 99–112.

Clarkson, J. J., Dodsworth, S., & Chase, M. W. (2017). Time-calibrated phylogenetic trees establish a lag between polyploidisation and diversification in Nicotiana. *Plant Systematics and Evolution, 303*, 1001–1012. https://doi.org/10.1007/s00606-017-1416-9

Clausen, R. E. (1928). Interspecific hybridization and the origin of species in *Nicotiana*. *Zeits Ind Abst Ver. Supbd., 1*, 547–553.

Clausen, R. E. (1932). Interspecific hybridization in *Nicotiana*. XIII. Further data as to the origin of *Nicotiana tabacum*. *Svensk Botanisk Tidskrift, 26*, 123–136.

Clausen, R. E. (1941). Polyploidy in *Nicotiana*. *The American Naturalist, 75*, 291–306.

D'Andrea, L., Sierro, N., Quadi, S., & Hasing, T. (2023). Polyploid *Nicotiana* section Suaveolentes originated by hybridization of two ancestral *Nicotiana* clades. *Frontiers in Plant Science, 14*, 999887. https://doi.org/10.3389/fpls.2023.999887

D'Arcy, W. G. (1976). New names and taxa in Solanaceae. *Annals of the Missouri Botanical Garden, 63*, 363–369.

Daly, K. (1959). On the origin and history of the red-flowered tobacco, *Nicotiana* x *Sanderae*. *Baileya, 7*, 37–41.

Descorbeth, M. P. (2004). *Genotype and phenotype of Nicotiana alata* × *Nicotiana rastroensis*. University of Missouri–Columbia. Office of Undergraduate Research.

Descorbeth, P., & McClure, B. (2005). Genotype and phenotype of *Nicotiana alata* X *Nicotiana rastroensis*. In vivo. *Cancer and the environment 37th annual MACUB conference, Long Island university, 26*(2), 12.

Dodsworth, S. A. (2015). *Genome skimming for phylogenomics*. PhD thesis, School of Biological and Chemical Sciences, Queen Mary University of London.

Doroszewska, T., Depta, A., & Czubacka, A. (2009). *Album of Nicotiana species*. Puławy. Instytut Uprawy Nawożenia i Gleboznawstwa – Państwowy Instytut Badawczy.

East, E. M. (1928). The genetics of the genus *Nicotiana*. *Bibliography Generator, 4*, 243–318.

Gerstel, D. U. (1960). Segregation in new allopolyploids of *Nicotiana*. Comparison of 6x (*N. tabacum* × *N. tomentosiformis*) and 6x (*N. tabacum* × *N. otophora*). *Genetics, 45*, 1713–1734.

Goodspeed, T. H. (1945). Cytotaxonomy of *Nicotiana*. *The Botanical Review, 11*, 533–592.

Goodspeed, T. H. (1954). *The genus Nicotiana*. Chronicle Bot.

Goodspeed, T., & Bradley, M. (1942). Amphidiploidy. *The Botanical Review, 8*, 272–316.

Goodspeed, T. H., & Clausen, R. E. (1928). Interspecific hybridization in *Nicotiana*. VIII. The *sylvestris-tomentosa-tabacum* triangle and its bearing on the origin of tobacco. *University of California Publications in botany, 11*, 127–140.

Gottula, J., Lewis, R., Saito, S., et al. (2014). Allopolyploidy and the evolution of plant virus resistance. *BMC Evolutionary Biology, 14*, 149. https://doi.org/10.1186/1471-2148-14-149

Gray, J. C. (1978). Serological reaction of fraction I proteins from interspecific hybrids in the genus *Nicotiana*. *Plant Systematics and Evolution, 129*, 177–183.

Gray, J. C., Kung, S. D., Wildman, S. G., & Sheen, S. J. (1974). Origin of *Nicotiana tabacum* L. detected by polypeptide composition of fraction I protein. *Nature, 252*, 226–227.

Greenleaf, W. H. (1941). Sterile amphidiploids: Their possible relation to the origin of *Nicotiana tabacum*. *American Naturalist, 75*, 394–399.

Greenleaf, W. H. (1942). Genic sterility in tabacum-like amphidiploids of *Nicotiana*. *Journal of Genetics, 43*, 69–96.

Hart, G. E. (1965). *Studies on extrachromosomal male sterility in Nicotiana*. Dissertation abstracts 26: 3595 (tobacco abstracts 10: 2108) Ph D thesis Univ California, Berkley USA.

He, H., Iizuka, T., Maekawa, M., Sadahisa, K., Morikawa, T., Yanase, M., Yokoi, S., Oda, M., & Tezuka, T. (2019). *Nicotiana suaveolens* accessions with different ploidy levels exhibit different reproductive isolation mechanisms in interspecific crosses with *Nicotiana tabacum*. *Journal of Plant Research, 132*, 461–471.

Horton, P. (1981). A taxonomic revision of *Nicotiana* (Solanaceae) in Australia. *Journal of Adelaide Botanic Gardens, 3*, 1–56.

Hussain, Z., Sun, Y., Shah, S. H., Khan, H., et al. (2023). The dynamics of genome size and GC contents evolution in genus *Nicotiana*. *Brazilian Journal of Biology, 83*, e245372. https://doi.org/10.1590/1519-6984.245372

Iwai, S., Tanabe, Y., & Kawashima, N. (1976). Origin of sequence heterogeneity of the small subunit of fraction 1 protein from Nicotiana tabacum. *Biochemical and Biophysical Research Communications, 73*, 993–996.

Jimenez-Duran, K., McClure, B., Garcia-Campusano, F., Rodriguez-Sotres, R., Cisneros, J., Busot, G., & Cruz-Garcia, F. (2013). NaStEP: A proteinase inhibitor essential to self-incompatibility and a positive regulator of HT-B stability in *Nicotiana alata* pollen tubes. *Plant Physiology, 161*, 97–107.

Julio, E. (2005). *Développement d'une carte génétique de Nicotiana tabacum et identification de QTLs liés à des caractères agronomiques et à la composition de la fumée*. PhD thesis, National polytechnique de Toulouse.

Kawashima, N., Tanabe, Y., & Iwai, S. (1976). Origin of Nieotiana *tabacum* detected by primary structure of fraction 1 protein. *Biochimica et Biophysica Acta, 427*, 70–77.

Kelly, L. J., Leitch, A. R., Clarkson, J. J., Hunter, R. B., Knapp, S., & Chase, M. W. (2010). Intragenic recombination events and evidence for hybrid speciation in *Nicotiana* (Solanaceae). *Molecular Biology and Evolution, 27*, 781–799. https://doi.org/10.1093/molbev/msp267

Kenton, A., Parokonny, A. S., Gleba, Y. Y., & Bennett, M. D. (1993). Characterization of the *Nicotiana tabacum* L. genome by molecular cytogenetics. *Molecular & General Genetics, 240*, 159–169.

Khan, M. Q., & Narayan, N. K. J. (2007). Phylogenetic diversity and relationships among species of the genus *Nicotiana* using RAPDs analysis. *African Journal of Biotechnology (AJB), 6*, 148–162.

Kitamura, S., Inoue, M., Shikazono, N., & Tanaka, A. (2001). Relationships among *Nicotiana* species revealed by the 5S rDNA spacer sequence and fluorescence in situ hybridizatio*N. Theoretical and Applied Genetics, 103*, 678–686.

Knapp, S. (2013). Nat. Hist. Museum NaturePLus. Seeking nightshades in South America. (*blog*) www.nhm.ac.uk

Knapp, S. (2020). Biodiversity of *Nicotiana* (Solanaceae). In N. V. Ivanov, N. Sierro, & Peitsch (Eds.), *The tobacco plant genome, compendium of plant genomes*. Springer. https://doi.org/10.1007/978-3-030-29493-9_2

Knapp, S., Chase, M. W., & Clarkson, J. J. (2004). Nomenclatural changes and a new section classification in *Nicotiana* (Solanaceae). *Taxon, 52*, 73–82.

Kostoff, D. (1936). The origin of cultivated tobacco. *Current Science, 4*, 872.

Kostoff, D. (1938). Studies on polyploid plants XVIII. Cytogenetic studies on *Nicotiana sylvestris* × *N. Tomentosiformis* hybrids and amphidiploids and their bearing on the problem of the origin of *N. Tabacum*. *CR Akad Nauk SSSR, 18*, 459–462.

Kostoff, D. (1939). The origin of the tetraploid *Nicotiana* from Backhurst. *Current Science*, 110–113.

Kostoff, D. (1943). *Cytogenetics of the genus Nicotiana*. State Printing House.

Kovarik, A., Dadejova, M., Lim, Y. K., Chase, M. W., Clarkson, J. J., Knapp, S., & Leisch, A. R. (2008). Evolution of rDNA in *Nicotiana* allopolyploids: A potential link between rDNA homogenization and epigenetics. *Annals of Botany, 101*, 815–823.

Kubo, T. (1985). Studies on hybrid breeding by means of cytoplasmic male sterility in tobacco. *Bulletin of the Iwata Tobacco Experiment Station, 17*, 69–138.

Kung, S. (1977). Expression of chloroplast genomes in higher plants. *Annual Review of Plant Physiology, 28*, 401–437.

Laskowska, D., & Berbeć, A. (2003). Preliminary study of the newly discovered tobacco species *Nicotiana wuttkei* Clarkson and Symon. *Genetic Resources and Crop Evolution, 50*, 835–839.

Lee, C. B., Page, L. E., McClure, B. A., & Holtsford, T. P. (2008). Post-pollination hybridization barriers in *Nicotiana* section *Alatae*. *Sexual Plant Reproduction, 21*, 183–195.

Leitch, I. J., Hanson, L., Lim, K. Y., Kovarik, A., & Chase, M. W. (2008). The ups and downs of genome size evolution in polyploid species of *Nicotiana* (Solanaceae). *Annals of Botany, 101*, 805–814.

Lewis, R. S. (2011). Nicotiana. In C. Kole (Ed.), *Wild crop relatives: Genomic and breeding resources, plantation and ornamental crops*. Springer. https://doi.org/10.1007/978-3-642-21201-7_10

Lewis, R. S., & Nicholson, J. S. (2007). Aspects of the evolution of *Nicotiana tabacum* L. and the status of the United States *Nicotiana* germplasm collection. *Genetic Resources and Crop Evolution, 54*, 727–740.

Liao, L. J., & Lai, K. L. (1994a). Fertility of somatic hybrids derived from protoplasts fusion between *Nicotiana sylvestris* and *N. Tomentosiformis*. *Agronomy Journal China, 4*, 75–78.

Liao, L. J., & Lai, K. L. (1994b). The growth characteristics of somatic hybrids between *Nicotiana sylvestris* and *N. Tomentosiformis*. *Agronomy Journal China, 4*, 59–74.

Liao, L. J., & Lai, K. L. (1994c). Chromosomal behavior of the somatic hybrids between *Nicotiana sylvestris* and *N. Tomentosiformis*. *Agronomy Journal China, 4*, 217–228.

Lilienfeld, F. A. (1952). Cytogenetical problems in the genus *Nicotiana*. *National Institute of Genetics Annual Report, 3*, 38–39.

Lilienfeld, F. A. (1953). Cytogenetical problems in *Nicotiana*. b. Continued investigation of synthesized *N. tabacum*. *National Institute of Genetics Annual Report, 4*, 33.

Lim, K. Y., Matyasek, R., Kovarik, A., & Leitch, A. R. (2004). Genome evolution in allotetraploid *Nicotiana*. *Biological Journal of the Linnean Society, 82*, 599–606.

Lim, K. Y., Matyasek, R., Kovarik, A., Fulnecek, J., & Leitch, A. R. (2005). Molecular cytogenetics and tandem repeat sequence evolution in the allopolyploid *Nicotiana rustica* compared with diploid progenitors *N. paniculata* and *N. undulata*. *Cytogenetic and Genome Research, 109*, 298–309.

Lim, K. Y., Souckova-Skalicka, K., Sarasan, V., Clarkson, J. J., Chase, M. W., Kovarik, A., & Leitch, A. R. (2006). A genetic appraisal of a new synthetic *Nicotiana tabacum* (Solanaceae) and the Kostoff synthetic tobacco. *American Journal of Botany, 93*, 875–883.

Liu, H., & Marubashi, W. (2014). Species origin of genomic factors in *Nicotiana nudicaulis* Watson controlling hybrid lethality in interspecific hybrids between *N. Nudicaulis* Watson and *N. Tabacum* L. *PLoS One, 9*, e97004. https://doi.org/10.1371/journal.pone.0097004

Macbride, J. F. (1930). Spermatophytes, mostly Peruvian. 3. Peruvian Solanaceae. *Publication Field Museum of Natural History Botanical Series, 8*(2), 105–112.

Macke, M. A., Gregor, W., Mette, M. F., Aufsatz, W., Kanno, T., Jakowitsch, J., & Matzke, A. J. (2004). Endogenous pararetroviruses of allotetraploid tobacco *Nicotiana tabacum* and its diploid progenitors, *N. Sylvestris* and *N. Tomentosiformis*. *Biological Journal of the Linnean Society, 82*, 627–638.

Marks, C. E. (2010). Definition of South Pacific taxa of *Nicotiana* section Suaveolentes (Solanaceae). *Muelleria, 28*, 74–84.

Marks, C. E., Ladiges, P. Y., & Newbigin, E. (2011). Karyotypic variation in *Nicotiana* section Suaveolentes. *Genetic Resources and Crop Evolution, 58*, 797–803.

Martinez-Croveto, R. (1978). Una nueve especie de *Nicotiana* de la flora Argentina. *Bonplandia, 5*, 7–10.

McCarthy, E. W., Arnold, S. E. J., Chittka, L., LeComber, S. C., Verity, R., Dodsworth, S., Knapp, S., Kelly, L., Chase, M. W., Baldwin, I. T., Kovarik, A., Mhiri, C., Taylor, L., & Leitch, A. R. (2015). The effect of polyploidy and hybridization on the evolution of flower colour in *Nicotiana* (Solanaceae). *Annals of Botany, 115*, 1117–1131.

Merxmueller, H., & Buttler, K. P. (1975). *Nicotiana* in der Afrikanischen Namib ein pflanzengeographisches und phytogenetisches Ratsel. *Mitteilungen der Botanischen Staatssammlung München, 12*, 91–104.

Moon, H. S., Nicholson, J. S., & Lewis, R. S. (2008). Use of transferable *Nicotiana tabacum* L. microsatellite markers for investigating genetic diversity in the genus *Nicotiana*. *Genome, 51*, 547–559.

Murad, L., Lim, K. Y., Christopodulou, V., Matyasek, R., Lichtenstein, C. P., Kovarik, A., & Leitch, A. R. (2002). The origin of tobacco's T genome is traced to a particular lineage within *Nicotiana tomentosiformis* (Solanaceae). *American Journal of Botany, 89*, 921–928.

Murfett, J., Lee, C., Page, L., Yates, A., et al. (2005). *Phylogenetics of Nicotiana section Alatae and description of N. 'Rastroensis'*. Botany home conference ID: 359.

Narayan, R. K. J., & Rees, H. (1974). Nuclear DNA, heterochromatin and phylogeny of *Nicotiana* amphidiploids. *Chromosoma, 47*, 75–83. https://doi.org/10.1007/BF00326273

Nikova, V., & Vladova, R. (2002). Wild *Nicotiana* species as a source of cytoplasmic male sterility in *Nicotiana tabacum*. *Beiträge zur Tabakforschung International, 20*, 301–311.

Ohashi, Y. (1976). *Nicotiana kawakamii: A new species of the genus Nicotiana*. Bul Spec CORESTA, Congress Tokyo: 71 (abstr) P001.

Okamuro, J. K., & Goldberg, R. B. (1985). Tobacco single-copy DNA is highly homologous to sequences present in the genomes of its diploid progenitors. *Molecular Genetics and Genomics (MGG), 198*, 290–298.

Petit, M., Lim, K. Y., Julio, E., Poncet, C., de Borne, F. D., Kovarik, A., Leitch, A. R., Grandbastien, M. A., & Mhiri, C. (2007). Differential impact of retrotransposon populations on the genome of allotetraploid tobacco (*Nicotiana tabacum*). *Molecular Genetics and Genomics, 278*, 1–15.

Raju, K. S., Sharma, R. K., & Murthy, T. G. K. (2012). Genetic and evolutionary relationship among Nicotiana species as elucidated by AFLP. *The Journal of Agricultural Science, 7*, 135–144.

Ren, N., & Timko, M. P. (2001). AFLP analysis of genetic polymorphism and evolutionary relationships among cultivated and wild *Nicotiana* species. *Genome, 44*, 559.

Renny Byfield, S., Chester, M., Kovarik, A., Le Comber, S., Grandbastien, M.-A., Deloger, M., Nichols, R. A., Macas, J., Novak, P., Chase, M. W., & Leitch, A. R. (2011). Next generation sequencing reveals genome downsizing in Allotetraploid *Nicotiana tabacum*, predominantly through the elimination of paternally derived repetitive DNAs. *Molecular Biology and Evolution, 28*, 2843–2854.

Riechers, D. E., & Timko, M. P. (1999). Structure and expression of the gene family encoding putrescine N-methyltransferase in *Nicotiana tabacum*: New clues to the evolutionary origin of cultivated tobacco. *Plant Molecular Biology, 41*, 387–401.

Runemark, A., Vallejo-Marin, M., & Meier, J. I. (2019). Eukaryote hybrid genomes. *PLoS Genetics, 15*, e1008404. https://doi.org/10.1371/journal.pgen.1008404

Särkinen, T., Bohs, L., Olmstead, R. G., et al. (2013). A phylogenetic framework for evolutionary study of the nightshades (Solanaceae): A dated 1000-tip tree. *BMC Evolutionary Biology, 13*, 214. https://doi.org/10.1186/1471-2148-13-214

Scarpa, G. F., & Rosso, C. N. (2011). Etnobotanica del "Coro", (*Nicotiana paa*, Solanaceae), un tabaco silvestre poco conocido del extremo sur de Sudamerica. *Bonplandia, 20*, 391–404.

Schiavinato, M., Marcet-Houben, M., Dohm, J. C., Gabaldón, T., & Himmelbauer, H. (2020). Parental origin of the allotetraploid tobacco *Nicotiana benthamiana*. *The Plant Journal, 102*, 541–554. https://doi.org/10.1111/tpj.14648

Sheen, S. J. (1972). Isozyme evidence bearing on the origin of *Nicotiana tabacum*. *Evolution, 26*, 143–154.

Sierro, N., Battey, J. N. D., Ouadi, S., Bakaher, N., Bovet, L., Willig, A., Goepfert, S., Peitsch, M. C., & Ivanov, N. V. (2014). The tobacco genome sequence and its comparison with those of tomato and potato. *Nature Communications, 5*, 3833.

Sierro, N., Battey, J. N. D., Bovet, L., et al. (2018). The impact of genome evolution on the allotetraploid *Nicotiana rustica*—An intriguing story of enhanced alkaloid production. *BMC Genomics, 19*, 855.

Sievert, R. C. (1972). Resistance to anthracnose in the genus *Nicotiana*. *Tobacco Science, 16*, 32–34.

Skalicka, K., Lim, K. Y., Matyasek, R., Koukalova, B., Leitch, A. R., & Kovarik, A. (2003). Rapid evolution of parental rDNA in a synthetic tobacco allotetraploid line. *American Journal of Botany, 90*, 988–996.

Skalicka, K., Lim, K. Y., Matyasek, R., Matzke, M., Leitch, A. R., & Kovarik, A. (2005). Preferential elimination of repeated DNA sequences from the paternal, *Nicotiana tomentosiformis* genome donor of a synthetic allotetraploid. *The New Phytologist, 166*, 291–303.

Slana, L. J., Stavely, J. R., Grosso, J. J., & Golden, A. M. (1977). Probable source of *Meloidogyne incognita* resistance in tobacco as indicated by reactions to five *Meloidogyne* isolates. *Phytopathology, 67*, 537–543.

Smith, L. B., & Downs, R. (1964). Solanaceas. Flora ilustrada Catarinense. 321. *Phytologia, 10*, 438.

Stavely, J. R. (1979). Disease resistance. (in) Durbin RD ed. *Nicotiana*. Procedures for experimental use. U.S. *Department of Agriculture Technical Bulletin, 1586*, 111–120.

Stavely, J. R., Kung, S. D., & Slana, L. J. (1977). Comparison of polypeptide composition of fraction I protein of certain *Nicotiana* species and hybrids related to root knot resistant tobacco. *Tobacco Science, 179*, 31–32.

Stehmann, J. R., Semir, J., & Ippolito, A. (2002). *Nicotiana mutabilis* (Solanaceae), a new species from southern Brazil. *Kew Bulletin, 57*, 639–646.

Symon, D. E. (1984). A new species of *Nicotiana* (Solanaceae) from Dalhousie Springs, South Australia. *Journal of the Adelaide Botanic Gardens, 7*, 117–121.

Symon, D. E. (1998). A new *Nicotiana* species from near Coober Pedy, South Australia. *Journal of the Adelaide Botanic Gardens, 18*, 1–4.

Symon, D. E. (2005). Native tobaccos (Solanaceae: *Nicotiana* spp.) in Australia and their use by aboriginal people. *The Beagle, 21*, 1–10.

Symon, D. E., & Keneally, K. F. (1994). A new species of *Nicotiana* (Solanaceae) from near Broome, Western Australia. *Nuytsia, 9*, 421–425.

Symon, D. E., & Lepschi, B. J. (2007). A new status in *Nicotiana* (Solanaceae): *N. Monoschizocarpa* (P. Horton) Symon & Lepschi. *Journal of the Adelaide Botanic Gardens*, 21, 92.

Tatemichi, Y. (Ed.). (1990). *Illustrated book of the genus Nicotiana*. Plant Breeding and Genetics Research Laboratory. Japan Tobacco Inc.

Wang, X., & Bennetzen, J. L. (2015). Current status and prospects for the study of *Nicotiana* genomics, genetics, and nicotine biosynthesis genes. *Molecular Genetics and Genomics, 290*, 11–21. https://doi.org/10.1007/s00438-015-0989-7

Wheeler, H. M. (1945). A contribution to the cytology of the Australian-South Pacific species of *Nicotiana*. *Proceedings of the National Academy of Sciences of the United States of America, 31*, 177–185.

Williams, E. (1975). Meiotic chromosome pairing in interspecific hybrids of *Nicotiana*. Comparative pairing of hybrids in section Suaveolentes. *New Zealand Journal of Botany, 13*, 601–609.

Yukawa, M., Tsudzuki, T., & Sugiura, M. (2006). The chloroplast genome of *Nicotiana sylvestris* and *Nicotiana tomentosiformis*: Complete sequencing confirms that the *Nicotiana sylvestris* progenitor is the maternal genome donor of *Nicotiana tabacum*. *Molecular Genetics and Genomics, 275*, 367–373.

Chapter 3
Experimental Interspecific Hybrids in *Nicotiana* and Barriers to Hybridization

3.1 Overview of History and Extent of Interspecific Hybridization in *Nicotiana*

Interspecific hybrids may arise spontaneously in laboratories that maintain their in-house collections of different *Nicotiana* species. Unlike these rare events, such as the hybrid 2x (*N. setchellii* × *N. otophora*) reported by Bawolska et al. (1978), the vast majority of the known hybrids within the genus were the result of purposeful manipulation. The first experimental crossing of two different species of *Nicotiana* is generally credited to the German scholar J. G. Koelreuter, who produced a hybrid between *N. paniculata* and *N. rustica* while working in St. Petersburg, Russia in 1760 (Kostoff, 1943; Ternovsky, 1962; Mayr, 1986). Koelreuter's classical work confirmed the existence of sexuality in plants and demonstrated the equivalency of maternal and paternal parents in contributing to their offspring. The first interspecific hybrid that involved *N. tabacum* was probably also created by Koelreuter, who backcrossed the hybrid *N. rustica* × *N. paniculata* to *N. paniculata* and mated the offspring with *N. tabacum* (Kostoff, 1943). One may speculate that the trispecific combination thus produced was probably composed of aneuploid plants possessing full haploid genomes of *N. paniculata* and *N. tabacum* plus varying numbers of chromosomes contributed by *N. rustica*. Koelreuter also produced the hybrid 2x (*N. glutinosa* x *N. rustica*) and created the first true amphihaploid hybrids between *N. tabacum* and other *Nicotiana* species: 2x (*N. glutinosa* × *N. tabacum*), 2x (*N. paniculata* x *N. tabacum*) and 2x (*N. rustica* x *N. tabacum*) (Kostoff, 1943; Mayr, 1986).

In the nineteenth century, several other interspecific hybrid combinations with *N. tabacum* were synthesized. According to the accounts by East (1928) and Kostoff (1943), Sageret crossed *N. suaveolens* with *N. tabacum* in the 1820s, Gaertner produced N. *rustica* × *N. tabacum* and *N. quadrivalvis* × *N. tabacum* in the 1840s, and in the 1850s, Naudin was the first to produce the hybrid N. *tabacum* × *N. alata*.

A. Berbeć, *A Century of Interspecific Hybridization and Introgression in Tobacco*,
https://doi.org/10.1007/978-3-031-54964-9_3

Brongniart and Gris (1861) were the first to report the hybrid N. *glauca* × *N. tabacum*.

East (1928), one of the early cytogeneticists of *Nicotiana*, divided artificially made interspecific hybrids of *Nicotiana* into 'pre-Mendelian' and 'post-Mendelian'. The former were produced by Koelreuter and by nineteenth century investigators such as Gartner, Focke, Gudron, Naudin, and Sageret. The post-Mendelian hybrids were those created by East himself and by his contemporaries. East's list of viable interspecific hybrids includes 22 'pre-Mendelian' hybrid combinations and 43 hybrids produced later, 65 interspecific hybrids altogether, not including reciprocals. Kostoff (1943) described 181 interspecific hybrids, and Goodspeed (1945) and Goodspeed and Thompson (1959) elevated their number to 243. Citing those authors, Smith (1968) estimated that more than 300 interspecific hybrids had been reported in the genus. Apparao and Ramavarma (1974) raised the estimated number to 340. The author of this volume found references to 455 interspecific hybrid combinations within the genus *Nicotiana* that have been produced by 2023, including several that were reported as nonsurviving seedlings or as mere viable seeds (see Chap. 7). The former, mostly included in the type II seedling death category, have been included on the grounds that hybrid seedling lethality, especially type II, was recently demonstrated as a remediable condition controlled by a simple genetic system. The hybrids, together with literature references limited mostly to the earliest ones, were compiled in Chap. 7 of this review. The supplement chapter lists hybrid combinations and the parentage of each including the reciprocal hybrid, if reported. Multispecies crosses have not been included (those that involve *N. tabacum* as one of the parents are treated separately in Chap. 4 and listed in Tables 4.5 and 4.6).

Table 3.1 lists *Nicotiana* species and the numbers of hybrid combinations with other *Nicotiana*e reported for each of them. It also includes information on whether a particular species was successfully crossed with the cultivated species *N. tabacum*. Included in Table 3.1 is also information about the degree of chromosome homology between *N. tabacum* and the other *Nicotiana* species. Structural similarity between the chromosomes of *N. tabacum* and those of its wild relatives can be treated as a preliminary element of the assessment of how much restriction to the genetic flow exists between the species of interest and cultivated tobacco. As can be readily seen from Table 3.1, this piece of information, essential from the standpoint of a potential interspecific breeder, is either lacking or difficult to locate for nearly half of the species that have been successfully hybridized with *N. tabacum*.

The majority of *Nicotiana* species (fifty-nine) listed in Table 3.1 have a record of having been hybridized with both *Nicotiana tabacum* and at least one other sister *Nicotiana*. Some (*N. mutabilis*, *N. petunioides*, *N. attenuata*, *N. corymbosa*, *N. linearis*, *N. burbidgeae*, *N. thyrsiflora* and *N. wigandioides*) were hybridized with at least one other *Nicotiana* but not with *N. tabacum*. Twenty-eight species listed in Table 3.1, the majority of which were discovered only recently (*N. gandarela*, *N. azambujae*, *N. acaulis*, *N. ameghinoi*, *N. cutleri*, *N. paa*, *N. longibracteata*, *N. spegazzini*, *N. bilybara*, *N. candelabra*, *N. fatuhivensis*, *N. faucicola*, *N. gascoynica*, *N. heterantha*, *N. hoskingii*, *N. insecticida*, *N. karijini*, *N. monoschizocarpa*, *N. murchinsonica*, *N. nota*, *N. paulineana*,

Table 3.1 Potential of *Nicotiana* species to hybridize with *Nicotiana tabacum* and with other sister *Nicotianae*

Species	Hybrid with *N. tabacum* Y = reported N = not reported	Chromosome pairing in F_1 hybrids with *N. tabacum*		Hybrids with species other than *N. tabacum* (number of species)[3]
		No. of bivalents: range and modal number (in parentheses)	Reported by	
Alatae				
N. alata	Y	5–8 (?) 1–6 (3–4) 0–10 (6)	Kostoff (1943), Patel, 1960 Takenaka (1954, 1956c) Berbeć (1987b)	24
N. azambujae	N			N
N. bonariensis	Y			12
N. forgetiana	Y	1–8 (3)	Takenaka (1963)	10
N. gandarela	N			
N. langsdorffii	Y	5–12 (11)[1] 0–5 (0)	Takenaka et al. (1955), Takenaka (1958), Hu (1956) Burk (1972)	21
N. longiflora	Y	5–7 (4–9) 0–4 (0–1)	Kostoff (1943) Takenaka (1957), Takenaka (1962a)	21
N. mutabilis	N			5
N. plumbaginifolia	Y	10–12 1–6 (3)	Patel (1960) Goodspeed (1954)	22
N. sp. ‘Rastroensis	N			6
N. x *sanderae*	Y	5–8 (6)	Kostoff (1943) Malecka (1977)	20
Nicotiana				
N.tabacum	N	x	x	x
Noctiflorae				
N. acaulis	N			N
N. ameghinoi	N			N

(continued)

Table 3.1 (continued)

Species	Hybrid with *N. tabacum* Y = reported N = not reported	Chromosome pairing in F_1 hybrids with *N. tabacum*		Hybrids with species other than *N. tabacum* (number of species)[3]
		No. of bivalents: range and modal number (in parentheses)	Reported by	
N. glauca	Y	9–12 0–7 (4) 9–10 7–12 0–8 (2–3)[2] 0–7 (0) 0–5 (0)	Kostoff (1943) Goodspeed (1954) Kehr and Smith (1952) Patel (1960) Berbeć (1971) Trojak-Goluch and Berbeć (2002)	34
N. noctiflora	Y			7
N. paa	N			N
N. petunioides	Y			3
Paniculatae				
N. benavidesii	Y	0–6 (3) 0–5 0–6 (0)	Goodspeed (1954) Takenaka (1958) Berbeć (1987a)	7
N. cordifolia	Y			4
N. cutleri	N			N
N. knightiana	Y	0–5 (0) 2–12 0–1, 0–3 (0)	Takenaka (1958, 1962c) Patel (1960) Berbeć et al. (1979), Berbeć (1987a)	8
N. paniculata	Y	0–4 (0) 0–4 (1) 2–5	Goodspeed (1954), Berbeć et al. (1979) Takenaka et al. (1956), Takenaka (1962a) Kostoff (1943)	24

N. raimondii	Y	3–5, 0–6 0–7 (1) 0–3 (0)	Kostoff (1939a, 1939b) Goodspeed (1954) Berbeć (1987a)	7
N. solanifolia	Y	0–7 (2) 1–5 (1)	Goodspeed (1954) Takenaka et al. (1955); Takenaka (1956b)	11
Polydicliae				
N. clevelandii	Y			9
N. quadrivalvis	Y	0–6 0–10 (4)	Patel (1960) Goodspeed (1954)	27
Petunioides				
N. acuminata	Y			3
N. attenuata	N			8
N. corymbosa	N			3
N. longibracteata	N			N
N. linearis	N			1
N. miersii	N			11
N. pauciflora	Y	0 0–4	Gentscheff (1931) Patel (1960)	3
Repandae				
N. nesophila	Y			4
N. nudicaulis	Y	0 1–5 (?) Max. 4	Gentscheff (1931) Kostoff (1943), Burk and Neas (1964)	15
N. repanda	Y			15
N. stocktonii	Y	0–4 (?)	Wong (1975)	4

(continued)

Table 3.1 (continued)

Species	Hybrid with *N. tabacum* Y = reported N = not reported	Chromosome pairing in F_1 hybrids with *N. tabacum*		Hybrids with species other than *N. tabacum* (number of species)[3]
		No. of bivalents: range and modal number (in parentheses)	Reported by	
Rusticae				
N. rustica	Y	0–7 (3) 1–10 (?) 0–6 (2–3)	Goodspeed (1954) Takenaka (1953) Moav and Cameron (1961)	30
Suaveolentes				
N. africana	Y	0–4 (?) 0–5 (0–1)	Gerstel et al. (1979) Doroszewska and Berbeć (1996)	4
N. amplexicaulis	Y	0–4 (0)	Berbeć et al. (1979)	12
N. benthamiana	Y			14
N. bilybara	N			N
N. burbidgeae	N			1
N. cavicola	Y			3
N. candelabra	N			N
N. debneyi	Y	0–5 (?)	Kostoff (1943)	28
N. eastii	Y			11
N. excelsior	Y			10
N. exigua	Y			13
N. fatuhivensis.	N			N
N. faucicola	N			N
N. fragrans	Y			5
N. gascoynica	N			N
N. goodspeedii	Y			7
N. gossei	Y	0–5 (1)	Takenaka (1962a)	20
N. hesperis	Y			1

N. heterantha	N			N
N. hoskingii	N			N
N. ingulba	Y			5
N. insecticida	N			N
N. karijini	N			N
N. maritima	Y			13
N. megalosiphon	Y	1–8 (3–4)	Takenaka (1962a)	19
N. monoschizocarpa	N			N
N. murchisonica	N			N
N. notha	N			N
N. occidentalis	Y	4–11 (?)	Wong (1975)	8
N. paulineana	N			N
N. pila	N			N
N. rosulata	Y			1
N. rotundifolia	Y			3
N. rupestris	N			N
N. salina	N			N
N. scopulorum	N			N
N. simulans	Y			4
N. stenocarpa	N			N
N. suaveolens	Y	0–3 0–4 (0)	Kostoff (after Goodspeed, 1954) Shinkareva (1979)	28
N. truncata	N			N
N. umbratica	Y			24
N. velutina	Y			18
N. walpa	N			N
N. wuttkei	Y	0–5 (0)	Laskowska and Berbeć (2012)	4
N. yandinga	N			N

(continued)

Table 3.1 (continued)

Species	Hybrid with *N. tabacum* Y = reported N = not reported	Chromosome pairing in F_1 hybrids with *N. tabacum*		Hybrids with species other than *N. tabacum* (number of species)[3]
		No. of bivalents: range and modal number (in parentheses)	Reported by	
Sylvestres				
N. sylvestris	Y	9–12 (12)	Goodspeed (1954)	22
Tomentosae				
N. kawakamii	Y			4
N. leguiana	N			N
N. otophora	Y	10–13 (12)	Goodspeed (1954), Takenaka (1957)	13
N. setchellii	Y	11-12 (12)	Goodspeed (1954)	6
N. tomentosa	Y	11-13 (12)	Goodspeed (1954)	13
N. tomentosiformis	Y	12 (12)	Goodspeed (1954)	19
Trigonophyllae				
N. obtusifolia	Y	2–10 (?) 0–11 (5–6)	Kostoff (1943) Takenaka (1953)	14
N. palmeri	Y	0–10 (4)	Goodspeed (1954)	8
Undulatae				
N. arentsii	Y			2
N. glutinosa	Y	0–11 (3) 2–6 (?)	Goodspeed (1954) Kehr and Smith (1952)	27
N. thyrsiflora	N			2
N. undulata	Y	0–8 (?) 0–8 (3–5)	Kostoff (1943) Takenaka (1953)	17
N. wigandioides	N			4

[1] data from the cross 4n *N. tabacum* x 2n *N. langsdorffii*
[2] data for F_1 *N. tabacum* x *N. glauca* (upper line) and its reciprocal (lower line)
[3] full list of interspecific hybrids (including those involving *N. tabacum*) is given in Chap. 7

N. pila, N. rupestris, N. salina N. stenocarpa, N. truncata, N. yandinga, N. walpa), were not found to have any hybridization record whatsoever.

3.2 Pre- and Post-fertilization Barriers to Production of Viable Hybrid Seeds in *Nicotiana*

3.2.1 *Manifestations of Pre- and Postzygotic Barriers to Hybridization*

The number of interspecific hybrids produced between various *Nicotiana* species, although impressive, is still only a small fraction of several thousand theoretically possible combinations. Not all hybrids are obtainable with equal ease. The process of speciation depends on the development of different blocks preventing plants of diverging taxa from crossing with each other and producing viable offspring. The barriers to hybridization can be roughly divided into prezygotic and postzygotic (He et al., 2019). Prezygotic barriers include inhibition of pollen adhesion to the stigma, inhibition of pollen germination, obstruction of pollen tube growth in the stigma and style, and failure of the pollen tube to penetrate the mycropyle. Postzygotic barriers manifest themselves as failure to produce a functional zygote, embryo abortion and failure of hybrid seeds to germinate, lethality of germinated seeds and seedlings, disturbed development of hybrid plants and their failure to develop flowers and, finally, failure of flowering hybrid plants to perpetuate themselves because of their inherent inability to produce offspring. In *Nicotiana*, blocks to hybridization have evolved relatively slowly, unlike gene and chromosome alterations, which have progressed more rapidly (Goodspeed, 1945). Barriers to hybridization were already encountered by early *Nicotiana* researchers. Numerous cases were recorded where upon crossing two different species, either no seeds were produced, the seeds were produced but they failed to germinate, the seeds germinated but the hybrid plants died, mostly at juvenile stages of growth (McCray, 1933; Kostoff, 1943) or mature flowering plants could not be perpetuated because of different forms of sterility (see Sect. 3.6).

3.2.2 *Cross Incompatibility*

The causes of cross incompatibility between *Nicotiana* species were discussed in detail by Kostoff (1943). Three major causes of the failure to obtain viable hybrid seeds were noted by him:

- failure of the pollen tube to reach the ovary
- failure of the sperm to fuse with the egg
- abortion of the hybrid embryo

The first of the causes may be attributed to various factors, such as differences in the length of styles between the maternal and paternal species (Swaminathan & Murthy, 1957; Stoyanova, 1979; Lee et al., 2008), physiological inhibition of pollen tube growth or mechanical obstruction to the growth of pollen tubes (Kostoff, 1943). Growth inhibition of the pollen tube (gametophytic incompatibility) was found to be conditioned genetically (Pandey, 1977). The growth-promoting substance (GPS) diffused in pollen walls becomes deactivated in incompatible stigmata by the genetically controlled system attached to GPS and prevents its normal function and pollen growth. Gametophytic incompatibility in the hybrid *N. tabacum* x *N. alata* was demonstrated to be controlled by differences in ribonuclease activities in self-compatible (SC) *N. tabacum* and self-incompatible (SI) *N. alata* (McClure et al., 1989; McClure, 1996). RNAase linked to the compatibility locus (S) resulting in the rejection of alien pollen was also implicated in the rejection of alien pollen in other intrasectional and intersectional crosses involving *N. alata* (Murfett et al., 1996). Mechanical obstruction may be caused by differences in cell size since somatic chromosome number is, at least within a certain range, correlated with cell size (Kostoff, 1943). By way of example, the thick pollen tubes of *N. tabacum* cannot make their way through the styles of *N. alata*, whereas reciprocal mating is possible (Kostoff, 1943). Another studied case in point was the cross between *N. tabacum* and *N. obtusifolia,* where the pollen tubes of the former species stopped growing in the styles of the latter (Chung et al., 1996). There are many other cases on record where a hybrid was successfully produced one way but not the other. It came to be known as unilateral incompatibility. Kostoff (1938a) found the hybrid *N. glauca* × *N. langsdorffii* to be produced with ease, whereas in reciprocal mating, the pollen tube of *N. glauca* was frequently arrested in the style of *N. langsdorffii*. Kuboyama et al. (1994) found that the pollen tubes of *N. repanda*, *N. obtusifolia*, and *N. rustica* were arrested at different phases of their growth through the styles of *N. tabacum.* The ability of *N. tabacum* pistils to inhibit foreign pollen tube elongation gradually increased as the flowers matured. A reverse situation was observed by Liao et al. (2017) in the hybrid *N. tabacum* cv. 'K326' x *N. alata*. Young flowers of *N. tabacum* inhibited the growth of *N. alata* pollen tubes and prevented them from reaching the ovary, but the inhibition was removed, and seeds could be obtained when senescent flowers were pollinated.

The Australian species of the section Suaveolentes are known to produce hybrids with *N. tabacum* when used as maternal parents, whereas reciprocal crosses are usually unsuccessful (Kostoff, 1943; Goodspeed, 1945). The hybrid of *N. tabacum* by the African species *N. africana* is a well-known exception to that rule (Gerstel et al., 1979; Doroszewska & Berbeć, 1996; Hancock et al., 2015), although successful reciprocal mating, *N. africana* × *N. tabacum*, was also reported (Nikova & Zagorska, 1990). Unilateral incompatibility is also absent when *N. fragrans* is crossed with *N. tabacum* (Tezuka et al., 2010). There are several cases on record when using the autotetraploid variant of a Suaveolentes species as the male parent resolved the problem of unilateral incompatibility with *N. tabacum* (see Sect. 3.4.1). Interestingly, Zaitlin and Mundell (2006) experienced no difficulty in obtaining viable hybrids by fertilizing female *N. tabacum* with pollen of *N. benthamiana,* whereas DeVerna et al. (1987) found that the mating was completely incompatible and they had to resort to ovule culture to overcome the incongruity.

Artificial pollination of ovules cultured *in vitro* was an effective approach to overcome unilateral incompatibility of *N. tabacum* crossed as females with several Suaveolentes species. In this way, viable hybrids of *N. tabacum* as the maternal parent were obtained with: *N. amplexicaulis* (Larkina, 1980), *N. benthamiana* (Subhashini et al., 1986; Mihaylova-Kroumova et al., 2020), *N. debneyi* (Butenko et al., 1970; Larkina, 2015, 2017), *N. excelsior* (DeVerna et al., 1987; Tezuka et al., 2010), *N. maritima* (Tezuka et al., 2010), *N. rosulata* (Ternovsky et al., 1976; Ternovsky & Larkina, 1978a; Larkina, 2015, 2017), N. *suaveolens* (Marubashi & Onosato, 2002, *N. velutina* (Tezuka et al., 2010) (for more information, see Sect. 3.4.3). Unilateral incompatibility between *N. tabacum* and the species of the section Suaveolentes also has important practical implications. Many of these species carry resistance to important diseases and pests of tobacco, and cytoplasmic factors from many of them produce cytoplasmic male sterility in cultivated tobacco. Since CMS is inherited maternally, it may obstruct the breeding process when the transfer of a desired trait from the Suaveolentes species to *N. tabacum* starts and proceeds in a matrilineal fashion (see Chap. 5).

The hybridization process between *N. tabacum* and another *Nicotiana* species that gets stalled at some stage, either pre- or postfertilization, cannot always be accurately classified in simple binary terms: cross compatible vs. cross-incompatible. There are many instances where success depends on whether the right genotype has been chosen within the parental species to be crossed. Kostoff (1943) cites several examples, e.g., *N. rustica* × *N. tabacum* or *N. tabacum* × N. *pauciflora*, where the wild species was successfully crossed with one *N. tabacum* variety but not with others. The crossability of *N. tabacum* with *N. alata*, generally very poor, was found to be largely affected by the genotype of *N. tabacum* (Kostoff, 1943), a phenomenon also confirmed by Ternovsky (1962) and Berbeć (1987b).

The ease with which hybrids of *N. tabacum* × *N. africana* could be obtained either directly from seeds or by resorting to an *in vitro* culture was substantially affected by the genotype of the *N. tabacum* parent in the study by Depta and Doroszewska (2019).

Different accessions of *N. suaveolens* also vary substantially in their ability to cross with *N. tabacum* (He et al., 2019).

The type of cytoplasm can also affect crossability. Liao et al. (2017) failed to obtain viable seeds by crossing regular male fertile *N. tabacum* cv. 'K326' with *N. alata,* but the mating was successful when a cytoplasmically male sterile lineage of K326 with the cytoplasm of *N. glauca* was used as the female parent.

3.2.3 Mechanisms of Pre- and Post Zygotic Incongruity

When the foreign pollen tube has reached the ovary but failed to fertilize the egg, the following phenomena may follow: cell division may occur in the nucellus without embryo or endosperm formation, resulting in no seeds, diploid endosperm may develop, resulting in seeds with no embryos, and induction of cell divisions in the egg may lead to the formation of parthenogenetic haploids or, rarely, diploid embryos. The phenomenon of haploid and diploid maternal soffspring of interspecific crosses will be addressed in one of the subsequent sections.

Various interspecific matings in *Nicotiana* result in the actual fusion of gametes, but the resulting zygote is dysfunctional, and the embryo is aborted soon after fertilization. An example case is the hybrid *N. obtusifolia* x *N. tabacum* (Chung et al., 1996), the hybrid embryo of which aborts soon after fertilization or develops poorly, resulting in inviable, nongerminating seeds. Disturbed and collapsed development of the hybrid embryo may result from the incongruity between the maternal endosperm and the "foreign" embryo as well as from the genetic factors inherent in the embryo itself. Brink and Cooper (1941) and Bannikova (1965a) studied seed development in the F_1 hybrid *N. rustica* × *N. tabacum*. The cross usually resulted in aborted seeds with less than 1% germination capability. Retarded growth of the endosperm, the failure of intergumentary cells to differentiate into conductive tissues and the overgrowth of the nucellus were the causes of the abortion of the hybrid seeds. Disturbed development of the embryo and the endosperm and deranged metabolic exchange between the embryo, endosperm and maternal tissue set in within a week after fertilization in the hybrid *N. rustica* × *N. paniculata* (Bannikova, 1965b). The development of the hybrid embryos was substantially slower compared to the embryo growth rate of self-fertilized *N. rustica*.

The nonviable seeds obtained from crossing *N. tabacum* with *N. alata* were void of both the embryo and the endosperm (Stoyanova, 1979). The embryo death by starvation in the hybrids *N. stocktonii* x *N. tabacum*, *N. nesophila* x *N. tabacum* and *N. repanda* x *N. tabacum* was attributed to the cessation of endosperm development (Reed & Collins, 1980a).

Recently, arrest of endosperm development leading to starvation and abortion of the embryo in the hybrid 4n *N. suaveolens* x 2n *N. tabacum* was attributed to disturbed EBN (Endosperm Balance Number) (He et al., 2020, 2022, 2023). According to the EBN hypothesis, the normal ratio of maternal vs. paternal ploidy level is approximately 2:1 due to the triploid genome of the endosperm. Disturbance of this ratio caused by different ploidy levels of parental genomes would lead to endosperm growth arrest. Specifically, excess maternal EBN in the cross 4n *N. suaveolens* x 2n *N. tabacum* led to precocious developmental transition and subsequent endosperm development failure accompanied by hypertrophy of the embryo arrested at the globular stage (He et al., 2020) or very early cessation of hybrid embryo growth, depending on the *N. suaveolens* accession used.

Chromosome elimination in postzygotic stages is yet another manifestation of postzygotic incongruity. Disturbed mitotic cell divisions in the hybrid embryo may lead to failure of some essential chromosomes to be included in the daughter cells and, consequently, result in embryo abortion. If uniparental chromosome elimination is the case, the affected embryos may develop into semiviable dysfunctional aneuploid hybrid plants or, in extreme cases, yield maternal or paternal haploids (see Sect. 3.5). Kramer and Reed (1988) and Hancock et al. (2015) reported the appearance of such irregular plants in the progeny of the hybrid *N. tabacum* x *N. africana*.

3.3 Seedling Lethality

3.3.1 *Occurrence of Premature Plant Death in* Nicotiana *Hybrids*

The incongruity between species of *Nicotiana* may extend beyond the survival of the embryos and affect the germinating seeds and newly emerged plantlets. Root necrosis, decay and death of juvenile plants are common among the hybrid offspring of different *Nicotianae.* In the first half of the twentieth century, East (1928) and Gentscheff (1931) listed *N. tabacum* × *N. alata, N. tabacum* × *N. langsdorffii, N. tabacum* × *N. longiflora*, and *N. tabacum* × *N. plumbaginifolia* among the hybrids that die at the cotyledonary stage or thereafter. McCray (1932) added *N. tabacum* × *N. glutinosa, N. nudicaulis* × *N. tabacum* and *N. suaveolens* × *N. tabacum* to the list of "weak" or prematurely dying hybrids. Kostoff (1943) also cited previously mentioned combinations and added *N. tabacum* × *N. pauciflora.* Foster (1943) was unable to raise the hybrid *N. repanda* x *N. tabacum* to maturity, and Clayton (1950) encountered that phenomenon in the hybrid *N. debneyi* × *N. tabacum.* Inviable hybrids of *N. eastii* × *N. tabacum* and *N. megalosiphon* × *N. tabacum* were reported by Chaplin and Mann (1961), and those of *N. gossei* × *N. tabacum* and *N. occidentalis* × *N. tabacum* were reported by Apparao and Ramavarma (1972) and Ternovsky et al. (1976), respectively. In later years, several other hybrid combinations were found to be lethal, especially those involving species of the section Suaveolentes and *N. tabacum* (Lloyd, 1975; Berbeć & Doroszewska, 1981; Nikova et al., 1991; Laskowska & Berbeć, 2012). There was one particular case in which the massive death of hybrid seedlings was advantageous. In the cross *N. tabacum* × *N. africana,* the seed set was good, and the seeds germinated well, but germination was followed by extensive root necrosis and death of the emerged seedlings (Gerstel et al., 1979). In that particular case, only ca. 1% of the plants survived beyond the cotyledonary stage, and the population of rare survivors was composed of true hybrids and occasional maternal haploids. The latter were used to produce completely homozygous lines in a one-step procedure that came to be deployed for both academic and practical purposes (Burk et al., 1979; Wernsman, 1992) as an alternative to the anther culture method (Nitsch & Nitsch, 1969).

3.3.2 *Types of Hybrid Lethality*

The mechanisms underlying the phenomenon of hybrid lethality remained obscure until the nineties of the last century. In-depth investigations into the causes of seedling unviability in interspecific hybrids of *Nicotiana* were undertaken primarily by Japanese researchers in the late 1980s. They were summed up by Tezuka et al. (2012) in an exhaustive review of the work that had been carried out by him and by his Japanese associates and colleagues.

According to the external symptoms exhibited by inviable seedlings, five types of hybrid lethality in *Nicotiana* have been distinguished to date:

- Type I: browning of shoot apex and root tip
- Type II: browning of hypocotyls and roots
- Type III: yellowing of true leaves
- Type IV: formation of multiple shoots
- Type V: fading of shoot color

Type II lethality seems to be prevalent in crosses involving *N. tabacum*. It is shown by 19 hybrids of *N. tabacum* with the species of the section Suaveolentes (Tezuka, 2012). The phenomenon occurred regardless of cross direction, with the exception of *N. wuttkei* × *N. tabacum*, the reciprocal of which was not successful (Laskowska & Berbeć, 2012). Further studies identified the causative factor of type II lethality in *N. debneyi* that interacts with the factors allegedly on chromosome Q in *N. tabacum* (Tezuka, 2012, see also Sect. 3.3.4).

Outside the hybrids with *N. tabacum*, Type II lethality was also observed in *N. paniculata* × *N. gossei*, *N. suaveolens* × *N. gossei* and *N. debneyi* × *N. repanda* (Tezuka, 2012). The lethality of the hybrid *N. tabacum* × *N. langsdorffii* (Watanabe & Marubashi, 2004) was not assigned to any of the lethality types. Based solely on the description by these authors but also from other accounts (Burk, 1972, Berbeć, unpublished), it could fit in with type II. However, from the account of Burk (1972), who used monosomics of *N. tabacum* to overcome the unviability of that hybrid, it appears that neither the use of Haplo H nor Haplo Q restored viability to the hybrid; some viable plants have been obtained from Haplos A, G, L and Z (see, however, also Sect. 3.3.4).

N. occidentalis produces type V lethal hybrids when crossed with *N. tabacum* (Tezuka, 2012).

Lethality types II and V are also peculiar in that the symptoms are temperature dependent, i.e. they develop at temperatures of 28 °C and below, are completely suppressed when the temperature is raised to 34–36 °C and recur once the temperature drops again below 30 °C. This phenomenon was first reported by Manabe et al. (1989) in the hybrid *N. suaveolens* × *N. tabacum*.

Of the other types of seedling lethality, type I was exhibited by *N. nudicaulis* × *N. tabacum*, type III by *N. repanda* × *N. tabacum,* (Iwai et al., 1984), *N. paniculata* × *N. nudicaulis* and *N. glutinosa* × *N. repanda, and* type IV by *N. paniculata* × *N. alata* and *N. paniculata* × *N. glutinosa* (Tezuka, 2012).

According to Iizuka et al. (2012), *N. benthamiana* and *N. fragrans* produce only viable hybrids with *N. tabacum*. However, Krusteva et al. (2003) and Nikova et al. (2006) reported serious survival issues among the emerging seedlings of the F_1 hybrid *N. benthamiana* x *N. tabacum*. Subhashini et al. (1986) and DeVerna et al. (1987) failed to obtain any seeds from mating male *N. benthamiana* to female *N. tabacum*.

3.3.3 *Causes of and Phenomena Associated with Hybrid Lethality*

The temperature-dependent lethality of hybrid seedlings observed in types II and V is called programmed cell death (PCD) and is accompanied by apoptotic phenomena such as condensation of chromatin, fragmentation of nuclei, disintegration of DNA and aggregation of insoluble proteins in dying cells (Ueno et al., 2019). It is one of the mechanisms of reproductive isolation and occurs in other plant genera (Tezuka & Marubashi, 2004; Hancock et al., 2015). Bomblies (2009) suggested that PCD and apoptotic hybrid lethality are caused by the hyperactivation of plant defense responses and thus may share a common mechanism with resistance to some viral pathogens (e.g., TMV or TSWV), which is also based on eliciting a similar hypersensitive reaction (HR). Hancock et al. (2015) drew attention to the fact that the gene conferring the hypersensitive necrotic response to TMV and the gene controlling apoptotic hybrid lethality in *N. tabacum* are both located on the same chromosome H contributed by the ancestral form of *N. tomentosiformis*. The idea of the common genetic background of PCD and HR plant defense response was also discussed by Chen and Lin (2016). The two authors thought it very likely that the genes that cause necrosis and death in interspecific hybrids are also involved in immune responses and that the appearance of apoptotic phenomena in wide hybrids can be related to the activation of pathogenesis-related genes. It was argued that the activation of the PCD-related processes is due to the mismatch between the delicately configured immune systems of individual species that make up an interspecific hybrid. This interpretation became more precise when the gene Nt6549g30 was identified to encode a CC-NRL protein in *N. tabacum*, the CC-NRL group being the largest class of plant defense proteins that recognize effectors from a pathogen (Ma, 2017; Ma et al., 2020). The same Nt6549g30 gene was also demonstrated to control seedling lethality in the hybrid *N. tabacum* x *N. africana* (Ma, 2017, Ma et al., 2020). More recently, Katsuyama et al. (2021) suggested that both NRL proteins and the associated chaperone protein complexes that aid NRLs in unfolding were involved in developing seedling lethality in the hybrid *N. gossei* x *N. tabacum*. Hence, hybrid seedling death is essentially the effect of autoimmunity-based responses caused by incorrect interactions between parent-of origin NRLs and their associated proteins. As the result of those PCD-type autolytic processes, insoluble protein progressively aggregates in dying hybrid cells, a process demonstrated for the hybrid *N. suaveolens* x *N. tabacum* (Ueno et al., 2019).

Development of teratological changes is a peculiar phenomenon that may be involved in the failure of certain hybrid combinations to grow to maturity. Apparently, in most interspecific hybrids, genetic tumors usually appear in senescing plants (see Sect. 4.5.9), although Burk (1972) reported their occurrence before flowering in a plant of *N. tabacum* x *N. langsdorffii*. Likewise, in some crosses that did not involve *N. tabacum* (*N. rustica* × *N. alata* and *N. rustica* × *N. langsdorffii*), tumors and teratomata were observed in juvenile plants of those hybrids and in the hybrid *N. rustica* × *N. alata* (Takanashi & Marubashi, 2017). The onset of these abnormalities was temperature dependent. The hybrid *N. obtusifolia* × *N. tabacum* developed vitrification and tumors as early as the seedling stage, and these abnormalities were the main cause of seedling death in that hybrid (Liu & Marubashi, 2014). In the latter case, the abnormalities were probably of an origin different than genetic tumors since neither parental species belonged to the "plus" group as classified by Näf (1958).

Nicotiana species that were found to produce inviable hybrids with *N. tabacum* are listed in Table 3.2.

Table 3.2 Lethality of juvenile hybrid plants of *Nicotiana tabacum* with other *Nicotiana* species

Section	Species	Lethality L=lethal V=viable
Alatae	*N. alata*	V/L
	N. langsdorffii	L/V
	N. longiflora	L
	N. plumbaginifolia	L
Repandae	*N. nudicaulis*[2]	L type I
	N. repanda	L type III
Rusticae	*N. rustica*	V/L
Suaveolentes	*N. africana*	L Type II
	N. amplexicaulis	L Type II
	N. cavicola	L Type II
	N. debneyi	L Type II
	N. eastii[1]	L
	N. excelsior	L Type II
	N. exigua	L Type II
	N. goodspeedii	L Type II
	N. gossei	L Type II
	N. hesperis	L Type II
	N. ingulba	L Type II
	N. maritima	L Type II
	N. megalosiphon	L Type II
	N. occidentalis	L Type II
	N. rosulata	L Type II
	N. rotundifolia	L Type V
	N. simulans	L Type II
	N. suaveolens	L Type II
	N. umbratica	L Type II
	N. velutina	L Type II
	N. wuttkei	L Type II
Trigonophyllae	*N. obtusifolia*	L/V
Undulatae	*N. glutinosa*	V/L

Adapted after Tezuka (2012)

[1]Sesquidiploid hybrid from crossing *N. eastii* with autotetraploid *N. tabacum* (Chaplin & Mann, 1961)

[2]Vitrification and tumors developed by seedlings (Liu & Marubashi, 2014)

3.3.4 Location and Identification of Genetic Factors Responsible for Lethality of Hybrid Seedlings

Chromosomal Location of Heritable Factors in *N. tabacum* by Monosomic Analysis The first monosomic line of *N. tabacum* experimentally produced was probably the one derived from backcrossing the hybrid *N. sylvestris* x *N. tabacum* to *N. tabacum*. It was reported by Clausen and Goodspeed (1926) and named 'corrugated'. The original complete series of 24 lines of *N. tabacum*, each of them monosomic for one of the 24 chromosomes of the species, was developed by Prof. R. E. Clausen with his associates at the University of California, Berkeley, starting from the mid-twenties of the last century (Olmo, 1935; Clausen & Cameron, 1944; Lewis, 2011). Generally, four sources of monosomic lines in *N. tabacum* were indicated by Clausen and Cameron (1944):

(a) Monosomic variants in segregating populations from backcrossing F_1 (*N. tabacum* x *N. tomentosa*[1]) and F_1 (*N. tabacum* x *N. sylvestris*) to *N. tabacum*. The method may have taken advantage of the 'Drosera' scheme pairing in the gametogenesis of the initial hybrids (Goodspeed, 1954) and of other irregularities in homoeologous pairing between the 48-chromosome *N. tabacum* and its ancestral 24-chromosome progenitors. Thus, the backcross F_1 (*N. tabacum* x *N. sylvestris*) x *tabacum* could occasionally yield some true *N. tabacum* individuals and aneuploids, including those monosomic for the T subgenome chromosomes. In a similar fashion, individuals monosomic for the S subgenome chromosomes could be picked among the survivors of the cross F1 (*N. tabacum* x *N. tomentosa*) x *N. tabacum*. In those early studies, several monosomics were obtained in this manner but only from backcross populations that involved *N. sylvestris* (Olmo, 1935). More recently, monosomic individuals of *N. tabacum* were also identified in hybrid populations derived from the cross F_1 (*N. tabacum* x *N. tomentosiformis*) x *N. tabacum* using SSR markers for specific linkage groups of *N. tabacum* (Liu et al., 2017). These plants were monosomic for linkage groups 3 and 6 according to *the N. tabacum* genome map developed by Bindler et al. (2011);
(b) Spontaneous genomic mutations in normal disomic tobacco populations;
(c) Segregating offspring of *N. tabacum* lines that carried the asynaptic pale-sterile mutation discovered by Clausen (1931). Pale sterile plants produce microsporocytes with, on average, only 11 bivalents per cell, and the rest of the chromosomes remain unpaired. The cross pale sterile x normal type produces highly diversified offspring that consist of trisomics, single, double and triple monosomics and various monosomic-trisomic combinations. Monosomics for a single chromosome pair are bred into the normal type by repeated backcrossing accompanied by morphological and cytological selection;

[1] At that time *N. tomentosa* was regarded as the progenitor species of *N. tabacum* and the contributor of its T subgenome

(d) Monosomics that arise spontaneously in the offspring of established ones due to the tendency of limited asynapsis in many monosomic types. Ultimately, those spontaneous variants proved to be the major source of monosomics developed in the genetic laboratory of the University of California in Berkeley (Clausen & Cameron, 1944).

The monosomics of *N. tabacum* and their morphological and developmental features in the background of the variety Red Russian were described by Prof. D. R. Cameron (Smith, 1968, 1979). At the same time, a labeling system of *N. tabacum* chromosomes was adopted in which chromosomes contributed by *N. tomentosiformis* are lettered from A through L and those by *N. sylvestris* originally from M through X (Olmo, 1935). The lettering was later changed from M through Z (Olmo, 1935; Smith, 1968) because the letters X and Y were traditionally assigned to the sex chromosomes.

Until recently, monosomic analysis was the only available tool with which the genes that control simply inherited traits could be assigned to particular chromosomes.

Location and Identification of the Lethality Factor in *N. tabacum* Responsible for Type II Hybrid Seedling Lethality Inoue et al. (1996) studied the survival of hybrids from crossing *N. suaveolens* with putative progenitor species of *N. tabacum,* i.e., *N. sylvestris* and *N. tomentosiformis*, which contributed subgenomes S and T, respectively, to the allopolyploid cultivated species. They found the hybrid plants *N. suaveolens* x *N. sylvestris* to develop typical lethality symptoms, whereas the plants from crossing *N. tomentosiformis* by *N. suaveolens* remained viable. The authors inferred from those results that it is the subgenome S of *N. tabacum* that carried the lethality factor. Based on an analogous study of the hybrids *N. debneyi* x *N. sylvestris* and *N. tomentosiformis* x *N. debneyi*, Tezuka et al. (2007) likewise concluded that the subgenome S of *N. tabacum* houses the factor/s that cause lethality in the hybrid *N. debneyi* x *N. tabacum* (Table 3.3,[2] column 4).

Based on the evidence discussed in the preceding paragraph, Marubashi and Onosato (2002) studied genetic causes of lethal hybrids by crossing ten monosomic *N. tabacum* lines for the sylvestroid chromosomes of *N. tabacum* (from Haplo N to Haplo Z except Haplo P and Haplo V) with *N. suaveolens*. When tested against *N. suaveolens* as the pollen parent, all but Haplo Q consistently yielded lethal hybrid offspring. In contrast, the progeny of Haplo Q x *N. suaveolens* segregated for viable and lethal seedlings, and this result was also confirmed in the study of Tezuka and Marubashi (2006a). Similarly, the progeny of the monosomic Haplo Q x *N. debneyi* segregated for lethal and viable individuals. Q chromosome-specific markers revealed the presence of the Q chromosome in lethal plants, whereas viable seedlings

[2] In this Table, as in those to follow across this book, mating direction of sexual hybrids is generally not indicated. That information should be sought through references to the hybrid of interest either in the tables or in Chap. 7 of this review where all known *Nicotiana* hybrids, including those with *N. tabacum*, are tabulated. Cross directions in the hybrids and in their reciprocals, if reported, are given in column IV of Table 7.1.

Table 3.3 Chromosomal and genomic location of the *N. tabacum* factor controlling type II lethality in hybrid seedlings from crossing *N. tabacum* with several species of the section Suaveolentes on chromosome H (subgenome T) vs. chromosome Q (subgenome S) as established by different research teams

Species producing lethal hybrids in crosses with N. tabacum	Chromosome location of type II lethality factor		Involvement of S subgenome of *N. tabacum* in conferring type II lethality to hybrid progeny
	Chromosome H	Chromosome Q	
N. africana	Gerstel et al. (1979)[1]	Tezuka et al. (2010)[5]	
	Hancock et al. (2015)[2]		
	Ma et al. (2020)[3,4]		
N. debneyi	Ma et al. (2020)[4]	Tezuka et al. (2007)[5]	Tezuka et al. (2007)[7]
N. excelsior	Ma et al. (2020)[4]	Tezuka et al. (2010)[5]	
N. goodspeedii	Ma et al. (2020)[4]	Tezuka et al. (2010)[5]	
N. gossei	Ma et al. (2020)[4]	Tezuka et al. (2010)[5]	
N. ingulba		Tezuka (2012)[6]	
N. maritima	Ma et al. (2020)[4]	Tezuka et al. (2010)[5]	
N. megalosiphon	Ma et al. (2020)[4]	Tezuka et al. (2010)[5]	
N. simulans	Ma et al. (2020)[4]		
N. suaveolens		Marubashi and Onosato (2002)[5]	Inoue et al. (1996)[8]
N. umbratica	Ma et al. (2020)[4]		
N. velutina	Ma et al. (2020)[4]	Tezuka et al. (2010)[5]	

[1]determined by monosomic analysis of the progeny from mating *N. tabacum* to *N. africana* across the whole set of *N. tabacum* monosomics (from Haplo-A to Haplo Z); only the progeny Haplo H x *N. africana* segregated for lethal and viable individuals

[2] determined by demonstrating association of chromosome H with linkage group 11 of *N.* tabacum (Bindler et al., 2011) and by microsatellite marker genotyping

[3] determined by demonstrating that the gene *Nt6549g30* at the NtHL1 locus on chromosome H of *N. tabacum* controls hybrid lethality in *N. tabacum* x *N. africana*; the authors assigned chromosome H to the subgenome contributed by *N. sylvestris* but did not elaborate on this revision of the recognized classification of *N. tabacum* chromosomes

[4] demonstrated by obtaining viable progeny from interspecific crosses in which the *N. tabacum* parent carried the *Nt6549g30* allele whose ethality-conferring function was disabled via CRISP-Cas9 technology

[5] determined by monosomic analysis: hybrid progeny that involved Haplo Q as the *tabacum* parent segregated for lethal and viable individuals

[6]no details

[7]indirectly demonstrated by lethality reaction in *N. sylvestris* x *N. debneyi* vs. viable offspring of *N. tomentosiformis* x *N. debneyi*

[8] indirectly demonstrated by lethality reaction in *N. suaveolens* x *N. sylvestris* vs. viable offspring of *N. tomentosiformis* x *N. suaveolens*

were deficient for the Q chromosome (Tezuka et al., 2007). In another study conducted in that laboratory, Tezuka et al. (2010) found that in the offspring from crossing Haplo-Q monosomics of *N. tabacum* with *N. africana*, *N. excelsior*, *N. goodspeedii*, *N. gossei*, *N. maritima*, *N. megalosiphon* and *N. velutina,* the seedlings that carried chromosome Q were lethal, whereas those deficient for chromosome Q were viable. Hence, the authors concluded that the factor/s that triggered the seedling death response in those hybrid combinations must reside on chromosome Q of *N. tabacum*. Additionally, Tezuka et al. (2012) found that the SSR (simple sequence repeats) markers present in the inviable hybrid seedlings of *N. tabacum x N. africana* mapped to a linkage group that had been coded as linkage group 11 in subgenome S of *N. tabacum* according to Bindler et al. (2011).

Recent investigations have indicated that rare spontaneous survivors among the dying offspring of type II lethality hybrids may arise due to the loss of the distal part of the Q chromosome as a result of homoeologous translocations during spermatogenesis of the male *N. tabacum* parent (Nakata et al., 2021).

Controversy Regarding the Identitification of the *N. tabacum* Chromosome That Carries the Lethality-Controlling Factor The results discussed in the preceding section are at odds with what was reported by Gerstel et al. (1979), Hancock et al. (2015) and Ma et al. (2020). Gerstel et al. (1979) tested the whole range of *N. tabacum* monosomics (from Haplo A through Haplo Z) by crossing them with *N. africana*. It is only in the Haplo H x *N. africana* progeny that he found segregation for viable and lethal individuals. The remaining combinations, including Haplo Q x *N. africana,* yielded predominantly lethal offspring. Their conclusion was that chromosome H of the T subgenome of *N. tabacum* controlled the lethality of the hybrid *N. tabacum* x *N. africana*.

The more recent report by Hancock et al. (2015) supported those old findings. By using the same or similar set of SSR markers as those used by Tezuka et al. (2012), the authors recognized the association of the seedling lethality factor with linkage group 11 of the *N. tabacum* genome but were firm in associating linkage group 11 with chromosome H rather than with chromosome Q. They supported their claim by citing evidence mainly from earlier studies on the inheritance of TMV resistance in *N. tabacum* (Lewis et al., 2005, other unpublished data) but also because one of the markers for linkage group 11 was detected on chromosome H.

In his thesis, Ma (2017) modified the position on the controversy by conceding that chromosome H originated from the S genome. In a subsequent paper (Ma et al., 2020), the authors reiterated that chromosome H (linkage group 11) of *N. tabacum* was contributed by the ancestral form of *N. sylvestris,* thereby effectively moving chromosome H from subgenome T to subgenome S. They did so in recognition of the facts (see the final paragraph of this section) but in contradiction to the existing nomenclature and classification of monosomics (Clausen & Cameron, 1944; Smith, 1968, 1979; Tezuka et al., 2010; Lewis, 2011).

Table 3.3 presents the results of the genetic control of type II hybrid lethality in hybrids involving *N. tabacum* reported by Japanese and American scientists.

A very intriguing and unanswered question is how this controversy may have come about. The monosomic stocks used by both Gerstel et al. (1979) and Marubashi and Onosato (2002) were bred into Red Russian. In all probability, both Japanese and American stocks descended from those developed and described by Dr. Clausen and Dr. Cameron. This was explicitly stated by Marubashi and Onosato (2002), and it is the best guess for the monosomics used in the USA. The provenance of the Haplo-Q stock used by Tezuka et al. (2007, 2010) was not specified, although in both studies, the monosomics were apparently also ultimately derived from the same source in Berkeley, California. Both chromosome Q, as reported in Japanese studies (Tezuka et al., 2012), and chromosome H, as studied by the Americans (Hancock et al., 2015), showed structural instability and were prone to breakage in the hybrid *N. tabacum* x *N. africana*. Marubashi and Onosato (2002) and Tezuka et al. (2007, 2010) reported the necessity to use Haplo-Q as the maternal parent in crosses with the Australian species because of the failure of their Haplo-Q to produce viable pollen, a feature characteristic of both Haplo-H and Haplo-Q (Smith, 1968). Although Haplo-H and Haplo Q seem to be discernible from each other by some morphological and growth features (Smith, 1968, 1979), morphology is not a wholly dependable criterion in selection for monosomic types, as these differences are strongly influenced by the environment (Liu et al., 2017). Clausen and Cameron (1944) wrote at length on how difficult it was to maintain the identity and integrity of their monosomics and that despite their efforts, doubts remained as to the validity of some of them. All these things put together, it is obvious that at least some monosomics of *N. tabacum* are not unlikely to be confused with one another. Hence, it is probable that Haplo-H used earlier by Professor Gerstel and more lately by the team headed by Professor Lewis and Haplo-Q used by Professor Tezuka, his associates and by other research teams from Japan in reality represented the same monosome type. It was tacitly conceded by Ma (2017) when, in support of the assertion that the hybrid lethality factor locus, N. tab_HL1, resides near the end of chromosome H, he quoted both Hancock et al. (2015) and Tezuka et al. (2012), even though the Japanese researchers had reported on chromosome Q rather than H. The fact was communicated in the very title of their study (Tezuka et al., 2012). The same interpretation of the controversy was recently also advanced by Nakata et al. (2021) and summarized by Mino et al. (2022).

With what we know, it is not wholly unjustified to assume that while Clausen and Cameron's collection of monosomics was being developed and/or maintained in American laboratories, monosome H was at some point wrongly named and wrongly assigned to the T (tomentosoid) genome, whichever came first. A very puzzling part of the story, however, is how the "rectification" of both the misnaming (from H to Q) and the misplacing (from T to S subgenome) came about in the "Japanese chapter" of the collection. The Japanese researchers seem to have ignored or been unaware of the inconsistency of their results with those published in the US, at least in the early stages of their studies. They did not question the validity of any of the collection stocks, nor did they comment on the discrepancy or report on any corrections thereof. Other things put aside, the series of Japanese reports concerning the issue under discussion has turned out to be consistent and logical, by coincidence or otherwise.

Pivotal to the success in resoving this controversy is to establish the association of the linkage groups in the gene map of *N. tabacum* developed by Bindler et al. (2011) to the physical carriers of those genes i.e. the chromosomes. In Bindler's map there are 24 linkage groups corresponding to 24 chromosomes of the amphidiploid *N. tabacum*. Eleven of these were assigned to the tomentosoid (T) genome, nine to the sylvestroid (S) genome, and the remaining four were mixed and consisted of groups of either T or S genetic markers. Neither Bindler (Bindler et al., 2011) nor anyone else has made any attempts to link those linkage groups to particular physical units (chromosomes). Generally, reports on associating linkage groups with chromosomes of *N. tabacum* have been scant as yet and limited to chromosomes A, H and Q. Vontimitta and Lewis (2012) assigned chromosome A to Bindler's linkage group 4 but later corrected the association in favor of group 15 (Ma, 2017). It is not clear, however, if the assignment of chromosome A was changed accordingly. The assignments of chromosomes H and Q were discussed above. Mapping linkage groups to the corresponding chromosomes has barely started, and nearly all the work is still to be done. Most likely, it is not until linkage groups and chromosomes become mutually identifiable that controversies such as the one above will finally be prevented.

In a recent development that may be regarded as a *sui generis* sequel to the chromosome Q vs. H controversy, the transposon-tagging method combined with CRISP-Cas9 gene-editing technology was used to identify the gene in the genome of *N. tabacum* that controls the death of juvenile hybrid plants of *N. tabacum* x *N. africana* (Ma et al., 2020). To this end, a strategy was deployed that involved the use of a maize-derived binary transposable *Ac*/*Ds* system consisting of the autonomous immobile activator (Ac) and the *Ac*-controlled transposon (*Ds*) plus selectable markers linked to the *Ac* and *Ds* elements. As a result, the association was established between the DNA sequence identified as *Nt6549g30* and the *NtHL1* locus previously demonstrated to house the factor responsible for triggering the series of apoptotic events that cause the death of seedlings of that hybrid (Hancock et al., 2015, see also Sect. 3.4.3). The authors stated that its chromosomal location was on chromosome H, or linkage group 11, contributed by *N. sylvestris*. They found a high degree of similarity of *Nt6549g30* to several homologous genes in *N. sylvestris*.

3.3.5 *Genetic Control of Type V Lethality in* N. occidentalis *x* N. tabacum

First, it was suggested that Type V lethality in the hybrid *N. occidentalis* x *N. tabacum* was related to factors located on both subgenomes, S and T, of *N. tabacum* (Tezuka & Marubashi, 2012). Although type V was phenotypically observed in the hybrid, the type II genetic system must have also been functioning alongside because the hybrids of *N. occidentalis* with the progenitor species of *N. tabacum*, *N. sylvestris* and *N. tomentosiformis* showed type II and type V lethality, respectively. Using two new accessions of *N. occidentalis* (PI555541 and PI555690) alongside the old one (JT), Kawaguchi et al. (2021) demonstrated that the genetic control of type V hybrid lethality in the hybrid *N. occidentalis* x *N. tabacum* differed depending on the *N. occidentalis* accession involved: in JT x *N. tabacum,* it was controlled by alleles in both S and T subgenomes, whereas in the hybrids involving the new *N. occidentalis* accessions (PI555541 and PI555690), it was only subgenome T of *N. tabacum* that housed the factor/s controlling type V lethality. Both PI555541 and PI555690 yielded viable offspring when crossed with *N. sylvestris,* and the hybrid progeny of analogous crosses with *N. tomentosiformis* showed type V lethality.

3.4 Methods to Overcome Cross-Incompatibility, Embryo Abortion and Mortality of Juvenile Hybrid Plants

3.4.1 Managing Cross Incompatibility

Numerous approaches have been used to overcome cross incompatibility between species of *Nicotiana*. Swaminathan and Murthy (1957) bypassed the incompatibility between *N. debneyi* and *N. tabacum* that results from the difference in style length of parental species by cutting off the style of the maternal *N. tabacum* to a length of 2–3 mm and smearing the cut surface with a drop of sucrose agar. The pollen of *N. debneyi* was placed on the thus-prepared styles of the egg parent. The problem with this particular cross was more efficiently resolved by resorting to culturing excised ovules of *N. tabacum* fertilized *in vitro* with the pollen of *N. debneyi* (Butenko et al., 1970; Ternovsky et al., 1976). By resorting to the same technique, Marubashi et al. (1988) and Marubashi and Onosato (2002) produced seedlings of the hybrid *N. tabacum* x *N. suaveolens*. The same method was adopted to overcome cross-incompatibility between the long-styled *N. tabacum* used as the female parent and the relatively short-styled *N. knightiana*, *N. rustica*, *N. benthamiana* and *N. rosulata* as the pollen parent (Slusarkiewicz-Jarzina & Zenkteler, 1983, Larkina, 1980, DeVerna et al., 1987, Ternovsky et al., 1976).

The problem of unilateral incompatibility of *N. gossei* and *N. tabacum* was overcome by using the autotetraploid variant of *N. tabacum* as the pollen parent (Valleau, 1952). In some other hybrid combinations, the use of the autotetraploid variant of wild *Nicotiana* ensured success. The first attempt to cross 4n *N. repanda* as a female with 2n *N. tabacum* was only partly successful since the hybrid plants failed to reach flowering (Foster, 1943). The same cross was repeated with success by Pittarelli and Stavely (1975). Surviving hybrid plants were also obtained from crossing 4n *N. palmeri* with *N. tabacum* (Berbeć et al., 1982). However, an opposite effect was recently reported for the cross 4n *N. suaveolens* x 2n *N. tabacum,* which was incompatible due to collapsed endosperm development (He et al., 2020, 2022), whereas the regular combination (2n *N. suaveolens* x 2n *tabacum)* was compatible. For more details on manipulating ploidy levels to bypass the incompatibility of interspecific *Nicotiana* crosses and the underlying EBN (Endosperm Balance Number) theory, the reader is referred back to Sect. 3.2.3 of this chapter.

Senescing flowers of *N. tabacum* were found to be more receptive to fertilization by pollen of *N. alata* than freshly developed flowers (Liao et al., 2017).

In the hybrid *N. gossei* × *N. tabacum,* the incongruity between the two species was overcome by inducing structural changes in the parental chromosomes through the exposure of the paternal pollen to helium ions or gamma rays (Kitamura et al., 2003). Chromosome loss and chromosome rearrangements were observed in rare surviving hybrids, and their survival may have been the result of the deletion or disablement of the gene/s controlling apoptotic seedling death in that hybrid (see also Sect. 3.4.3).

3.4.2 Embryo Rescue by Culturing Ovules

Inviable embryos can be rescued, and hybrid plants can be regenerated by pollinating field- or greenhouse-grown plants *in situ* and culturing the excised fertilized ovules *in vitro*. Such an approach was adopted to overcome incongruity between *N. tabacum* and, e.g., *N. plumbaginifolia*, *N. sanderae*, *N. acuminata*, *N. nesophila*, and *N. stocktonii* (Nikova et al., 2006; Iwai et al., 1986; Reed & Collins, 1978).

Alternatively, *in vitro* cultured ovules can also be pollinated *in vitro*. Pollinating the cultured ovaries of *N. tabacum* with the pollen of *N. rustica* Marubashi & Nakajima, 1985 rescued the lethal hybrid *N. tabacum* × *N. rustica*. Liu et al. (2017) used this approach to obtain progeny from backcrossing the amphihaploid hybrid *N. tabacum* x *N. tomentosiformis* to *N. tabacum* used as the male parent.

In another successful embryo rescue experiment, the ovaries of *N. repanda* were X-irradiated prior to fertilization with pollen of *N. tabacum* (Shintaku et al., 1985). The two surviving hybrid plants were similar to *N. tabacum* in habit and growth type, but their flower morphology resembled that of *N. repanda*. Both showed chromosome deficiency (for 1 and 2 chromosomes, respectively) as judged by the expected amphihaploid chromosome number. This prompted the authors to surmise that in those plants, the lethality factor/s were eliminated with the missing chromosome/s. A similar approach based on partial chromosome elimination in the male parent through the use of irradiation was used by Shizukuda et al. (1983) to obtain "partial" or asymmetric hybrids of *N. tabacum* x *N. rustica*.

More examples of culturing fertilized ovules as a means to raise other incongruous interspecific hybrids to maturity are given in Table 3.4. Where embryo abortion is the result of a malfunctioning endosperm, the beneficial effect of aseptic culture is that it substitutes for the endosperm in providing nutrients to the embryo, thus assuring its continued development (Reed & Collins, 1980a).

Table 3.4 In vitro cultures used to overcome the death of embryos and juvenile plants in incongruous hybrids of *N. tabacum* with other *Nicotiana* species

Section	Species involved in the hybrid with *N. tabacum* [1]	Manifestation of incongruity	Method adopted	Reported by
Alatae	*N. alata* (M)	LS	OC *in vitro*	Papadopoulou et al. (1997)
	N. longiflora (M)	LS	OC *in situ*	Venkateswarlu et al. (1998)
	N. plumbaginifolia (F)	IS/ LS	SC OC *in situ*	Nikova et al. (2004) Nikova et al. (2006)
	N. plumbaginifolia (M)		OC *in vitro*	Prasad et al. (1985)
	N. sanderae (M)	IS	OC *in situ*	Nikova et al. (2003, 2006)
Paniculatae	*N. knightiana* (M)	NS	OC *in vitro*	Slusarkiewicz-Jarzina and Zenkteler (1983)
Petunioides	*N. acuminata* (M)	IS (embryo abortion)	OC *in situ*	Iwai et al. (1986)
Repandae	*N. nesophila* (F)	IS (embryo abortion)	OC *in situ*	Reed and Collins (1978)[1,2], Evans et al. (1982)
	N. nudicaulis (M)	LS	CC	Yamada et al. (1999)
	N. repanda (F)	LS type II	CL OC *in situ* OC *in vitro* + OI	Iwai et al. (1985) Iwai et al. (1985) Shintaku et al. (1985, 1986), Pontes et al. (2005)
	N. repanda (M)	IS, LS	OC *in vitro*, OC *in vitro* followed by treatment by IAA of germinating seedlings Culture of ovules pollinated in vitro + addition of IAA to potting medium to prevent dieback at later growth stages	Choi et al. (1998) Zhou et al. (1991)
	N. stocktonii (F)	IS (embryo abortion)	OC *in situ*	Reed and Collins (1978)[1,2], Muraida and Marubashi (2015)

(continued)

Table 3.4 (continued)

Section	Species involved in the hybrid with *N. tabacum* [1]	Manifestation of incongruity	Method adopted	Reported by
Rusticae	*N. rustica* (F)	NS	OC *in situ* OC *in vitro* by X-irradiated pollen OC *in vitro*	Douglas et al. (1983), Choi and Lee (1991), Sarala et al. (2023) Shizukuda and Nakajima (1982), Shizukuda et al. (1983) Choi and Hong (1992)
	N. rustica (M)	NS	OC *in vitro*	Marubashi and Nakajima (1985)
Suaveolentes	*N. africana* (F) *N. africana* (M) *N. africana* (M)	LS type II	GSC CC OC *in situ*	Nikova et al. (1988), Nikova and Zagorska (1990), Doroszewska and Berbeć (1990) Tezuka et al. (2010)
	N. amplexicaulis (M) *N. amplexicaulis* (F)	NS ?	OC *in vitro* OC *in situ*	Larkina (1980), DeVerna et al. (1987), Venkateswarlu et al. (1998), Sarala et al. (2023)
	N. benthamiana (M) *N. benthamiana* (F)	NS IS	OC *in situ*	Subhashini et al. (1986), DeVerna et al. (1987), Nikova et al. (2006)
	N. cavicola (F)	LS type II	OC *in situ*	Nikova et al. (2006)
	N. excelsior (M)	LS type II	OC *in situ*	Tezuka et al. (2010)
	N. debneyi (M)	NS	OC *in vitro* followed by CC OC *in situ*, OC *in vitro*	Butenko et al. (1970), Ternovsky et al., 1976, Tezuka et al. (2007)
	N. debneyi (M) *N. debneyi* (F)	LS	OC *in vitro* OC *in vitro* followed by CC OC *in situ*	Ternovsky and Larkina (1978b), Larkina (2015, 2017) Tezuka and Marubashi (2006b), Tezuka et al. (2010) Sarala et al. (2023)
	N. goodspeedii (F)	NS	OC *in vitro* followed by CC	Tezuka et al. (2010)

	N. gossei (F)	LS type II	OC *in situ* CC	Adachi and Inoue (1995), Sarala et al. (2023) Mino et al. (2002)
	N. ingulba	LS ? IS LS	GSC OC *in situ* OC *in vitro*	Butenko et al. (1970) Nikova, Palakarcheva, et al. (1998a) Tezuka et al. (2012)
	N. maritima (M)	NS	OC *in vitro*	Tezuka et al. (2010)
	N. megalosiphon (M) (F)	NS LS type II	OC *in situ* GSC OC *in situ*	Tezuka et al. (2010) García Cruz et al. (2008) Sarala et al. (2023)
	N. occidentalis (F)	LS type II	GSC	Butenko et al., 1970, Ternovsky et al. (1972), Semenova and Ivanova (1973)
	N. rosulata (M)	NS	OC *in vitro*	Ternovsky et al. (1976), Ternovsky and Larkina (1978a), Larkina (1980, 2015, 2017)
	N. suaveolens (M)	NS	OC *in vitro*	Marubashi and Onosato (2002), Tezuka and Marubashi (2004, 2006b)
	N. suaveolens (F)	LS	GSC, ISC	Lloyd (1975), Shinkareva (1979), Inoue et al. (1994)
	N. suaveolens (F)	LS type II	ISC in liquid media	Inoue et al. (1997)
	N. umbratica (M)	?	OC *in situ*	Sarala et al. (2023)
	N. velutina (F) *N. velutina* (M)	NS	OC *in vitro* OC *in vitro*	Tezuka et al. (2010)
	N. wuttkei (F)	LS type II	CC	Laskowska and Berbeć (2012)
Trigonophyllae	*N. obtusifolia* (F)	NS, IS	OC *in situ*	Chung et al. (1988), Choi and Lee (1991), Liu and Marubashi (2014)

Abbreviations used: *LS* lethal seedlings, *LS type II* seedling lethality type II, *IS* inviable seeds, *NS* no seeds, *OC in situ* culture of excised ovules from flowers pollinated in situ, *OC in vitro* culture of excised ovules pollinated in vitro, *CC* culture of cotyledons, *CL* culture of true leaf explants, *OI* irradiation of ovules, *GSC* culture of germinating seeds, *ISC* culture of intact seedlings; Footnotes marked by superscript numerals (arranged columnwise and downwards within columns); parent of the non-*tabacum* species in the hybrid: *F*= female, *M*=male

[1] Fertile amphidiploids recovered from culture

[2] Followed by cotyledon culture of ovule culture-raised seedlings

3.4.3 Managing Lethality of Juvenile Plants

Overcoming Hybrid Lethality Through Natural Processes Up to the mid-sixties of the last century and in some laboratories even much later, the *Nicotiana* investigators and breeders tried to manage the lethality of hybrid seedlings by the simple expedient of pollinating as many plants and flowers and sowing as many seeds as possible in the hope of obtaining rare phenotypes that would be able to survive to maturity. In quite a few cases, the policy bore fruit (e.g., Tsikov, 1966; Clayton et al., 1967; Gerstel et al., 1979; Berbeć & Doroszewska, 1981; Hancock et al., 2015). Since one of the major obstacles in obtaining hybrid seeds was premature flower drop after pollination, Burk and Chaplin (1979) gave a general recommendation to apply indole acetic acid (IAA) to the pedicels of pollinated flowers to prevent abscission. To the same end, Wark (1970) applied IAA to calyx sectors of maternal flowers. Recently, He et al. (2022), in a study on the causes of hybrid lethality in *Nicotiana,* confirmed the effectiveness of IAA in suppressing abscission in the interspecific cross *N. suaveolens* x *N. tabacum*, although the treatment *per se* had no effect on the lethality of hybrid seeds.

At least in the case of *N. tabacum* × *N. africana,* the survival of rare hybrid phenotypes was found to be related to the somatic instability of the hybrid embryos. As a result of chromosome loss or fragmentation, the specific loci contributed by *N. africana* and/or by *N. tabacum* and responsible for the development of apoptotic plant death syndrome in hybrid seedlings were eliminated, and rare hybrid genotypes deficient for lethality factors could be grown to maturity (Hancock et al., 2015). In another hybrid that showed PCD syndrome, *N. amplexicaulis* × *N. tabacum* (Berbeć & Doroszewska, 1981), different chromosomal constitutions were found among rare surviving hybrids, including an aneuploid deficient for 3 chromosomes, 6 aneuploids with two to four supernumerary chromosomes 12, and apparently regular amphihaploids ($2x = 42$) plus one apparent amphidiploid ($4x = 84$).

Overcoming Hybrid Lethality Through *In Vitro* Cultures The methods to overcome hybrid unviability that depended on fortuitous processes were inefficient, time and labor consuming and final success was a matter of luck. For plant breeders, another trade-off of this approach is the possibility that the surviving aneuploids may be deficient for the gene/genes of their specific interest.

Butenko and Luneva (1966) were probably the first to overcome hybrid lethality in *Nicotiana* by resorting to *in vitro* culture. They placed aseptically produced seeds of the nonsurviving hybrid *N. alata* × *N. glauca* on a nutrient medium supplemented with kinetin and produced viable hybrid plants. With time, different variants of *in vitro* cultures on solidified media to rescue dying hybrid seedlings became a widely used practice. A compilation of cases where interspecific hybrids involving *N. tabacum* were obtained by resorting to *in vitro* culture, regardless of method applied and cause of incongruity, is presented on a species-by-species basis in Table 3.4. By comparing Tables 3.4 and 3.10, it is notable that the majority of the hybrids listed in Table 3.4 were also obtained by earlier hybridizers in the natural way by direct pollination without the aid of *in vitro* techniques. Nonetheless, the gains of saved time and labor resulting from the improved seedling survival rate (Tezuka et al., 2010) caused the traditional approaches to have been all but displaced by aseptic cultures.

The role of tissue culture in overcoming lethality of *Nicotiana* hybrids is still not wholly understood, and the extent to which particular mechanisms are involved may differ among individual cases. The process of organ culture and plant regeneration is known to exacerbate somaclonal variation. According to Tezuka et al. (2012), the deletion of a chromosome or a chromosome fragment may lead to the elimination of the causative agent for interspecific lethality in hybrids rescued by tissue culture. The validity of that supposition was confirmed by Nakata et al. (2021). The authors demonstrated that the cultured hybrid seedlings of *N. suaveolens* x *N. tabacum* that had overcome lethality lacked the distal part of chromosome Q of *N. tabacum* that carries the Nt6549g30 gene responsible for the development of lethality syndrome (Ma et al., 2020). Such spontaneous chromosomal alterations that led to the cancellation of the PCD syndrome were convincingly demonstrated for the hybrid *N. tabacum* × *N. africana* by Hancock et al. (2015), see also Sect. 3.2.3). The preponderant class of lethal *N. africana* × *N. tabacum* hybrid plants obtained by Nikova and Zagorska (1990) that were rescued by tissue culture had fewer (44) mitotic chromosomes than the theoretically expected number (47). On the other hand, the *N. tabacum* × *N. africana* hybrid plants recovered from cotyledon culture by Doroszewska and Berbeć (1996) all had a regular number of 47 chromosomes. In yet another study of the same hybrid (Depta & Doroszewska, 2019), hybrid plants regenerated from cultured cotyledons were also classified as amphihaploids. In the latter case, flow cytometry was used to determine the ploidy level of viable regenerants. However, this relatively fast and convenient method lacks the resolution power to discriminate at the single chromosome level, as can be clearly seen from the data supplied by Hancock et al. (2015). Nonetheless, the chromosome counts of one of the first lethal hybrids rescued by cotyledon culture, *N. suaveolens* × *N. tabacum* (Lloyd, 1975), also revealed a regular amphihaploid number of 40.

Culture conditions and medium components may induce mutations that result in disabling the function of the lethality genes in otherwise chromosomally regular hybrid genotypes (Tezuka, 2012). What all those and some other tissue-culture rescued hybrids (Ternovsky et al., 1972, 1976; Yamada et al., 1999) had in common was that they were cultured and regenerated on media that included cytokinins in their composition. Inoue et al. (1994, 1997) demonstrated that the addition of cytokinins to the culture medium had a decisive effect on the survival of the lethal hybrid plants of *N. suaveolens* × *N. tabacum,* but the efficacy of the treatment varied with the type and concentration of the cytokinin used. Nakata et al. (2021) surmised that the increased rate of survival of *N. suaveolens* × *N. tabacum* hybrid seedlings on cytokinin-enriched medium was due to the presence of cytokinin-induced reactive oxygen species (ROS) in the studied hybrid explants. Since ROSs are known to cause chromosome breakage, they may have induced the observed fracture and loss of the distal part of chromosome Q that carried the Nt6549g30 gene essential for triggering the plant death response. This mechanism was assumed to account for the restoration of viability to at least a large part of the potentially apoptotic hybrid plants of *N. suaveolens* x *N. tabacum*. Hancock et al. (2015) also linked the rare survival of *N. tabacum* x *N. africana* seedlings grown *in vivo* to chromosome instability and to the deletion of the Nt6549g30 gene, but they associated the phenomenon with the innate behavior of chromosome H in that particular hybrid milieu rather than with the influence of any external factors (see, however, section 3.3.4 for comments on the identity of chromosome H and chromosome Q).

Temperature may be another factor related to overcoming hybrid lethality by tissue culture. High temperature-dependent suppression of hybrid lethality of types II and V is known to become ineffective once the hybrid plant is returned to normal growth conditions (Tezuka, 2012). However, in many cases, the temperatures prevailing inside culture containers are likely to be above the critical threshold of ca. 30 °C and thus may facilitate the operation of other, not temperature-related, factors in the *in vitro* environment. Incidentally, the survival *in vivo* of hybrid seedlings of *N. amplexicaulis* × *N. tabacum* was substantially improved when seed germination and seedling culture were moved from the greenhouse to a growth chamber fitted with fluorescent tubes that kept the ambient temperatures ca. 30 °C, and the process was conducted in covered glass containers (Berbeć & Doroszewska, 1981).

Shiragaki et al. (2020) reported that treatment of juvenile plants of the hybrid *N. suaveolens* x *N. tabacum* with L-2-aminooxy-3-phenylpropionic acid (AOPP) suppressed the development of type II lethality by inhibiting the phenylalanine ammonia-lyase (PAL) responsible for the production of phenolic compounds involved in seedling death.

Overcoming Hybrid Lethality Through the Use of *N. tabacum* Monosomics Burk (1972) was the first to resort to a monosomic parent to overcome hybrid lethality in an interspecific hybrid of *Nicotiana*. By crossing *N. langsdorffii* as the male parent to each of the 24 monosomies of *N. tabacum* as females, he was able to produce several viable hybrid plants, most of which had been obtained from Haplo A plus a few single survivors from Haplos G, L, and Z. A few years later, Gerstel et al. (1979) demonstrated the involvement of the *N. tabacum* H chromosome in the lethality of the hybrid N. *africana* x *N. tabacum* by producing viable and inviable plants in the progeny of *N. africana* x Haplo H. Tezuka et al. (2010) repeated the success of Gerstel et al. (1979) with the hybrid N. *afric*ana x *N. tabacum* and with hybrids of *N. tabacum* with several other species of the section Suaveolentes, reporting Haplo Q as the monosomic *N. tabacum* parent (see, however, Sect. 6.4.3 on the identity of chromosomes H and Q).

Even though the monosomic method seems to be easy and dependable for the production of viable hybrids, especially those exhibiting type II lethality, few, if any, researchers, apart from those cited above, have tried this approach. Most of those who did had goals in mind other than bypassing lethality obstacles. The most likely reason is that the monosomic stocks are relatively hard to access from external sources, and their in-house development and perpetuation are both costly and troublesome.

Overcoming Hybrid Lethality Through Gene Editing Ma et al. (2020) identified the gene of *N. tabacum* that controls the seedling death of *N. tabacum* x *N. africana* hybrid plants (see also Sect. 3.3.3). As the confirmation part of the study, the authors used CRISPR–Cas9 gene-editing technology to insert frameshift mutations within the *Nt6549g30* region. A transformant homozygous for a mutation of the *Nt6549g30* gene was found that conditioned nearly 100% survival of the hybrid seedlings that resulted from pollinating the engineered *N. tabacum* by *N. africana*. Thus, it was demonstrated that the mutation disabled the lethality-conferring function of *Nt6549g30,* thereby reconfirming the identity of *Nt6549g30* as the gene controlling hybrid lethality in *N. tabacum* x *N. africana*.

In yet another part of the same study (Ma et al., 2020), several crosses were made of the mutant line of *N. tabacum* with some other species of the section Suaveolentes (*N. amplexicaulis*, *N. debneyi*, *N. excelsior*, *N. gossei*, *N. megalosiphon*, *N. occidentalis*, *N. simulans*, *N. umbratica*, *N. velutina*) to determine whether the same factor was responsible for seedling death of those hybrids that previously had been known to exhibit type II seedling lethality (Tezuka, 2012). The viability of all but one of those hybrids was restored. The exception was the hybrid from crossing *N. occidentalis* with the Nt6549g30 mutant of *N. tabacum*. The results were consistent with lethality types exhibited by the studied hybrids. While all the surviving hybrids represented type II lethality, the hybrid *N. occidentalis* x *N. tabacum* that failed to respond to the Nt6549g30-mutant gene had been classified as type V (Tezuka, 2012). Obviously, a different mechanism of hybrid lethality is involved in the latter hybrid. The authors of the study (Ma et al., 2020) pointed out that their novel approach may expand the gene pool available for tobacco breeding. The current practical disadvantage of the method is that the CRISPR–Cas technology involves what in many countries is termed genetic modification and as such is subject to legal restrictions.

3.5 Maternal and Paternal Plants in the Offspring of Interspecific Crosses in *Nicotiana*

3.5.1 *Gynogenic and Androgenetic Haploids*

Both gynogenic and androgenetic haploids were reported in the offspring of interspecific crosses in *Nicotiana*. Gynogenic (maternal) haploids of several *Nicotiana* species (*N. alata, N. rustica, N. paniculata, N quadrivalvis*) stimulated by the pollen of an alien *Nicotiana* species were probably first reported by Wellington (1913). Single haploid plants of *N. rustica* were found among the progeny from crossing *N. tabacum* as the pollen parent with the former species as the female (Savelli cited by Kostoff, 1943) and with *N. gossei* (Apparao & Ramavarma, 1972). Clausen and

Mann (1924) discovered a haploid plant of *N. tabacum* among the offspring of the cross *N. tabacum* × *N. sylvestris*. Haploids of *N. tabacum* were also reported from crossing *N. tabacum* with *N. alata* and *N. quadrivalvis* (Wellington, 1913), *N. glutinosa* (McCray, 1932), and *N. longiflora* and *N. glauca* (Ternovsky, 1936a, 1936b). Haploid plants were induced in the offspring of *N. tabacum* plants whose stigmata were dusted with X-irradiated pollen of *N. alata* (Tanaka & Kurihara, 1968). The thus obtained gynogenic haploids were compared with those from anther culture (Kumashiro & Oinuma, 1985).

Androgenetic haploids among the offspring of interspecific crosses seem to occur much more rarely than their gynogenic counterparts. A supposedly androgenetic haploid of *N. tabacum* was found in the progeny from crossing maternal amphidiploid 4x (*N. glutinosa* × *N. tabacum*) with male *N. tabacum* (Clausen & Goodspeed, 1925; Clausen & Lammerts, 1929). A single androgenetic haploid of *N. africana* was detected among the offspring from mating *N. tabacum* with genetically transformed *N. africana* (Hancock et al., 2015). The divisions of the sperm of *N. tabacum* in the nucleus of *N. eastii* followed by chromosome doubling in the developing embryo may have accounted for the appearance of a cytoplasmically male sterile *N. tabacum* plant as a result of crossing maternal *N. eastii* with *N. tabacum* as the pollen parent (Berbeć & Berbeć, 1992). A case that bears a strong resemblance to the story of cms *eastii* was reported by Nikova et al. (1997). The authors attempted to transfer cms factors from *N. excelsior* to *N. tabacum*. Following the F_1 generation, the transfer seemed to proceed smoothly, but the very nature of the F_1 plants cannot be readily explained based on what the authors themselves reported. In the account by Nikova and her associates, their F_1 plants were similar to the male *N. tabacum* save for flower malformations, and they could be backcrossed with ease to the male recurrent parent. Based solely on morphological and fertility evidence supplied by the authors, it is verging on impossible that those F_1's were regular amphihaploid hybrids and were very unlikely to be spontaneous amphidiploids. Based on what the investigators themselves reported, it can be hypothesized that the male sterile *N. tabacum*-like phenotypes in the F_1 offspring may have originated as a result of unilateral elimination of *N. excelsior* chromosomes from the hybrid zygotes followed by chromosome doubling (compare the reports on gynogenic haploids by Chimoyo and Pupert (1988) and by Hancock et al. (2015) referred to in the paragraph next but one below.

Spontaneous androgenetic haploids in *N. tabacum* were proposed to be a convenient vehicle to transfer cytoplasmic male sterility from one variety to another in a single step by using a rootless mutation to discriminate between rare androgenetic haploids and true intervarietal hybrids (Horlow et al., 1993).

One of the interspecific crosses, *N. tabacum* × *N. africana*, produces maternal haploids very regularly, albeit normally at a very low rate. The phenomenon was deployed to develop a method by which haploid plants of *N. tabacum* can be generated for experimental and breeding purposes (Burk et al., 1979, Nielsen & Collins, 1989, see also Sect. 3.3.1). Gerstel and Wernsman (1979) assumed that gynogenic haploids arise spontaneously during the reproductive process without direct involvement of the pollinating species. The massive death of true interspecific hybrids at the cotyledonary stage gives a selective advantage to rare gynogenic haploids that are otherwise difficult to discern and pick up among regular progeny. One must note, however, that not necessarily all the haploids found in the progenies of mating *N. tabacum* to *N. africana* are purely gynogenic. At least some of them may have arisen as products of selective elimination of *N. africana* chromosomes from originally hybrid embryos (Chimoyo & Pupert, 1988; Hancock et al., 2015).

3.5.2 Maternal Diploids

Various Cases of Maternal Diploid Induction in *Nicotiana* Maternal diploid plants in the offspring of interspecific matings in *Nicotiana* have long been a controversial topic. At first, Goodspeed (1915) dismissed diploid maternals, apomictic or otherwise, as products of experimental errors in disagreement with East (1930) but later abandoned his former view, citing *N. paniculata* as the prime example of maternal diploidy following hybridization with other species (Goodspeed, 1954). Maternal plants continued to be reported as byproducts or even as sole products of mating maternal *N. tabacum* with *N. rustica* (Lehmann, 1936), *N. repanda* (Pittarelli & Stavely, 1975), *N. amplexicaulis* (Berbeć & Doroszewska, 1981), *N. wuttkei* (Laskowska et al., 2015), *N. africana* (Hancock et al., 2015), *N. sylvestris* (Eghis 1930) and various other interspecific matings (Murthy & Subbarao, 2004). Alleged apomictics were also reported in the progeny of *N. alata* and *N. forgetiana* as induced by pollination with X ray-irradiated pollen of *N. langsdorffii* (Pandey, 1974). Maternal plants were reported as the preponderant category of offspring when *N. tabacum* was mated to the amphidiploid 4x (*N. wuttkei* x *N. tabacum*) (Laskowska et al., 2015).

A singular case was reported by Kostoff (1935, 1938a), who described a parthenogenetically produced amphidiploid 4x (*N. glauca* × *N. langsdorffii*) obtained by fertilizing an amphihaploid plant 2x (*N. glauca* × *N. langsdorffii*) with pollen of *N. langsdorffii*. Kostoff explained the phenomenon by the presence of an unreduced monad cell with the doubled chromosome complement of the maternal hybrid that had been stimulated by pollen of *N. langsdorffii* to develop parthenogenetically into an amphidiploid embryo (Kostoff, 1935). Another case of that type that also involved *N. langsdorffii* was the production of amphidiploid seeds by the amphihaploid *N. knightiana x N. tabacum* upon pollination with pollen of F_1 *N. alata x N. langsdorffii* (Berbeć et al., 1982). In the latter case, along with maternal origin, another plausible course of events is the fusion of the unreduced egg cell produced by the amphihaploid with the sperm of *N. alata* x *N. langsdorffii* followed by selective elimination of the alatoid chromosomes during embryogenesis or/and embryo development. The plausibility of such an explanation is further supported by the fact that the same amphihaploid 2n (*N. knightiana* x *N. tabacum*) was prone to spontaneous seed setting (Berbeć et al., 1982). In a related phenomenon, Apparao et al. (1980) observed a massive selective loss of *N. gossei* chromosomes from the hybrid 4n *N. tabacum* × 2n *N. gossei,* resulting in F_1 progeny composed of plants phenotypically close to *N. tabacum*, each of which contained a full diploid complement of the maternal parent and two chromosomes from *N. gossei*.

Induction of Maternal Plants by Pollen of *N. alata* The mating of *N. tabacum* with *N. alata* is noteworthy, as it seems to have yielded maternal phenotypes more frequently than any other interspecific cross that involved *N. tabacum* as the female parent. Both maternal haploids and diploids of *N. tabacum* were induced by X-ray-irradiated pollen of *N. alata* (Tanaka & Kurihara, 1968; Pandey & Phung, 1982; Kumashiro & Oinuma, 1985). Spontaneous diploid maternal phenotypes were observed in the offspring of the cross *N. tabacum* × *N. alata* by Stoyanova (1979), Sarychev (1987), Berbeć (1987b), Naumenko (2012), and Liao et al. (2017). In the study of Stoyanova (1979), diploid maternals were practically the only class of progeny obtained from mating an unspecified Virginia variety of *N. tabacum* to *N. alata*, whereas crosses between other varieties of *N. tabacum* and *N. alata* yielded no maternal plants. In the latter instance, the author herself suspected experimental error, but she did not exclude other causes.

In one of his early experiments with interspecific hybrids, the author of this review studied the hybrid *N. tabacum* x *N. benavidesii* (Berbeć, 1978). In the course of his study, he obtained a number of subsesquidiploids (TTB) with chromosome numbers ranging from 55 to 58. Under open self-pollination, these sesquidiploids did not produce seeds by selfing but could be easily backcrossed to *N. tabacum*. A surprising part of the study was that they could also set seeds when their flowers were pollinated with pollen of *N. alata*. Generally, the offspring from regular backcrosses were morphologically similar to those resulting from pollinating the sesquidiploid with *N. alata*, with no traces of introgression from the pollinator species. Likewise, both classes of offspring showed a similar behavior in meiosis with approximately 24 bivalents and varying numbers of univalents, but the back-cross products involving *N. alata,* univalents were generally higher in number. All these observations pointed to induction of parthenogenetic development of unreduced female gametes of the BC1 plants by pollen of *N. alata* without gametic fusion having actually taken place. Alternatively, fusion of gametes may have taken place, but the *N. alata* chromosomes were selectively eliminated from the developing embryo. However, another plausible but, under these circumstances, not very likely explanation is an experimental error, i.e. inadvertent spontaneous self-pollination in BC_1 plants. Interestingly, in the previously mentioned study on the hybrid *N. tabacum* x N. *alata* by the same author (Berbeć, 1987b), N. *tabacum* (TT) plants well protected against uncontrolled pollen contamination were mated to the sesquidiploid (TTA) as pollen parent. The mating resulted in 36 surviving plants, of which 29 closely resembled the maternal plants and the remaining seven, while departing from the maternal phenotype, did not show any traces of introgression from *N. alata*.

When emasculated but unprotected flowers of *N. tabacum* were fertilized with pollen of *N. alata* (Berbeć, 1987b), some genotypes regularly yielded viable and fertile *N. tabacum*-like phenotypes as the preponderant class of offspring, the rest being inviable or poorly viable hybrid seedlings. When carried out alongside and under the same conditions, interspecific matings of female *N. tabacum* with male *N. langsdorffii* and *N. benavidesii* only occasionally produced surviving maternal plants (unpublished observations of the author of this volume).

A phenomenon very similar to if not identical with those described in the preceding two paragraphs was reported by Naumenko (2010, 2012). Diploid maternals described as 'pseudogamic' and resulting from fertilizing intraspecific F_1 hybrids of *N. tabacum* with pollen of *N. alata* could be generated regularly and were not much different from the regular selfed progeny of those plants. Apomictic populations from crossing F_1 intervarietal hybrid plants showed segregation for plant height, leaf number and leaf size that was characteristic of the F_2 pulations obtained from the same plants by selfing. According to Naumenko, the difference between the alleged apomictics and the regular hybrids was that the former became stabilized already in the F_2 generation whereas the regular hybrids kept segregating in subsequent generations. This description recalls the behaviour of populations derived from gynogenic or androgenetic doubled haploid hybrid plants. In this particular instance, *N. alata* sperms may have induced chromosome doubling and embryo development in the maternal plants without actual gamete fusion. Alternatively, gamete fusion may have been followed by massive elimination of *N. alata* chromosomes and chromosome doubling, whichever is more likely. This way or the other, the most puzzling feature of Naumenko's account is the massive and regular incidence of the described phenomenon, the fact also reported earlier by Sarychev (1987).

Liao et al. (2017) recovered self-fertile maternal phenotypes along with apparently regular interspecific hybrids by fertilizing senescent flowers of cytoplasmically male sterile *N. tabacum* with pollen of *N. alata*. The puzzling part of the latter report was that the *N. tabacum*-like plants showed restored normal stamen morphology and self-fertility, and according to flow cytometry measurements, their genome was considerably smaller than that of their maternal parent. According to Liao et al. (2017), the maternal phenotypes in the offspring were actually hybrids rather than true maternals since *N. alata*-specific fragments, including the putative male fertility-restoring genes, were amplified in them with selected SSR probes. The two accounts, by Naumenko (2012) and by Liao et al. (2017), bear intriguing similarities, but Naumenko obviously used male fertile maternal plants in her study since she propagated her apomictics by selfing.

Some of the results described in this section are loosely reminiscent of the experiments reported by Pandey on what he called egg transformation without gametic fusion (Pandey, 1974, 1975, Pandey & Phung, 1982, see Sect. 4.4.4).

Overall, due to its elusive and inconsistent nature as well as different manifestations, the phenomenon of maternal genotypes in the progeny of some interspecific crosses is not readily amenable to systematic study and, therefore, is habitually dismissed as resulting from inadequate protection against inadvertent pollen contamination. This notwithstanding, maybe it deserves more attention than it has hitherto received.

3.6 Sterility of Interspecific Hybrids

3.6.1 Causes of Sterility in Amphihaploid Hybrids of Nicotiana

In *Nicotiana*, once all previously discussed pre- and postfertilization barriers have been successfully overcome or circumvented and the hybrid plant has been brought to flowering, it can be reasonably expected to be completely sterile. Actually, the very first interspecific *Nicotiana* hybrid ever reported, *N. paniculata* x *N. rustica*, was described by its creator J. G. Koelreuter as sterile, which was in stark contrast to the full fertility of the parental forms (Mayr, 1986).

In tobacco, just as in other organisms that perpetuate themselves by sexual reproduction, the fusion of two viable gametes, male and female, is a crucial event in the reproductive process. An orderly and undisturbed reductional division in meiocytes whereby the number of somatic chromosomes of an individual is reduced by half is the key element in assuring the formation of viable and functional gametes. In interspecific hybrids, this regular process can be upset because the chromosomes contributed by the two parental species are structurally different, which prevents their normal pairing, and/or the numbers of chromosomes are different, which leaves some of the chromosomes without a partner with which to pair.

The genus *Nicotiana* is unique among other plant genera in that its species show a high degree of cross-compatibility (approximately 450 hybrid combinations have been produced thus far (see Chap. 7 of this review), and at the same time, the vast majority of those hybrids are practically self- and cross-sterile, i.e., they are not capable of producing seeds either if pollinated by their own pollen or by the pollen of another species. In other genera, approximately 75% of successful interspecific hybrids show at least some degree of self-fertility (Stebbins, 1950). According to Goodspeed (1954), in the genus *Nicotiana,* chromosomal rearrangements and other structural changes accumulated faster than barriers to hybridization. This leads to deranged micro- and macrosporogenesis in hybrids, resulting in chromosomally imbalanced and thus inviable or dysfunctional gametes and, consequently, aborted or inviable ovules and pollen. Some intrasectional hybrids, e.g., in the sections Alatae, Trigonophyllae, Tomentosae and Suaveolentes, with equal numbers of chromosomes and chromosome homology high enough to ensure normal pairing regularly produce selfed offspring and can be backcrossed to their parental species. A puzzling exception to this rule was the intersectional hybrid between *N. alata* (section Alatae, chromosome number $n = 9$) and *N. amplexicaulis* (section Suaveolentes, chromosome number $n = 18$) reported by Gopinath et al. (1970). Although difficult to produce, once obtained, the amphihaploid hybrid (27 somatic chromosomes) was reportedly self- and cross-fertile. Unfortunately, no one else is known to have reproduced and studied that hybrid.

All known interspecific hybrids involving the cultivated tobacco *N. tabacum* with their number approaching 60 are sterile, although those originating from crossing it with its direct ancestors or their close relatives occasionally yield viable offspring when backcrossed to the *N. tabacum* parent (Clausen & Cameron, 1944, 1957). Upon backcrossing to *N. tabacum*, the hybrid *N. tabacum* × *N. tomentosiformis* produced offspring among which fertile 48-chromosome *N. tabacum*-like segregants were found (Brieger (1928). This is not surprising bearing in mind the 'Drosera' chromosome pairing in the hybrid parent of the cross (compare Sect. 3.3.4). Other exceptions and reservations concerning the sterility of interspecific hybrids with *N. tabacum* are indicated in subsequent sections. The sterility of an interspecific hybrid makes it of little use, especially if further generations are considered, e.g. for breeding purposes.

3.6.2 Bioconfinement

In some cases, hybrid sterility may be of potential advantage. Chambers et al. (2011), Ling et al. (2012), and Rice et al. (2013) considered the applicability of the hybrid *N. tabacum* × *N. glauca* for transgene-controlled production of pharmaceuticals because of the hybrid's capacity for biomass production and its other merits. They pointed to the sterility of the hybrid as the safeguard against unintended escape of the transgene to the environment, although they were aware that the sterility of the hybrid was not complete (Chambers et al., 2011; Rice et al., 2013), an issue that will be briefly discussed in the next paragraph of this section. The bioconfinement effect of hybrid sterility was also established for two closely related species, *Nicotiana tabacum* and *N. sylvestris* (Ahl-Ahmad et al., 2006). The hybrid of the two species was reported to produce no offspring either by selfing or by backcrossing to *N. sylvestris*. The authors concluded that the cultivation of a transgene-carrying *N. tabacum* in close proximity to native or ornamental *N. sylvestris* poses no significant risk of the unintended release of some novel genetically modified genes to the environment. Because of the high degree of self- and cross sterility plus the added benefits of perennial growth and the ease of clonal propagation, Lim et al. (2006) envisaged the usefulness of genetically modified first-generation 'synthetic tobacco' (4x (*N. sylvestris* × N. *tomentosiformis*)) for biopharmacy. However, the latter hybrid, although practically sterile, may not be entirely gene escape-proof (see Chap. 2), especially when commercial tobacco is grown nearby. Cost-effective production of seeding material may also be an issue.

3.6.3 Conversion of Sterile Amphihaploids to Fertile Alloploids by Making Use of Natural Processes in Hybrid Plants

Partial or Occasional Self- and Cross-Fertility of Otherwise Sterile Allohaploid Hybrids In most intended or actually implemented practical uses of interspecific *Nicotiana* hybrids, their sterility is a liability rather than an asset. The desired goal has been a hybrid that can be sexually perpetuated by selfing and, preferably, one that is also able to produce offspring upon backcrossing to the cultivated species. In their efforts to restore fertility to their interspecific hybrids, the early breeders were, consciously or otherwise, heavily dependent on the production of restitution gametes by their experimental materials. They thus made use of the fact that during the aberrant events of gametogenesis, one or both reductional divisions may fail, resulting in the formation of unreduced gametes that are usually chromosomally balanced and viable, as they retain the genomic integrity of their parental plants. The rate of formation of such restitution gametes may vary from negligible to quite substantial depending on parental genotypes, environment, plant age, etc. (Kostoff, 1943, Goodspeed, 1954, Doroszewska & Berbeć, 1996).

Kostoff (1936, 1938a) heavily relied on a high rate of restitution gametes in his attempt to develop an amphidiploid N. *sylvestris* x *N. tomentosiformis* (SSTT) by crossing the sterile F1 hybrid (ST) first to *N. sylvestris* and the resulting SST hybrid to N. *tomentosiformis*, the form that came to be known as "Kostoff's hybrid" (see also Sect. 2.2.2). Kostoff (1943) observed that the production of restitution nuclei by *N. sylvestris* × *N. tomentosiformis* was further intensified when the hybrid was exposed to high temperatures accompanied by restricted water supply. Similarly, Rybin (1927) and Eghis (1927) crossed a tetraploid variant of *N. tabacum* (TTTT) with diploid *N. rustica* (RR). Upon backcrossing the resulting sesquidiploid (TTR) to the diploid parent (RR), they obtained occasional plants that had the amphidiploid (TTRR) genomic constitution. The above procedure did not always prove successful. In an attempt to produce amphidiploid 4x (*N. tabacum* x *N. benavidesii*), the author of this review backcrossed the sesquidiploid *N. tabacum-N. tabacum-N. benavidesii* to *N. benavidesii*. All but one of the offspring thus obtained were poorly viable amphihaploids with extra univalents, probably from *N. benavidesii,* and the remaining single plant was an unstable 68-chromosome subamphidiploid (Berbeć, 1978).

In a sterile amphihaploid, if an unreduced female gamete fuses with its unreduced male counterpart, a fertile amphidiploid may be produced by selfing. The best illustration of this fortuitous process is the case of the first artificial fertile *Nicotiana* amphidiploid ever made. After many years of unsuccessful attempts to self-pollinate the sterile hybrid 2n (*N. glutinosa* × *N. tabacum*), Clausen and Goodspeed (1925) obtained three seeds, out of which one germinated and grew to an amphidiploid. A few years later, that amphidiploid was used by Holmes (1938) to transfer resistance to tobacco mosaic virus (TMV) from *N. glutinosa* to *N. tabacum* (see Sect. 4.1). That resistance continues to be deployed in contemporary cultivars and thus, at least in several cases, its origin can be traced back to those remote but fateful events recorded by Clausen and Holmes many decades ago. There are several cases on record in which otherwise sterile hybrids of different *Nicotiana* species with *N. tabacum* yielded viable offspring through assiduous effort of selfing the amphihaploid plants. The key to success consisted in self-pollinating on a scale large enough to increase the small odds of an unreduced male and female gamete fusing together into an allopolyploid zygote. In exceptional cases, amphidiploids or sesquidiploids can be produced directly from mating two diploid species. Such a chance fusion of two restitution gametes produced by two parental species may give rise to a direct allopolyploid hybrid. Ternovsky (1962) reported spontaneous amphidiploids directly from the crosses *N. glutinosa* × *N. tabacum*, *N. tabacum* × *N. glauca* and *N. tabacum* × *N. sylvestris*. Another case of this kind is the appearance of the sesquidiploid plant in the hybrid progeny of the cross *N. tabacum* × *N. sylvestris* that appears to have arisen from the union of diploid restitution gametes of *N. tabacum* with a normal haploid gamete of *N. sylvestris* (Webber, 1930). If an amphihaploid was backcrossed to the *N. tabacum* parent, partly fertile sesquidiploids or near sesquidiploid plants were the usual outcome. Clayton (1954) observed a high rate of restitution gametes in the hybrid *N. debneyi* × *N. tabacum,* resulting in partially restored self-fertility and the production of allopolyploid progeny.

Low temperatures and excess moisture led to partially restored fertility in the hybrids *N. debneyi* × *N. tabacum* and *N. tabacum* × *N. glauca* (communicated to the author of this review by J. Berbeć). The production of restitution nuclei and viable pollen grains increased in the hybrid 2x (*N. tabacum* × *N. africana*) as the plants grew older (Doroszewska & Berbeć, 1996). In the latter case, the hybrid plants failed to yield any spontaneous seeds, notwirhstanding.

In some other reports (Berbeć & Opoka, 1966; Berbeć, 1971), the hybrid *N. tabacum* × *N. glauca* and its reciprocal retained some vestigial self-fertility, and the two hybrids could also be used both as pollen and egg parents to produce backcross offspring with *N. tabacum* from which further selfed generations were obtained. This simple expedient to overcome the sterility barrier by backcrossing the sterile hybrid to one of the parents was probably fairly frequently resorted to and in several cases proved successful (East, 1928; Ternovsky, 1936a, 1936b; Burk, 1967; Wichert-Kobus, 1967, 1971; Berbeć, 1980; Berbeć et al., 1982; Nikova et al., 1997).

Aneuploid plants having from 28 to 34 chromosomes obtained by culturing the anthers of the sterile hybrid *N. tabacum* × *N. sylvestris* probably also arose from the restitution gametes produced by that hybrid (Takahashi, 1973). Another plausible mechanism was the production of partly functional generative nuclei by the hybrid as a result of the 'Drosera' pairing process.

The citations of cases where spontaneous amphidiploids or sesquidiploids involving *N. tabacum* were produced are given in Table 3.5 on a species-by-species basis. The allopolyploids were obtained either directly from crossing diploid parental species or by selfing or cross-pollination of amphihaploid hybrids.

Use of Autotetraploid Forms of One or Both Parental Species Sterility barriers can also be circumvented if one or both parental species are used in the autotetraploid form. Clayton (1947) crossed autotetraploid *N. tabacum* with autotetraploid *N. longiflora* and obtained a partially fertile amphidiploid that was used in the interspecific transfer of a disease resistance factor. Similarly, an autotetraploid accession of *N. knightiana* was mated as the female to an autotetraploid variant of *N. tabacum* ($4n = 96$) to produce a fertile amphidiploid 4x (*N. knightiana* × *N. tabacum*) (Berbeć & Doroszewska, 1992). After a series of successful backcrosses to *N. tabacum* as the recurrent male parent, an alloplasmic lineage

Table 3.5 Spontaneous production of allopolyploid offspring by direct crosses of *N. tabacum* with an alien species or by amphihaploid hybrids involving *N. tabacum* as a result of self-fertilization (S) or cross-pollination (CP) with the *N. tabacum* parent

Species involved in the hybrid with *N. tabacum*	Cytological status of offspring	Fertilisation	Author
N. alata	TTAA	S	Ternovsky (1936), Stoyanova, 1979
	TTA	S, C_{TxA}	Stoyanova (1978, 1979)
	TTSanSan	S	Ternovsky (1936)
N. sanderae	TTSan	C_{TxA}	Ternovsky (1936, 1962)
	TTGlaGla	C_{TxA}	Ternovsky (1936)
N. glauca	TTGla	S, CP	Ternovsky, 1936a, 1936b, Stoyanova and Konotop (1975)
	GlaGlaTT, GlaTT	S, CP	Berbeć and Opoka (1966), Berbeć (1971)
N. rustica	TTR	S	Ternovsky (1936a, 1936b), Zhukov (1939)
	TTRR	CP	Eghis (1927)[2]
N. benavidesii	TTB	CP	Berbeć (1978, 1980)
N. knightiana	KKTT	S, CP	Berbeć et al. (1982)[3]
N. paniculata	PTT (?)	CP	Holmes (1937a, b)
N. glutinosa	GluGluTT	S, C_{TxA}	Clausen and Goodspeed (1925), Ternovsky (1962)
	TGluGlu	C_{TxA}	Ternovsky (1962)
N. sylvestris	TTSS	C_{TxA}	Ternovsky (1962)
	TTS	CP	East (1928)
	TTS (?)[1]	CP	Burk (1967)
N. amplexicaulis	AmAmTT AmTT	C_{AxT} CP	Berbeć and Doroszewska (1981) Nikova et al. (1997)
N. debneyi	DebDebTT (?)	S	Clayton (1950, 1954), Berbeć (1964)
N. exigua	ETT	CP, S	Wichert-Kobus (1971)
N. goodspeedii	GGTT	S	Palakarcheva et al. (1978)
N. megalosiphon	MMT (?)	CP	García Cruz et al. (2008)[4]
N. setchellii x *N. otophora*	TTSeO	C	Berbeć et al. (1982)
N. tomentosa	TTTom	CP	East (1928)
N. tomentosiformis	TTTmf	CP	East (1928)

Abbreviations and symbols used: haploid genomes: *T N. tabacum*, *A N. alata*, *Am N. amplexicaulis*, *San N. sanderae*, *Gla N. glauca*, *R N. rustica*, *B N. benavidesii*, *K N. knightiana*, *Glu N. glutinosa*, *S N. sylvestris*, *E N. exigua*, *G N. goodspeedii*, *M N. megalosiphon*, *Tom N. tomentosa*, *Tmf N. tomentosiformis*; *Se N.* setchellii; *O N.* otophora; TxA or AxT typed in subscript stand for T (*N.tabacum*) and A (alien species) and indicate the direction of crossing

[1]Allo-aneuploids of *N. sylvestris* x *N. tabacum* possessing full haploid genomes from both species plus unspecified number of extra chromosomes from *N. repanda* were backcrossed to *N. tabacum* (for more details see Sect. 4.4.1);

[2] Amphidiploids TTRR were obtained by crossing a sesquidiploid TTR with diploid N. tabacum;

[3] amphidiploids KKTT were obtained by spontaneous seed set by the amphihaploid KT and by induction of amphidiploid seeds by the amphihaploid KT with pollen of *N. alata* x *N. langsdorffii*;

[4]cytological status of the backcross hybrids not reported

N. tabacum cms knightiana was developed that showed vestigial self-fertility. The negative aspect of this "tetraploid-tetraploid" approach is that autotetraploids usually show a certain percentage of imbalanced gametes due to frequently occurring polyvalent associations in meiosis. The resulting offspring of such tetra-tetra crosses are thus very likely to contain aneuploids along with regular amphidiploids and may happen to be deficient for a desired genetic factor from the wild species.

More frequently, autotetraploid variants of *N. tabacum* were crossed with diploid *Nicotiana* species to produce sesquidiploids. Since sesquidiploids contain, at least in theory, a full diploid chromosome complement of the tetraploid parent, they are able to produce some chromosomally balanced, viable gametes and are usually at least partially fertile. The first sesquidiploids obtained in this manner were produced by crossing autotetraploid *N. tabacum* with *N. gossei* (Valleau, 1952) and with *N. plumbaginifolia* (Chaplin, 1954; Ar-Rushdi, 1957). Discussion on the use of autotetraploid forms of Nicotiana as components of interspecific hybrids is further expanded in part 4.4.3 of this discourse.

Known interspecific combinations involving autotetraploid *N. tabacum* are listed in Table 3.6.

3.6.4 Conversion of Sterile Hybrid Plants to Fertility by Means of External Agents

Use of Diverse Chemical or Physical Agents Since conversion to fertility that relies on the vagaries of nature was both time consuming and highly unreliable, external agents that might induce the chromosome doubling process were tried. Eghis (1930) applied chloroform to obtain the allopolyploid 4x (N. *tabacum* × *N. sylvestris*).[3] In another early attempt, Ternovsky (1939) restored fertility to the F_1 hybrid *N. tabacum* × *N. sylvestris* by exposing it to high temperature. Another approach to polyploidization was attempted by Kostoff (1937), who centrifuged the allohaploid germinating seeds of *N. tabacum* x *N. rustica* and, as a result, a fertile branch with the doubled chromosome complement was obtained.

Use of Antimitotic Drugs The search for efficient methods to induce chromosome doubling resulted in the identification of substances that act as antimitotic drugs. Such substances interfere with the formation of the spindle during cell division. The chromosomes lag at the equatorial plate of the dividing cell, and the newly synthetized nuclear membrane surrounds the chromosomes, which are now double in number. If the anti-mitotic agent is removed in a timely manner, the cells continue to divide, thus giving rise to polyploid tissues. Acenaphtene was probably the first antimitotic drug used by some *Nicotiana* investigators for its chromosome doubling

[3] some authors consider that hybrid an allohexaploid (6x) on account of the amphidiploid origin of *N. tabacum* and *N. sylvestris* being one of the parental species of *N. tabacum*.

Table 3.6 Instances of sesquidiploid hybrids obtained by direct crossing autotetraploid forms of *N. tabacum* with other *Nicotiana* species

Species involved with the hybrid with *N. tabacum*	Author
N. alata	Kostoff (1930), Chaplin and Mann (1961), Chaplin (1962), Takenaka (1960, 1962b), Takenaka and Yoneda (1964), Ivancheva-Gabrovska and Manolov (1982), Berbeć (1987b), Laskowska and Berbeć (2005)
N. forgetiana	Burk (1972)
N. langsdorffii	Takenaka et al. (1955), Takenaka (1958, 1962b), Burk (1972)
N. longiflora	Takenaka (1962a)
N. plumbaginifolia	Clausen in 1952 (after Ar-Rushdi, 1957), Chaplin (1954), Moav (1958), Moav and Cameron (1960), Chaplin and Mann (1961), Chen (1971), Baalawy and Fox (1971), Dang et al. (2019)
N. sanderae	Ivancheva-Gabrovska and Manolov (1982)
N. glauca	Chaplin and Mann (1961), Chaplin (1962), Wichert-Kobus (1971), Kobus (1971)
N. benavidesii	Takenaka (1962b), Berbeć (1986)
N. knightiana	Chaplin and Mann (1961), Berbeć et al. (1982)
N. paniculata	Chaplin & Mann, 1961, Baalawy and Fox (1971)
N. raimondii	Berbeć et al., 1982
N. repanda	Valleau (1952)[1]
N. quadrivalvis	Chaplin and Mann (1961)
N. pauciflora	Chaplin & Mann, 1961
N. rustica	Chaplin and Mann (1961), Legg and Mann (1961), Pandeya and White (1981, 1984); Chaplin and Sisson (1984), Pittarelli and Sisson (1989), Nifong (2008)
N. africana	Doroszewska and Berbeć (1990)
N. amplexicaulis	Berbeć et al. (1982)
N. debneyi	Clayton (1950)
N. exigua	Wichert-Kobus (1967), Wichert-Kobus (1971), Kobus (1971)
N. gossei	Valleau (1952), Moav and Cameron (1960) Apparao et al. (1980)[2]
N. megalosiphon	Manolov et al. (1978)
N. suaveolens	Chaplin (1959), Chaplin and Mann (1961)
N. sylvestris	Ar-Rushdi (1955), Chaplin and Mann (1961), Wichert-Kobus (1971), Kobus (1971)
N. otophora	Ar-Rushdi (1955), Chaplin and Mann (1961)
N. setchellii	Ar-Rushdi (1955)[3]
N. tomentosa	Ar-Rushdi (1955)
N. tomentosa var. 'Acomayo'	Ar-Rushdi (1955)[3]
N. tomentosiformis	Ar-Rushdi (1955) , Chaplin and Mann (1961)
N. glutinosa	Clausen and Cameron (1957), Chaplin and Mann (1961), Baalawy and Fox (1971), Pirrie and Power (1986), Giddings and Rees (1992)[4]

[1]a slow growing hybrid plant that died before flowering;
[2]*N. gossei* used as male parent, selective loss of *N. gossei* chromosomes observed;
[3]female sterile and male fertile sesquidiploids;
[4]sesqudiploids were obtained by fusing tetrad protoplasts of *N. glutinosa* with mesophyll leaf protoplasts of *N. tabacum* (gametosomatic hybrids)

effects. Using that agent, Ternovsky (1962) obtained the amphidiploid 4x (*N. sylvestris* × *N. tomentosiformis*). Bolsunov used acenaphtene to produce amphidiploids from F_1 (*N. rustica* x *N. tabacum*) (Bolsunov, 1963) and from F_1 (*N. rustica* × *N. exigua*) (Bolsunov, 1970).

More recently, oryzalin, an herbicidal substance but also a very powerful antimitotic, was used to induce chromosome doubling in the amphihaploid *N. sylvestris* × *N. tomentosiformis* (Lim et al., 2006).

Starting with the first experiments by Warmke and Blakeslee (1939) that involved the hybrid *N. tabacum* × *N. glutinosa,* colchicine became the antimitotic drug of choice to restore fertility to sterile *Nicotiana* hybrids. Colchicine is used in various formulations, e.g., as water solutions of different strengths, water solutions with agar, mixtures with lanolin, etc. It was applied both *in situ*, on field- or greenhouse-grown plants and *in vitro* in various types of aseptic cultures. Chromosome doubling is induced at various growth stages and in different plant parts, in cultured embryos, germinating seeds, seedlings and growing plants. The dividing meristems to which the drug is applied include apices and ancillary buds. Some examples are given in Table 3.7.

3.6.5 Regeneration of Hybrid Plants from In Vivo *and* In Vitro *Cultures*

Regeneration from Callus *In Situ* This oldest and probably long-forgotten method made use of polyploid cells already present in the plant by inducing the growth of calli *in vivo*. Polyploid cells originate during the tissue differentiation process, mostly through endomitosis, which involves the mitotic division of chromosomes within an intact nuclear membrane. Using this approach, Protassenya (1935) obtained an allopolyploid hybrid from 2x (*N. rustica* × *N. tabacum* and Greenleaf (1938) doubled the chromosome complements of 2x (*N. sylvestris* × *N. tomentosiformis*), 2x (*N. sylvestris* × *N. setchellii*), 2x (*N. sylvestris* × *N. tomentosa*) and 2x (*N. glutinosa* × *N. sylvestris*). The callus growth in the amphidiploids produced by Greenleaf was induced by the application of heteroauxin (IAA).

Regeneration from Calli *In Vitro* As discussed in a previous section, aseptic cultures were found to be helpful in overcoming incongruity of certain interspecific combinations that resulted in premature death of hybrid seedlings. Since the regeneration of viable plants from the explants of lethal hybrids passes through the callus phase and sometimes requires several passages of culture to take effect, it also offers an opportunity for the preexisting endomitotic cells to develop into allopolyploid along with amphihaploid shoots. The reader is referred back to Table 3.4 for rare instances of hybrids involving *N. tabacum,* the culture of which resulted in restoration of both viability and fertility. More frequently, chromosome number was

Table 3.7 Induction of alloploidy in some interspecific hybrids involving *N. tabacum* by using colchicine

Species or hybrid involved with the hybrid with *N. tabacum*	Phase of treatment/treated organ	Author
N. alata	seedlings, apical meristem immature embryo	Gajos (1975, 1981) Patrascu et al. (1999)
N. longiflora	apical meristem	Venkateswarlu et al. (1998)
N. plumbaginifolia	seedlings	Moav and Cameron (1960)
N. glauca	germinating seeds	Smith (1939), Valleau (1952), Trojak-Goluch and Berbeć (2007)
N. raimondii	seedlings, germinating seeds	Berbeć (1988)
N. rustica[1]	apical meristem seedlings plant cuttings axillary buds	Smith (1939) Furusato (1960) Moav and Cameron (1961) Takenaka (1963), Marubashi and Nakajima (1985)
N. nudicaulis	axillary buds plus inflorescence	Burk and Neas (1964)
N. amplexicaulis	axillary buds	Wark (1970)
	germinating seeds, apical meristem	Berbeć and Doroszewska (1981) Berbeć and Doroszewska (1992)
N. benthamiana	immature embryo	Subhashini et al. (1986)
N. benthamiana × *N. glutinosa*	seedlings, apical meristem	Ramavarma et al. (1977)
N. debneyi	apical meristem	Smith (1941), Sand (1968), Ternovsky et al. (1976)
N. exigua	seedlings	Wichert-Kobus (1967)[3], Kobus (1971)[3]
N. goodspeedii	axillary buds	Wark (1970)
N. gossei	axillary buds	Burk and Dean (1975)
N. maritima	not reported	Wark (1970)
N. occidentalis	axillary buds	Ternovsky et al. (1972)
N. rosulata	apical meristem	Ternovsky and Larkina (1978a)
N. suaveolens[2]	apical buds	Lloyd (1975)
N. velutina	axillary buds	Wark (1970)
N. otophora	seedlings apical meristems	Gerstel (1960) Larkina (2015, 2017)
N. setchellii	apical meristems germinating seeds	Larkina (2015, 2017) Berbeć (unpublished)
N. setchellii x N. otophora	germinating seeds	Berbeć (1982)[4]
N. tomentosiformis	seedlings	Gerstel (1960)
N. sylvestris	apical meristem	Smith (1939)
N. obtusifolia	seedlings	Chung et al. (1988, 1996)

(continued)

Table 3.7 (continued)

Species or hybrid involved with the hybrid with *N. tabacum*	Phase of treatment/treated organ	Author
N. glutinosa	apical meristem, axillary buds	Blakeslee and Avery (1937), Warmke and Blakeslee (1939)
N. glutinosa x N. obtusifolia	seedlings	Appa Rao and Krishna Murthy (1963)

[1]amphidiploid *N. rustica* x *N. tabacum* also induced by acenaphtene (Bolsunov, 1963);
[2]amphidiploid also induced by treatment with acenaphtene by Izard and Hitier (1955);
[3]induction of polyploidy by joint action of colchicine and gibberellin;
[4]induction of chromosome doubling in trispecific allohaploid (allotriploid) *N. tabacum* x *N. setchellii* x *N. otophora*), the resulting allotriploid (allohexaploid) was male fertile but female sterile

doubled, and fertility was restored by taking explants from sterile hybrid plants that had grown past the stage critical for survival (Table 3.8).

3.6.6 'Synthetic Species'

Synthetized amphidiploids in *Nicotiana* are known to differ from one another in the extent of variation they exhibit in successive selfed generations. Some lineages of the new amphidiploid 4x (*N. wuttkei* × *N. tabacum*) (Laskowska et al., 2015) were stable enough to deserve the name of a 'synthetic species'. Such stable, self-perpetuating 'synthetic species' were previously developed within the section Suaveolentes by Krishnamurthy and Gopinath (1969): 4x (*N. velutina* × *N. amplexicaulis*), 4x (*N. occidentalis* × *N. amplexicaulis*), including the nullisomic lineage 4n = 76 of the latter amphidiploid. Each of those artificial amphidiploids was given a regular botanical description and treated, as the authors put it, "equal in rank with the existing *Nicotiana* species". Some of such 'synthetic species' were even given specific names: *N.* × *obtusiata* for 4x (*N. obtusifolia* × *N. attenuata*) (Anssour et al., 2009; Krügel, 2010; McCarthy et al., 2015) and *N.* × *mierata* for 4x (*N. miersii x N. attenuata*) (Pearse et al., 2006; Krügel, 2010), *N. edwardsonii* for 4x (*N. glutinosa* × *N. clevelandii*) (Christie, 1969), *N. vavilovii* for 4x (*N. glauca* × *N. langsdorffii*) (Kostoff, 1938b; Kostoff, 1939a, 1939b), *N.* × *diruex* for 4x (N *rustica* × *N. exigua*) (Bolsunov, 1970), *N.* × *didebta* for 4x (*N. debneyi* × *N. tabacum*) (Clayton et al., 1967, He et al., 2019, misnamed by the latter authors '*N.* x *didepta*), *N.* × *digluta* for 4x (*N. glutinosa* × *N. tabacum*) (Clausen & Goodspeed, 1925; Clausen, 1928); N. x *disualovii* for *N. suaveolens* x *N. quadrivalvis* (*bigelovii*) (Modilevsky, 1939); *N.* x *flindersiensis* for 4x (*N. suaveolens* x *N. glauca*) (Smith & Abashian, 1963); *N.* × *ditagla* for 4x (*N. tabacum* × *N. glauca*) (Ternovsky, 1934; Modilevsky, 1936). Other lineages

Table 3.8 Instances of restoring fertility to amphihaploid hybrids involving *N. tabacum* by means of culturing explants from viable hybrid plants at advanced growth stages

Section	Species involved in the amphihaploid hybrid with *N. tabacum*	Explant	Cytological status of regenerants	Reported by
Alatae	*N. alata*	ST ST	amphidiploids subamphidiploids, mixoploids[1]	Skucińska et al. (1977), Dorossiev et al. (1978) Nikova et al. (1999)
	N. plumbaginifolia	ST	aneuploids, mixoploids	Nikova et al. (2004)
	N. longiflora	ST	Allopolyploids[2]	Nikova et al. (2001)
	N. sanderae	ST	near-amphidiploids	Skucińska et al. (1977)
		ST	amphidiploids	Dorossiev et al. (1990)
			self-fertile regenerants of different ploidy level	Nikova et al. (2003, 2006)
Paniculatae	*N. paniculata*	ST	mixoploids	Nikova et al., 1991, Nikova and Vladova (2002)
Noctiflorae	*N. noctiflora*	ST	amphidiploids	Stanoeva and Petkova (1978), Dorossiev et al. (1978, 1990)
	N. glauca	ST	mixoploids	Raicu et al. (1978)
Suaveolentes	*N. africana*	ST C	amphidiploids amphidiploids	Keum et al. (1994), Nikova et al. (1988), Nikova and Zagorska (1990) Doroszewska and Berbeć (1990, 2000)
	N. amplexicaulis	LM	amphidiploids	DeVerna et al. (1987)
	N. benthamiana	LM ST	amphidiploids subamphidiploids/ mixoploids	DeVerna et al. (1987) Nikova et al. (1991), Krusteva et al. (2003)
	N. goodspeedii	ST	amphidiploids	Zagorska and Palakarcheva (1978) Dorossiev et al. (1990)
	N. gossei	ST	amphidiploids (?)	Dorossiev and Palakarcheva (1990)
			mixoploids	Nikova, Palakarcheva, et al. (1998a)
		ST	amphidiploids	Palakarcheva et al. (1995)
	N. ingulba	ST	amphidiploids	Nikova, Vladova, et al. (1998b)
	N. maritima	ST	amphidiploids	Dorossiev et al. (1978, 1990)
	N. velutina	ST	mixoploids	Nikova et al. (1991)

Abbreviations in column 3: *ST* stem pith, *LM* leaf midrib, *C* cotyledons/cotyledon segments

[1] Aneuploids of different ploidy level (44-93 chromosomes), mixoploids

[2] Male sterile, partly female fertile;

of the latter amphidiploid were found to be highly unstable (Szilagyi, 1975). Unstable amphidiploids will be discussed in Sect. 4.5.4.

The relatively recently produced synthetic species *N. excelsiana* for 4x (*N. excelsior* × *N. benthamiana*) (Fitzmaurice, 2002) shows that there is still a large unexploited potential, both academic and commercial, in interspecific combinations in *Nicotiana*. Due to its biological properties, *N. excelsiana* gained the status of a "proprietary species" protected by patent rights as a convenient tool in a newly developed protein production technology. In a recent study, *N. x excelsiana* was demonstrated as a valuable and agronomically exploitable source of griffithsin, an anti-HIV drug (Eapen et al., 2020) and was also studied for other secondary metabolites (Mihaylova-Kroumova et al., 2020).

3.7 Parasexual Hybrids

3.7.1 Parasexual Hybrids by Fusion of Isolated Somatic Protoplasts

The idea of circumventing prefertilization barriers to crossability that had evolved at the gametic level prompted the attempts to fuse the somatic cells that lacked such obstructions. It was also envisaged that fusion of unreduced somatic cells would directly produce fertile amphidiploids, thereby bypassing the sterility of sexually produced hybrids. In its basics, the procedure has remained unchanged from its inception and is divided into three stages:

- isolation of protoplasts by enzymatic degradation of cell walls
- inducing the naked protoplasts to fuse by the presence of chemical agents or by electric fields
- selective culture of fused protoplasts based on their physical properties, the presence of fluorescent markers or genetic complementation, e.g., resistance to antibiotics

Since spontaneous fusion of naked protoplasts is a rare phenomenon, several agents, both chemical and physical, were tested for fusion-inducing action. Among the physical agents, the electric-field mediated method, also called electrofusion, is most frequently used. In this method, the protoplasts are brought into close contact by the application of an alternating electric field (AC) followed by exposure to direct current (DC) pulses (Davey, 2017).

Despite the high efficiency of electrofusion, polyethylene glycol (PEG), a chemical agent, has been most popular in creating interspecific somatic hybrids in *Nicotiana*, although the compound is toxic to plant cells. Another frequently used approach is the combination of high pH and high concentration of Ca^{2+} cations (Ilcheva & San, 1997; Davey, 2017).

After the fusion treatment, the medium contains both fused heterokaryons and unfused parental protoplasts. Different methods were applied to discriminate against the latter in the selection process. They included simple visual identification (Bates, 1985; Nagao, 1978; Hamill et al., 1984) and the use of various genetic markers, such as chlorophyll mutations (Evans et al., 1981, 1982, 1983; Aviv & Galun, 1986, 1987), resistance to antibiotics, both spontaneous (Medgyesy et al., 1980) and transgenic (Bates, 1990; Pental et al., 1988, 1989; Lu & Yang, 1996). Transgenic resistance to two different antibiotics provided double complementary selective systems under which only heterokaryons could survive in a growth medium containing both markers (Sproule et al., 1991; Donaldson et al., 1993, 1995; Ilcheva et al., 2000, 2001).

Although Zheng et al. (2018) wrote that the cytoplasmically male sterile line of *N. tabacum* 'cms-sua' used in their study originated from the somatic fusion between *N. suaveolens* and *N. tabacum* made in the 1950s (sic!) the first well-documented interspecific somatic hybrid in *Nicotiana* was reported by Carlson et al. (1972). The hybrid *N. langsdorffii* + *N. glauca* was a regular fertile amphidiploid ($4x = 42$) and thus seemed to confirm the hopes attached to interspecific hybridization at the somatic level. Soon thereafter, however, it became apparent that the method had serious constraints. Over the years, the number of interspecific somatic combinations, mostly those involving *N. tabacum*, that yielded genetically stable, fertile amphidiploids was disappointingly small and limited to hybrids of *N. tabacum* with *N. glauca, N. nesophila, N. debneyi, N. megalosiphon*, and *N. otophora* (Table 3.9). The other hybrids synthesized by protoplast fusion that involved *N. tabacum* showed high variability in external morphology and mostly aneuploid chromosome numbers due to chromosome elimination, either random or preferential, resulting in asymmetric hybrids, i.e., those with predominance of genetic material from one parental species (e.g., Donaldson et al., 1995; Ilcheva et al., 1997, 2000). Nonetheless, the irregularity and imbalance of interspecific protoplast fusion products may carry inherent advantages, which is explained in the paragraph to follow.

Asymmetric hybrids have also been synthetized consciously using chemical agents (iodoacetate) or irradiation (gamma or X-rays) to inactivate all or part of the nuclear genome of one parent prior to fusion (e.g., Bates, 1990). Highly asymmetric hybrids to which the cytoplasmic DNA was contributed almost exclusively by the irradiated parent whereas the other parent donated both cytoplasmic and nuclear DNA (the so-called cybrids obtained by the donor-recipient method) proved to be a useful tool to produce interspecific mitochondrial recombinants and provided a fast method to transfer whole plasmons or selected cytoplasmic traits from one species to another (see also Sect. 5.3.1). Cytoplasmic recombinants are practically impossible to obtain by sexual hybridization since in *Nicotiana*, as in most other genera, the cytoplasmic DNA is inherited unilaterally through maternal lineage, save for some rare exceptions (Medgyesy et al., 1985; Horlow et al., 1990; Svab & Maliga, 2007). In this context, one may also note that such unilaterally incomplete or asymmetric hybrids can also be obtained from sexual matings using irradiated pollen (see Sect. 3.4.2). An extreme case of that latter approach was the highly contested "egg transformation" (see Sect. 4.4.4).

Table 3.9 Interspecific parasexual hybrids involving *N. tabacum*

Species involved in the hybrid with *N. tabacum*	Methods of fusion induction and selection of heterokaryons	Hybrid type/ Cytological status	Author	Notes
N. alata	PEG; visual	sterile hyper-aneuploids	Nagao (1979)	66-71 chromosomes
N. bonariensis	? ; visual	not determined	DeVerna (1984)	regenerated shoots
N. plumbaginifolia	EF; visual identification	Not determined	Bates (1985), Bates and Hasenkampf (1985)	
	EF, transgenic kanamycin resistance	Donor-recipient cybrids	Bates et al. (1987)	48 and 49-chromosome asymmetric male sterile *N. tabacum*-like hybrids, self- and cross-sterile aneuploid hybrids of intermediate morphology (58-70 chromosomes)
	Terbutryn resistance, X-irradiation	Donor-recipient cybrids	Menczel et al. (1986)	
	?, ?	not determined	Medgyesy et al. (1985)	morphology intermediate between parental species
	EF, ?	Not determined	Hamill et al. (1987)	
	PEG; elimination of non-fused protoplasts by inability to divide and form colonies	various types	Desprez et al. (1992)	sterile gametosomatic allotriploids and allotetraploids showing mixoploidy, chromosome numbers ranging from 44 to 68 including asymmetric cybrid (48 chromosomes)
N. sanderae	PEG; selective markers (kanamycin resistance and nitrate reductase deficiency)	sterile true amphidiploids	Dragoeva et al. (1977)	
N. glauca	PEG; chloroplast-deficient mutations	amphidiploids (fertile)	Evans et al. (1980), Gleba et al. (1984)	
	?, ?		Sun et al. (2007)	
			Fuentes et al. (2014)	vegetative hybrid obtained by grafting

(continued)

Table 3.9 (continued)

Species involved in the hybrid with *N. tabacum*	Methods of fusion induction and selection of heterokaryons	Hybrid type/ Cytological status	Author	Notes
		amphidiploids (fertile)		
N. knightiana	PEG; shoot inducibility and albino mutation	sterile mixoploids and aneuploids	Maliga et al. (1978)	44 to 126 chromosomes
	PEG; visual	amphidiploids (fertile) and aneuploids	Menczel et al. (1981)	68 to 115 chromosomes
N. paniculata	?; *Agrobacterium* T-DNA markers, nitrate reductase deficiency	cellular lines	Mueller-Gensert and Schieder (1987)	
N. quadrivalvis	PEG; chlorophyll mutations	fertile and sterile cybrids	Aviv and Galun (1986, 1987)	
N. nesophila	PEG; chlorophyll mutations	fertile amphidiploids	Evans et al. (1981, 1982)	
N. repanda	PEG; visual	sterile aneuploids	Nagao (1982)	58 to 64 chromosomes
	EF; nopaline gene, transgenic kanamycin resistance	aneuploids and mixoploids	Bates (1990)	obtained by gamma irradiation of *N. repanda* protoplasts
N. stocktonii	PEG; chlorophyll mutations	fertile hybrids	Evans et al. (1981)	amphidiploids (?)
	EF; ?		Chen et al. (2013)	partly fertile hybrids

N. rustica	PEG; visual	aneuploids	Nagao (1978)	60 to 91 chromosomes
	?; visual	aneuploids	Hamill et al. (1984)	63 to 87 chromosomes, poor to good pollen stainability accompanied by prevalence of self-sterile regenerants
	?; ?; PG, double transgenic markers (resistance to kanamycin and methotrexate)	rare amphidiploids and aneuploids	Naton et al. (1992), Donaldson et al. (1993)	56 to 96 chromosomes, massive chromosome elimination, PG as the fusing agent and selection of heterokaryons by double selective markers used by Donaldson et al. (1993)
	PEG; visual	fertile hybrids	Choi et al. (1993)	
	high pH/Ca^{++} method; transgenic resistance to kanamycin	gametosomatic hybrids	Pental et al. (1988, 1989)	unknown cytological status, transmission of mitochondria but not chloroplasts from the microspore parent (*N. tabacum*) found in fusion products
	high pH/Ca^{++} method; transgenic resistance to kanamycin		Lu and Yang (1996)	sesquidiploid chromosome number reported
N. africana	?; ?	donor-recipient cybrids	Kumashiro et al. (1988)	*N. tabacum* like cms phenotypes
N. benthamiana	?; ?	fertile amphidiploids (?)	Hagimori et al. (1993)	reported as 'true hybrids'
N. debneyi	PG; double transgenic markers (resistance to kanamycin and methotrexate)	fertile amphidiploids and aneuploids	Sproule et al. (1991), Dijak et al. (1991)	
	PEG+ high pH/Ca^{++}; ?	donor-recipient cybrids	Kumashiro and Kubo (1986)	*N. tabacum* like cms phenotypes

(continued)

Table 3.9 (continued)

Species involved in the hybrid with *N. tabacum*	Methods of fusion induction and selection of heterokaryons	Hybrid type/ Cytological status	Author	Notes
N. megalosiphon	?;?	?	Chupeau (1987)	Self-sterile hybrids
	?; ?	amphidiploids and aneuploids (?)	Brandle et al. (1992)	variable intermediate morphology
	PEG; double transgenic markers (resistance to kanamycin and methotrexate)	amphidiploids and aneuploids (?)	Donaldson et al. (1995)	variable intermediate morphology
	PEG, EF; double transgenic markers (resistance to kanamycin and bialaphos)	aneuploids	Ilcheva et al. (1997, 2000)	highly variable morphology, highly asymmetric hybrids resulting from non-preferential loss of chromosomes of one or the other species
	: ?; X-irradiation followed by double transgenic markers (resistance to kanamycin and methotrexate)	donor recipient cybrids	Kasza and Kandra (1990)	cms forms characteristc of cytoplasm recipient and cytoplasmic recombinants
N. rotundifolia	PEG; double transgenic markers (resistance to kanamycin and bialaphos)	aneuploids	Ilcheva et al. (2001)	non-preferential chromosome loss
N. suaveolens	PEG; centrifugation using Percoll flotation protocol	Donor-recipient cybrids	Fitter et al. (2005)	Mitochondrial recombinants obtained with stigmatoid and carpeloid stamens characteristic of *N. tabacum* cms *suaveolens* but showing split corolla
	?	?	Zheng et al. (2018)	cms suaveolens alloplasmics derived from *N. tabacum* + *N. suaveolens* somatic hybrid reportedly obtained in the 1950s

N. sylvestris	PEG; no selection	donor-recipient cybrids	Zelcer et al. (1978), Aviv et al. (1980)	
	PEG; streptomycin resistance	amphidiploids and aneuploids	Medgyesy et al. (1980)	highly variable morphology
	PEG; Su mutation	amphidiploids	Evans et al. (1983)	
N. otophora	PEG; Su mutation	amphidiploids	Flick and Evans (1982), Evans et al. (1983)	
N. arentsii		not determined	DeVerna (1984)	regenerated non-flowering shoots
N. glutinosa	PEG; visual; PG; double transgenic markers (resistance to kanamycin and methotrexate)	amphidiploids and aneuploids	Nagao (1979), Donaldson et al. (1994)	selection of heterokaryons by double seletion markers by Donaldson et al. (1994)
	EF; chlorophyll mutation, nitrate reductase deficiency	not determined	Kim and Choi (1991)	reported as somatic hybrids based on morphological and biochemical evidence
	PEG; no selection	not determined	Uchimiya (1982)	hybrid character determined based on intermediate morphology and cms trait
		gametosomatic sesquidiploids (pentaploids)TTG	Pirrie and Power (1986), Giddings and Rees (1992)	
	dextran method; amino acid resistance and plant regeneration ability	aneuploids	Horn et al. (1983)	from 34 to 60 chromosomes, massive chromosome loss observed
N. undulata	PEG; no selection	not determined, intermediate morphology	Uchimiya (1982)	various mitochondrial recombinants expressing different manifestations of male sterility
	PEG; no selection	donor-recipient cybrids	Aviv and Galun (1986)	mitochondrial recombinants, including restored male fertility to fusion products
	PEG; ?	donor-recipient cybrids (aneuploids)	Liu-Bao et al. (1995)	highly asymmetric hybrids (51 to 66 chromosomes)

Column 2 contains abbreviated method of protoplast fusion followed by brief description of how heterokaryons were selected; question mark is put instead if the method applied is not known

Abbreviations: *PEG* polyethylene glycol, *EF* electrofusion

3.7.2 Gametosomatic Hybrids

A fusion of diploid mesophyll species of *N. tabacum* with haploid gametophyte protoplasts of an alien species was devised with an expectation to obtain sesquidiploid plants as a starting material for interspecific gene transfer equivalent to sesquidiploids from sexual matings (Davey et al., 1996, see also Sects. 3.6.3, 4.5.6). The method was experimentally tested by Pirrie and Power (1986) and by Giddings and Rees (1992). Their gametosomatic hybrids 2n *N. tabacum* + 1n *N. glutinosa* actually mimicked sesquidiploids (pentaploids) from sexual matings. In those experiments, haploid protoplasts were isolated at the tetrad stage. Alternatively, protoplasts isolated from mature pollen grains can be used for gametosomatic fusion (Desprez et al., 1992, Lu & Yang, 1996, Ping et al., 1996). Another benefit of gametosomatic fusion is that haploid gametophyte protoplasts fail to divide and do not form colonies in culture, which simplifies the selection of heterokaryons (Davey et al., 1996).

However, not all products of gametosomatic fusion represent true pentaploids. Gametosomatic regenerants (1n *N. tabacum* + 2n *N. plumbaginifolia*) made by Desprez et al. (1992) represented an array of sterile aneuploid forms that also showed mixoploidy plus a 48-chromosome asymmetric cybrid. According to those investigators, androgenetic regenerants obtained by asymmetric gametosomatic fusion can be used to transfer mitochondrial genomes separately from chloroplast genomes.

Despite their potential value, few interspecific gametosomatic hybrids have been reported. They include, apart from the previously mentioned 2n *N. tabacum* + 1n *N. glutinosa*, 1n *N. tabacum* + 2n *N. plumbaginifolia* (Desprez et al., 1992) and 1n *N. tabacum* + 2n *N rustica* (Pental et al., 1988, 1989; Mukhopadhyay et al., 1991; Lu & Yang, 1996; Ping et al., 1996).

A concise and informative review of parasexual hybridization by protoplast fusion in *Nicotiana* was prepared by Ilcheva and San (1997).

3.7.3 Graft Hybrids and Horizontal Gene Transfer

Grafting occurs when segments of two different plants come into close contact and fuse together into one separate and independent plant organism. In such a union of two plants or plant pieces, one of them provides the root of the new plant and is called rootstock or simply stock, while the other serves as the shoot with leaves and is referred to as 'scion'. Grafting as a fusion of two or more separate plant organisms occurs spontaneously in nature but has also been practiced by agriculturists.

In fruit farming practice, grafts have been used for two different purposes: for better growth and desirable performance and for genetic effects (Ohta, 1991).

The former function of grafting has been widely practiced by horticulturists from early antiquity, and its major objective is to improve the growth and performance of fruit plants or ornamentals by combining the superior qualities of the stock with those of the scion, e.g., winter hardiness with palatable fruit or other edible parts. The latter, much less popular and more controversial, was about to impose heritable changes in the scion by transferring genetic information from the stock. In the 1940s and 1950s, the concept gave rise to the so-called Michurinist genetics developed by the Soviet horticulturist I. V. Michurin, and the idea was officially proclaimed in the Soviet Union as an essential part of the so-called "revolutionary agrobiology", the movement advanced by T. Lysenko and his followers throughout the Eastern bloc in deliberate and blatant opposition to the Mendelian principles of heredity (Goldschmidt, 2014; Zhou & Liu, 2015). The 'graft hybrids' theory was deeply mistrusted by scientists in the West and repeatedly proven to lack a scientific basis (Goldschmidt, 2014). Ultimately, it was abandoned both in the Soviet union and in other Eastern bloc countries. However, the recent decades have witnessed a revived interest in epigenetic and hereditary effects of grafting (Ohta, 1991; Goldschmidt, 2014).

Nicotiana species are easily amenable to grafting both among themselves or with other related solanaceous plants, but the technique has, until very recently, attracted little attention from Mendelian tobacco geneticists and breeders. It is only at the beginning of this century that R. Bock and his team from the Max Planck Institute came up with evidence that the entire chloroplast genomes could be transferred through the graft junction from *N. tabacum* to two other *Nicotiana* species: *N. glauca* and *N. benthamiana* (Stegemann et al., 2012). In their experiment, the authors demonstrated that the transfer was restricted entirely to chloroplast DNA and did not involve any nuclear DNA fragments.

However, in another experiment reported 2 years later (Fuentes et al., 2014), the same team demonstrated the interspecific fusion of the whole genomes of two grafting partners, *N. glauca* and *N. tabacum.* After the fusion of stock and scion had taken place, fragments of tissue of the fusion zone were excised and cultured *in vitro.* Callus culture and plant regeneration were performed by following the genetic complementation protocol based on double selectable markers, a technique routinely applied in somatic hybridization by protoplast fusion. As a result, a fully fertile, regular 72-chromosome amphidiploid 4x (*N. glauca* × *N. tabacum*) was obtained and given the name '*Nicotiana tabauca*', a new species that arose by natural fusion of somatic cells.

A few observations can be made in connection with this unusual finding. The authors' argument that their discovery supports the likelihood of spontaneous asexual hybridization taking place in nature seems to be essentially valid, although in their experiments they had to resort to artificial tools, unknown to nature, to make it happen. This notwithstanding, when put in their long-term evolutionary perspective, even extremely rare and least likely events can and most likely do occur. The authors also presented their discovery as a new tool for crop improvement mostly because, as they argued, grafting is technically less demanding than protoplast fusion. Indeed, it is, but the sexual method is even less demanding. Precisely the allopolyploids *N. tabacum* × *N. glauca* and their reciprocals have been repeatedly obtained by conventional crossing for nearly a century, and even the name for that 'artificial species' was invented ('*Nicotiana* ditagla') preceding '*N.* tabauca' coined by the authors of the report by several decades (see Sect. 3.6.6). This notwithstanding, the significance of that discovery cannot be overestimated and fully deserves to be followed by other experiments involving other *Nicotiana* species, including those that show a high degree of mutual incompatibility. Regrettably, no new reports in that area, theoretical or practical, seem to have emerged thus far.

The demonstration of the feasibility of obtaining an interspecific hybrid by grafting is obviously reminiscent of the graft hybrids in the former Soviet Union mentioned in the introductory remarks to this section. The recent report by scientists from the Max Planck Institute may shed somewhat different light on the work of Mitschurin in Russia but also similar horticultural experiments by Burbank in the United States. The historical context to the achievement of Ignacia Fuentes and her colleagues was recalled by Zhou and Liu (2015).

One should also add that an analogous interspecific horizontal transfer through grafting was recently reported for mitochondrial genes from *N. sylvestris* responsible for restoring male fertility to the alloplasmic line of *N. tabacum* with the mitochondrial genome of *N. undulata* (Gurdon et al., 2016, see Sect. 5.3.2 on restoring male fertility to cms lines).

3.8 Ending Notes on Sexual and Asexual Interspecific Hybrids Involving *N. tabacum*

Somatic and gametosomatic hybrids that involve *N. tabacum* are listed in Table 3.9. A compilation of all interspecific hybrids involving cultivated tobacco, regardless of the method by which they were obtained, is presented in Table 3.10. Information was found on a total of 59 hybrids involving *N. tabacum*, and reciprocals were not included in the count.

In spite of various barriers to crossability discussed in the previous sections, the vast majority of those hybrids could be obtained by conventional crossing. Actually, only very few of those combinations may be considered to have been made possible owing to the use of advanced technologies. *N. nesophila* x *N. tabacum* (Reed & Collins, 1978), *N. occidentalis* x *N. tabacum* (Butenko et al., 1970), *N. rosulata* x *N. tabacum* (Ternovsky et al., 1976), *N. stocktonii* x *N. tabacum* (Reed & Collins, 1978) are known only from reports where hybridization was aided by tissue culture. *N. rotundifolia* + *N. tabacum* was reported as a somatic hybrid only (Ilcheva et al., 2001). Genetic engineering was deployed to obtain viable hybrids of *N. simulans* x *N. tabacum* and *N. umbratica* x *N. tabacum* (Ma et al., 2020). However, the latter two hybrids were also reported or hinted at by other authors (Kubo, 1985; Murthy et al., 2014), and *N. stocktonii* x *N. tabacum* was reported by Wong (1975). The reports on the latter three hybrids lacked details on how they were produced. Of two reports on *N. tabacum* x *N. bonariensis*, no details are known on the hybrid reported by Busconi et al. (2010).

Table 3.10 All interspecific hybrids involving the cultivated species *Nicotiana tabacum* (after Berbeć and Doroszewska 2020) with minor additions and modifications)

Section	Species involved in the hybrid with *N. tabacum*	F_1 hybrid (amphihaploid) reported by:[1]	Amphipolyploid (sexual or somatic) reported by:[1]
Alatae	*N. alata*	premendelian (Naudin (after Kostoff (1943)) East and Hayes (1912), East, 1928)	Ahuja (1962)[14], Gajos (1975, 1981), Dorossiev et al. (1978); Stoyanova (1978, 1980)[15]; Nikova et al. (1999)[16]; Nagao (1979)[17]
	N. bonariensis	Stavely (1979), Busconi et al. (2010)	Busconi et al. (2010) (?)[11] [18] triple hybrid 4x (*N. undulata-tabacum*) × *N. bonariensis* (Ahuja (1962))
	N. forgetiana	Takenaka (1963), Burk (1972)[2]	
	N. langsdorffii	East and Hayes (1912), East (1928), Hu (1956), Takenaka (1962b)[2], Burk (1972)[3]	Ahuja (1962)[14]
	N. longiflora	Malloch and Malloch (1924), Gentscheff (1931), Ternovsky (1936b)	Clayton (1947), Ahuja (1962), Morgan (1964), Smith et al. (1970), Sievert (1972a)
	N. plumbaginifolia	Gentscheff (1931), Pal and Nath (1936), Kincaid (1949), Chobanova (1977), Nikova et al. (2004)	Ar-Rushdi (1957), Moav and Cameron (1960), Apple (1962); Nikova et al. (2004)[20]
	N. sanderae	Christoff (1928), Kostoff (1930), Whitaker (1934), East (1935), Ternovsky (1962), Malecka (1977)	Ternovsky (1936), Skucińska et al. (1977), Malecka (1977), Dragoeva et al. (1977) (somatic),
Noctiflorae	*N. glauca*	premendelian (East (1928)): reported by Brongniart and Gris (1861) and Naudin in 1865 (according to Ternovsky (1936b)	Sarana (1934), Ternovsky (1934), Modilevsky (1936), etc.[19] many others, first ever interspecific graft amphidiploid reported by Fuentes et al. (2014)
	N. noctiflora	Palakarcheva (1975, 1992); Stanoeva and Petkova (1978), Dorossiev et al. (1978)	Dorossiev et al. (1978), Stanoeva and Petkova (1978)
	N. petunioides	Gisquet et al. (1940)	

(continued)

Table 3.10 (continued)

Section	Species involved in the hybrid with *N. tabacum*	F_1 hybrid (amphihaploid) reported by:[1]	Amphipolyploid (sexual or somatic) reported by:[1]
Paniculatae	*N. benavidesii*	Goodspeed (1945), Grebenkin (1968), Durbin and Uchytil (1977), Berbeć (1978a, 1980, 1987a)	Berbeć (1978a)[20], (1986)[20]
	N. cordifolia	Durbin and Uchytil (1977), Burk and Durbin (1978)	Burk and Durbin (1978)
	N. knightiana	Goodspeed (1945), Tanaka (1961), Takenaka (1962a), Slusarkiewicz-Jarzina and Zenkteler (1983), Berbeć (1987a)	Morgan (1964), Sievert (1972b), Berbeć et al. (1982)[21], Berbeć and Doroszewska (1992)
	N. paniculata	premendelian East (1928), Kostoff (1932), Holmes (1937a, b)	Nikova et al. (1991), Nikova and Vladova (2002)[22]
	N. raimondii	Kostoff (1943), Grebenkin (1968), Berbeć (1987a)	Burk et al. (1982), Berbeć (1988)
	N. solanifolia	Goodspeed (1945), Takenaka (1956b), Grebenkin (1968)	
Acuminatae	*N. acuminata*	Kostoff (1943), Iwai et al. (1986)	
	N. pauciflora	Gentscheff (1931), Kostoff (1943)[4], Goodspeed (1945)	
Polydicliae	*N. clevelandii*	Kehr and Smith (1952), Kaul (1988)	
	N. quadrivalvis	premendelian (East, 1928)	Fardy and Hitier (1945, 1947), Burk (1960), Calitz and Milne (1962), Morgan (1964), Smith et al. (1970), Gerstel and Burns (1983)
Repandae	*N. nesophila*	Reed and Collins (1978, 1980), Huesing et al. (1989)	Reed and Collins (1978),
	N. nudicaulis	Gentscheff (1931), Kostoff (1943), Burk and Neas (1964)	Burk and Neas (1964), Sievert (1972a)
	N. repanda	Foster (1943)[5], Kincaid (1949), Pittarelli and Stavely (1975), Nagao (1982[26]), Shintaku et al. (1985), Iwai et al. (1985), Choi et al. (1998)	Pittarelli and Stavely (1975), Zhou et al. (1991), Pontes et al. (2005)[23]
	N. stocktonii	Wong (1975), Reed and Collins (1978)	Reed and Collins (1978)

Rusticae	*N. rustica*	*N. rustica* × *N. tabacum*: premendelian (East (1928)), first reported by Koelreuter (Mayr, 1986) *N. tabacum* × *N. rustica*: (Eghis (1927))	Eghis (1927), Rybin (1927), Protassenya (1935), Ternovsky and Khudina (1938), Kostoff (1937, 1943), Furusato (1960), Moav and Cameron (1961), Takenaka (1963)
Suaveolentes	*N. africana*	Gerstel et al. (1979), Burk et al. (1979), Kandra (1984)[6] Keum et al. (1991, 1994), Doroszewska and Berbeć (1996)[7] Nikova and Zagorska (1990)[8], Ma et al. (2020)[9]	Doroszewska and Berbeć (1990, 1996)[7] Nikova and Zagorska (1990)[8] Keum et al. (1991, 1994)
	N. amplexicaulis	Wark (1970), Berbeć and Doroszewska (1981), DeVerna (1984), DeVerna et al. (1987)	Wark (1970), Berbeć and Doroszewska (1981), DeVerna et al. (1987)
	N. benthamiana	Subhashini et al. (1986), DeVerna (1984), DeVerna et al. (1987), Dorossiev et al. (1990), Nikova et al. (1991), Zaitlin and Mundell (2006), Iizuka et al. (2012), Mihaylova-Kroumova and Wagner (2016)	Subhashini et al. (1986), DeVerna et al. (1987), Dorossiev et al. (1990), Nikova et al. (1991), Krusteva et al. (2003)
	N. cavicola	Nikova et al. (2006)	Nikova et al. (2006) (?)
	N. debneyi	Kostoff (1943), Valleau (1952), Clayton (1958) (reciprocal), Grebenkin (1970), Larkina (2015) (reciprocal) Palakarcheva (1978) (reciprocal), Ma et al. (2020)[9]	Berbeć (1964), Bailov et al. (1964), Clayton (1968)[3], Grebenkin (1970) (4x (*N. tabacum* × *N. debneyi*))
	N. excelsior	Wark (1970), Gillham et al. (1977), Nikova (1986), Ma et al. (2020)[9]	Wark (1970)
	N. exigua	Wichert-Kobus (1967), Kobus (1971), Wark (1970), Manolov et al. (1978)	Wichert-Kobus (1967), Kobus (1971); Wark (1970), Manolov et al. (1978)
	N. fragrans	Durbin and Uchytil (1977), Tezuka et al. (2010)	
	N. goodspeedii	Butenko et al. (1970), Wark (1970), Palakarcheva (1974), Palakarcheva et al. (1978)	Wark (1970), Palakarcheva (1974), Zagorska and Palakarcheva (1978), Palakarcheva et al. (1978)
	N. gossei	Valleau (1952)[10] Takenaka (1962a), Parr and Thurston (1968), Dean (after Burk and Dean (1975)), Wark (1970), Ma et al. (2020)[9]	Wark (1970), Burk and Dean (1975), Tsikov and Tsikova (1986), Palakarcheva and Dorossiev (1992)
	N. hesperis	Kubo (1985)[11]	Kubo (1985)[11]

(continued)

Table 3.10 (continued)

Section	Species involved in the hybrid with *N. tabacum*	F_1 hybrid (amphihaploid) reported by:[1]	Amphipolyploid (sexual or somatic) reported by:[1]
	N. ingulba	Butenko et al. (1970), Nikova, Vladova, et al. (1998b), Tezuka et al. (2012)[12]	Nikova, Vladova, et al. (1998b)
	N. maritima	Wark (1970), Palakarcheva (1975), Dorossiev et al. (1978)	Wark (1970), Stanoeva and Petkova (1978), Dorossiev et al. (1978), Nikova et al. (1991)
	N. megalosiphon	Clayton (1950), Takenaka (1962a), Hranov (1970), Manolov et al. (1978), Ma et al. (2020)[9]	Palakarcheva and Bailov (1976), Manolov et al. (1978) (subamphidiploids)
	N. occidentalis	Butenko et al. (1970), Ternovsky et al. (1972), Wong (1975)	Ternovsky et al. (1972, 1973)
	N. rosulata	Ternovsky et al. (1976)	Ternovsky and Larkina (1978a)
	N. rotundifolia		Ilcheva et al. (2001)
	N. simulans	Kubo (1985)[11], Ma et al. (2020)[9]	Kubo (1985)[11]
	N. suaveolens	Premendelian (East (1928)), Izard and Hitier (1955), Lloyd (1975)	Izard and Hitier (1955), Morgan (1964)[24] Wark (1970), Lloyd (1975), Shinkareva (1979), Stavely (1979)
	N. umbratica	Murthy et al. (2014)	Murthy et al. (2014)[25]
	N. velutina	Wark (1970), Powell (1979), Ma et al. (2020)[9]	Wark (1970), Nikova et al. (1991) (fertile alloploids of undefined status)
	N. wuttkei	Laskowska and Berbeć (2012)	Laskowska et al. (2015)
	N. eastii	Chaplin and Mann (1961) (direct sesquidiploid EaTT)	
Sylvestres	*N. sylvestris*	East and Hayes (1912), Bellair (1913), Malinowski (1916), Goodspeed and Clausen (1917), Sachs-Skalińska (1917)	Rybin (1929), Eghis (1930), Ternovsky (1936a)

Tomentosae	*N. kawakamii*	triple hybrid only (*sylvestris* × *kawakamii*) × *tabacum* (Ohashi (1985))	Ohashi (1985)
	N. otophora	Goodspeed (1945), Ar-Rushdi (1955), Takenaka (1962a), Grebenkin (1968)	Goodspeed and Bradley (1942) Yang (1960), Gajos (1979)
	N. setchellii	Greenleaf (1941), Goodspeed (1945), Grebenkin, 1968	Larkina (1983), Berbeć (female sterile, unpublished)
	N. tomentosa	Goodspeed and Clausen (1928), McCray (1932)	Goodspeed and Bradley (1942)
	tomentosiformis	Brieger (1928), Breisser (1934), Lehmann (1936)	Fardy and Hitier (1945), Yang (1960), Gerstel (1960)
Trigonophyllae	*N. obtusifolia*	Takenaka (1956a), Tanaka (1961)	Chung et al. (1996)
	N. palmeri	Goodspeed (1945), Krishnamurthy et al. (1960), Berbeć et al. (1982)[13]	
Undulatae	*N. arentsii*		DeVerna (1984) (somatic)
	N. glutinosa	premendelian (East (1928)) first reported by Koelreuter (Mayr (1986))	Clausen and Goodspeed (1925), Clausen and Lammerts (1929), Ternovsky (after Ternovsky and Khudina (1938))
	N. undulata	Kehr and Smith (1952), Takenaka (1953), Takenaka (after Goodspeed and Thompson, 1959), Cameron (after Chaplin, 1964)	Kehr and Smith (1952), Ahuja (1962), Morgan (1964)

[1] only a maximum of six reports was referred to selected primarily by their order of appearance or by importance, in several cases many more investigators obtained or studied particular hybrid combinations;
[2] from crossing female autotetraploid *N. tabacum,* [2a]by mating *N. langsdorffii* to monosomic strains of *N. tabacum;*
[3] Clayton (1968) reported the amphidiploid 4x (*N. debneyi* × *N. tabacum*) to have existed already in 1938;
[4] crossability highly dependent on *N. tabacum* variety;
[5] F_1 hybrid 4n *N. repanda* x 2n *N. tabacum* died before flowering;
[6] F_1 hybrids between cms lines of *N. tabacum* and *N. africana* reported with restored male fertility;
[7] true amphihaploids (Doroszewska & Berbeć, 1996) or aneuploid near-amphihaploids (Gerstel et al. 1979) were obtained from mating female *N. tabacum* to male *N. Africana*;
[8] mixoploids, near-amphihaploids ($2n = 44$) and a near-amphidiploid obtained from mating female *N. africana* to male *N. tabacum;*

(continued)

[9] hybrid lethality was overcome and viable hybrid obtained by inactivating the gene present in the $NtHL_1$ locus of *N. tabacum* and responsible for the apoptotic death of hybrids with the species of the section Suaveolentes. The inactivation was achieved through editing of the $NtHL_1$ locus (Ma et al. 2020, see also sections 6.4.3 and 7.3.4);
[10] by using autotetraploid *N. tabacum* as the female parent (see Table 3.6);
[11] only circumstantial evidence exists for the hybrids *N. hesperis* × *N. tabacum* and *N. simulans* × *N. tabacum* having been synthesized based on the report on new cytoplasmically sterile lines cms *hesperis* and cms *simulans* (Kubo 1985);
[12] F_1 hybrids *N. tabacum* x *N. ingulba* obtained via ovule pollination in vitro by mating Haplo-Q monosomics of *N. tabacum* as females to *N. ingulba*;
[13] autotetraploid *N. palmeri* × *N. tabacum* (sesquidiploid PPT);
[14] failed to grow beyond seedling stage;
[15] apparently spontaneous amphidiploids derived from F_1 hybrids;
[16] male sterile subamphidiploids with partly restored female fertility;
[17] somatic hybrid;
[18] the account of converting several F1 hybrids, including *N. tabacum* × *N. bonariensis*, to fertile amphidiploids is ambiguous;
[19] subamphidiploids with partly restored male and female fertility;
[20] 68-chromosome allopolyploid obtained by backcrossing the sesquidiploid 3x (*N. tabacum* × *N. benavidesii*) (TTB) to *N. tabacum* plausibly by fusing an aneuploid unreduced gamete of TTB with a haploid gamete of *N. tabacum*;
[21] the amphidiploid was arrived at by three approaches: spontaneous seed set by amphihaploid *N. knightiana* x *N. tabacum*, direct crossing of tetraploid *N. tabacum* x tetraploid *N. knightiana*, induction of seeds by pollinating amphihaploid *N. knightiana* x *N. tabacum* with pollen of *N. langsdorffii* x *N. alata*;
[22] partly fertile unstable aneuploids;
[23] reported recovery of a fertile hybrid plant;
[24] the order of species in the listed amphidiploid suggests *N. suaveolens* as the maternal parent;
[25] there is a hint to fertile hybrids having been obtained but not explicit enough;
[26] somatic sterile aneuploids

Thirty-six *Nicotiana* amphihaploids that involved *N. tabacum* amphidiploids or near amphidiploids were produced by using various approaches. Most of those diploidized F_1 hybrids showed at least partially restored self-fertility: one was a female sterile amphidiploid (*N. tabacum* × *N. setchellii*), and the other was a nearly female sterile amphidiploid (*N. obtusifolia* × *N. tabacum*). Female sterility was also observed in the trigenomic allohexaploid 6x (*N. tabacum* × (*N. setchellii* × *N. otophora*) (Berbeć et al., 1982).

The author of this book found information on 26 *Nicotiana* species that were hybridized with *N. tabacum* by protoplast fusion, fewer than half the number of reported sexual hybrids (compare Tables 3.9 and 3.10). Among those 26 somatic hybrids, only a handful represented those in which whole genomes of both parents became united (*N. glauca* + *N. tabacum*, including a graft hybrid, *N. nesophila* + *N. tabacum*, *N. rustica* + *N. tabacum*, *N. debneyi* + *N. tabacum*, *N. otophora* + *N. tabacum*, *N. glutinosa* + *N. tabacum*). To the authors' knowledge, only two somatic hybrids with *N. tabacum* obtained by somatic fusion, *N. rotundifolia* + *N. tabacum* and possibly also *N. arentsii* + *N. tabacum,* have not been obtained by conventional sexual methods.

It appears that while the parasexual approach circumvents the prefertilization barriers that separate species from one another, its role in alleviating the incongruities existing between the fused genomes is of far less importance. The experience with somatic hybrids has confirmed an early observation by Zenkteler and Melchers (1978) that protoplast fusion contributes but little to expanding the crossability of different species within a genus. Chromosome loss, genome instability and other manifestations of intergenomic incongruities are among the common consequences of alloploidization and have been documented for both natural and synthetic allopolyploids that have arisen through the sexual process, a subject discussed in one of the previous sections. From numerous accounts, it appears that parasexual hybridization seems to exacerbate rather than mitigate those inherent incongruities since they are obviously far more numerous and more intense in somatic hybrids than those encountered in analogous hybrids synthesized via the sexual process. This said, asymmetry, a frequent phenomenon in most asexually produced hybrids, may actually facilitate gene flow between the fused genomes by bypassing the so-called bottle-necks of sexual introgression and by eliminating many deleterious linkage or epistatic effects that plague the sexual routes of gene transfer.

Last but not least, from the perspective of practical breeding issues to be discussed in the subsequent chapters, parasexual methods share the same disadvantage with genetic transformations at the molecular level. To wit, lawmakers of at least some countries have listed protoplast fusion been among the technologies used to generate genetically modified organisms (GMOs) thus effectively banning it in the development of commercially exploitable cultivars.

References

Adachi, A., & Inoue, M. (1995). Temperature-response of interspecific hybrid plants between *Nicotiana gossei* Domin and *N. tabacum* L., obtained from ovule culture. *Japanese Journal of Breeding, 2*, 287.

Ahl-Ahmad, H., Galili, S., & Gressel, J. (2006). Infertile interspecific hybrids between transgenically mitigated *Nicotiana tabacum* and *Nicotiana sylvestris* did not backcross to *N. sylvestris*. *Plant Science, 170*, 953–961.

Ahuja, M. R. (1962). A cytogenetic study of heritable tumors in *Nicotiana* species hybrids. *Genetics, 47*, 865–880.

Anssour, S., Krugel, T., Sharbel, T. F., Saluz, H. P., Bonaventure, G., & Baldwin, I. T. (2009). Phenotypic, genetic and genomic consequences of natural and synthetic polyploidization of *Nicotiana attenuata* and *Nicotiana obtusifolia*. *Annals of Botany (London), 103*, 1207–1217. https://doi.org/10.1093/aob/mcp058

Appa Rao, K., & Krishna Murthy, K. V. (1963). Studies on multiple polyploids in *Nicotiana*. *Genetica, 34*, 66–78.

Apparao, K., & Ramavarma, K. T. (1972). A gynogenic haploid plant in *N. gossei*. *Current Science, 41*, 645–646.

Apparao, K., & Ramavarma, K. T. (1974). Cytogenetics of the amphidiploid *Nicotiana benthamiana* × *N. glutinosa*. *Cytologia, 39*, 379–382.

Apparao, K., Ramavarma, K. T., & Joshi, B. G. (1980). Interspecific hybridization and breeding for pest resistance in tobacco. *Tobacco Science, 24*, 46–48.

Apple, J. L. (1962). Transfer of resistance to black shank (Phytophthora parasitica var. *Nicotiana*e) from *Nicotiana*. *Phytopathology, 52*, 1.

Ar-Rushdi, A. H. (1955). Inheritance in *Nicotiana tabacum*. XXVI. Sterility genes from *tomentosa*e species. *Journal of Genetics, 54*, 9–22.

Ar-Rushdi, A. H. (1957). The cytogenetics of variegation in a species hybrid in *Nicotiana*. *Genetics, 42*, 312–325.

Aviv, D., & Galun, E. (1986). Restoration of male fertile *Nicotiana* by fusion of protoplasts derived from two different male sterile cybrids. *Plant Molecular Biology, 7*, 411–417.

Aviv, D., & Galun, E. (1987). Chondriome analysis in sexual progenies of *Nicotiana* cybrids. *Theoretical and Applied Genetics, 73*, 821–826.

Aviv, D., Fluhr, R., Edelman, M., & Galun, E. (1980). Progeny analysis of interspecific hybrids: *Nicotiana tabacum* (CMS) + *N.sylvestris* with respect to nuclear and chloroplast markers. *Theoretical and Applied Genetics, 56*, 145–150.

Baalawy, H. A., & Fox, J. A. (1971). Resistance to Osborne's cyst nematode in selected *Nicotiana* species. *Journal of Nematology, 3*, 395–398.

Bailov, D., Palakarcheva, M., & Daskalov, S. (1964). Novi amfidiplodi *Nicotiana tabacum* x *N. debneyi* Dom. (New amphidiploids of *Nicotiana tabacum* × *N. debneyi* Dom). *Rastienievod Nauki, 7*, 3–16.

Bannikova, V. P. (1965a). Cytoembryology of distant hybrids. II. Disturbances in the fertilization process on crossing *Nicotiana glutinosa* L. with *N. rustica*. *Ukrainian Botanical Journal, 22*, 40–46.

Bannikova, V. P. (1965b). Comparative study on the development of seed resulting from self-pollination and hybridization of *N. rustica* with *N. paniculata*. *Ukrainian Botanical Journal, 22*, 58–66.

Bates, G. W. (1985). Electrical fusion for optimal formation of protoplast heterokaryons in *Nicotiana*. *Planta, 165*, 217–224.

Bates, G. W. (1990). Asymmetric hybridization between *Nicotiana tabacum* and *N. repanda* by donor recipient protoplast fusion: Transfer of TMV resistance. *Theoretical and Applied Genetics, 80*, 481–487.

Bates, G. W., & Hasenkampf, C. A. (1985). Culture of plant somatic hybrids following electrical fusion. *Theoretical and Applied Genetics, 70*, 227–233.

Bates, G. W., Hasenkampf, C. A., Contolini, C. F., & Piastuch, W. C. (1987). Asymmetric hybridization in *Nicotiana* by fusion of irradiated protoplasts. *Theoretical and Applied Genetics, 74*, 718–726.

Bawolska, M., Kobus, I., & Jurzysta, A. (1978). A characteristic of spontaneous hybrid *Nicotiana setchellii* × *N. otophora*. *Genetica Polonica, 19*, 49–59.

Bellair, G. (1913). Recroissees entre elles deux species qui sont degagees d'un hybride n'obeissent plus a la loi Mendelienne de la dominance. *Comptes Rendus IV Conference Internationale de Genetique*, 201–203.

Berbeć, J. (1964). Experience with developing domestic tobacco varieties resistant to blue mold at Puławy (in Polish). *Wiad Tyt, 11*, 170–172.

Berbeć, J. (1971). Investigations on reciprocal F_1 hybrids of *N. tabacum* var. Virginica Comes with *N. glauca* Grah (in Polish). *Pamiętnik Puławski*, (43), 5–32.

Berbeć, A. (1978). *Investigations of the interspecific hybrid Nicotiana tabacum L x N. benavidesii L.* Communicate delivered at the CORESTA Symposium, Sofia, Bulgaria, October 1–5.

Berbeć, A. (1980). Chromosome substitution and gene mutation concurring to produce a white-flower derivative of the interspecific hybrid *Nicotiana tabacum* L. × *N. benavidesii* Goodspeed. *Genetica Polonica, 21*, 283–289.

Berbeć, A. (1986). Genetic and breeding potential studies of hybrid populations derived from crossing cultivated tobacco *N. tabacum* L. with the species *N. benavidesii* Goodspeed (in Polish). *Biul CLTP, 1–4*, 29–43.

Berbeć, A. (1987a). Chromosome pairing and pollen fertility in the interspecific F_1 hybrids *Nicotiana tabacum* L. × *N. benavidesii* Goodspeed, *N. knightiana* Goodspeed × *N. tabacum* and *N. raimondii* Macbride × *N. tabacum*. *Genetica Polonica, 28*, 263–269.

Berbeć, A. (1987b). Cytogenetical study on *Nicotiana tabacum* L. cv. Nadwiślański Mały (2x and 4x) × *Nicotiana alata* link et Otto hybrids. *Genetica Polonica, 28*, 251–261.

Berbeć, A. (1988). Morphology, cytogenetics and resistance to potato virus Y of the amphidiploid *Nicotiana raimondii* Macbride × *N. tabacum* L. *Genetica Polonica, 29*, 41–52.

Berbeć, J., & Berbeć, A. (1992). Męska jałowość u tytoniu (*Nicotiana tabacum* L.) uzyskana drogą jednoetapowego podstawienia cytoplazmy gatunku *Nicotiana eastii* Kostoff. (Male sterility in tobacco obtained by one-step substitution of the cytoplasm from the wild species *Nicotiana eastii* Kostoff). *Pamiętnik Puławski*, (100), 136–139. (in Polish).

Berbeć, A., & Doroszewska, T. (1981). Investigations of the interspecific hybrid *Nicotiana amplexicaulis* Burbidge × *Nicotiana tabacum*. *Genetica Polonica, 22*, 197–207.

Berbeć, A., & Doroszewska, T. (1992). Alloplazmatyczne formy tytoniu uprawnego (*Nicotiana tabacum* L.) z podstawioną cytoplazmą gatunków *Nicotiana amplexicaulis*, *N. knightiana* i *N. raimondii*. *Pamiętnik Puławski, 100*, 141–150.

Berbeć, J., & Opoka, B. (1966). Badania nad krzyżówką *Nicotiana glauca* x *N. tabacum* (Resistance of the interspecific hybrids *Nicotiana tabacum* L. × *N. glauca* Grah. to *Lycopersicum virus*). *Pamiętnik Puławski, 21*, 195–209. (Inf. PAN 7:213-340) (in Polish).

Berbeć, A., & Doroszewska, T. (2020). The use of *Nicotiana* Species in tobacco Improvement. In N. Ivanov, N. Sierro, & M. Peitsch (Eds.), *The tobacco plant genome. Compendium of plant genomes*. Springer.

Berbeć, J., Berbeć, A., & Czop-Doroszewska, T. (1979). *Investigations on obtaining new varieties and breeding lines of Nicotiana tabacum resistant to virus leaf rib necrosis (Potato Virus Y) and virus degeneration of plants by Lycopersicum virus 3*. USDA Grant no. FG-Po-341 (JB-6) Annual Report 2 (1978-1979) (archived in IUNG-PIB, Puławy, Poland).

Berbeć, J., Berbeć, A., Glazewska, Z., Opoka, B., Kobus, I., & Czop, T. (1982). *Investigations on obtaining new varieties and breeding lines of Nicotiana tabacum resistant to virus leaf rib necrosis (potato virus Y) and virus degeneration of plants by Lycopersicum virus 3* (USDA Grant no. FG-Po-341 (JB-6) final report). Institute of Soils Science and Plant Cultivation, Pulawy, Poland (archived in IUNG-PIB, Puławy, Poland).

Bindler, G., Plieske, J., Bakaher, N., et al. (2011). A high density genetic map of tobacco (*Nicotiana tabacum* L.) obtained from large scale microsatellite marker development. *Theoretical and Applied Genetics, 123*, 219–230. https://doi.org/10.1007/s00122-011-1578-8

Blakeslee, A. F., & Avery, A. G. (1937). Methods of inducing doubling of chromosomes in plants by treatment with colchicine. *The Journal of Heredity, 28*, 393–411.

Bolsunov, I. (1963). *Amphihaploid obtained by hybridization between aurea forms of Nicotiana rustica L. and Nicotiana tabacum L.* Congr Intl Genet La Haye (The Hague) Abstract 11 (CORESTA Bull Inf 1964, 0104).

Bolsunov I (1970) New *Nicotiana* amphidiploid as a valuable starting material for the control of virulent blue mould lines. Intl Tob Sci Congr Proc Hamburg:160.

Bomblies, K. (2009). Too much of a good thing? Hybrid necrosis as a by-product of plant immune system diversification. *Botany, 87*, 1013–1022.

Brandle, J. E., Court, W. A., & Gleddie, S. (1992). The application of somatic hybridization to a tobacco breeding program. *Tobacco Chemical Research Conference, 46: 30*, abstr. 17.

Breisser, E. (1934). Zytologisch-genetische Untersuchungen an der Bastarden *Nicotiana* tabacum × *N. rusbyi* und ihren Rueckkreuzungen zu *N. rusbyi* und *N. sylvestris* I. Die Nachcommen aus reduzierten Eizellen. *Zeitschrift fuer indukt. Abstamm. u. Vererbungsl., 67*, 115–151.

Brieger, F. (1928). Uber die Vermehrung der Chromosomenzahl bei dem Bastard *Nicotiana tabacum* L. x *N. Rusbyi* Britt. *Zeitschrift Inductive Abstam u Vererbungslehre, 47*, 1–53.

Brink, R. A., & Cooper, D. C. (1941). Incomplete seed failure as a result of somatoplastic sterility. *Genetics, 26*, 487–505.

Brongniart, A., & Gris, A. (1861). Note sur le genre *Joinvillea* de Gaudichaud et sur la famille des Flagellariées. *Bulletin de la Société botanique de France, 8*(5), 264–269. https://doi.org/10.1080/00378941.1861.10829565

Burk, L. G. (1960). Male-sterile flower anomalies in interspecific tobacco hybrids. *The Journal of Heredity, 51*, 27–31.

Burk, L. G. (1967). An interspecific bridge-cross—*Nicotiana repanda* through *N. sylvestris* to *N. tabacum*. *The Journal of Heredity, 58*, 215–218.

Burk. (1972). Viable hybrids from monosomics *N. tabacum* by *N. langsdorffii*. *Tobacco Science, 16*, 43–45.

Burk, L. G., & Chaplin, J. F. (1979). Hybridization (in) *Nicotiana*.Procedures for experimental use. (ed. RD Durbin). *USDA Department of Agriculture Technical Bulletin, 1586*, 23–27.

Burk, L. G., & Dean, C. E. (1975). Hybrid fertility and aphid resistance in the cross *Nicotiana tabacum* × *N. gossei*. *Euphytica, 24*, 59–63.

Burk, L. G., & Durbin, R. D. (1978). Reaction of *Nicotiana* species to tentoxin. *The Journal of Heredity, 69*, 117–120.

Burk, L. G., & Neas, M. O. (1964). 4n(*N. tabacum* × *N. nudicaulis*), a colchicine induced fertile hybrid. *Tobacco Science, 8*, 65–66.

Burk, L. G., Gerstel, D. U., & Wernsman, E. A. (1979). Maternal haploids of *Nicotiana tabacum* L. from seed. *Science, 206*, 585.

Burk, L. G., Gooding, G. V., & Chaplin, J. F. (1982). Reaction of *Nicotiana* species and cultivars or breeding lines of *Nicotiana tabacum* to three strains of Potato Virus Y. *Tobacco Science, 26*, 85–88.

Busconi, M., Reggi, S., Lorenzoni, C., & Fogher, C. (2010). Interspecific crosses in the genus *Nicotiana*. *Proceedings of the 56th Italian Society of Agricultural Genetics Annual Congress*. Post 7.02.

Butenko, R. G., & Luneva, M. Z. (1966). Применение метода стерильных культур для выращивания отдаленных гибридов *Nicotiana*. Use of the sterile culture technique for the culturing of distant hybrids of Nicotiana (in Russian). *Fizjologija Rastenij, 13*, 733–736.

Butenko, R. G., Ternovsky, M. F., Semenova, I. V., & Moiseeva, M. E. (1970). The significance of tissue culture in the creation of initial material on the immunity to tobacco diseases, *CORESTA 5th Intl Tobacco Scientific Congress Inf Bul Abstr, A102*, 12.

Calitz, P. C., & Milne, D. L. (1962). Reaction of *Nicotiana* species and species crosses to the root knot nematode, Meloidogyne javanica. *South African Journal of Agricultural Science, 5*, 123–126.

Carlson, P. S., Smith, H. H., & Dearing, R. D. (1972). Parasexual interspecific plant hybridization. *Proceedings of the National Academy of Sciences of the United States of America, 69*, 2292–2294.

Chambers, O., Rice, J. H., Mundell, R., Millwood, R. J., Stewart, C. N., & Davies, H. M. (2011). *Field evaluation of an interspecific hybrid as a containment strategy for plant-made pharmaceutical applications in tobacco*. TSRC, Tob Sci Res Conf 65, abstr 62.

Chaplin, J. F. (1954). Use of tetraploid *Nicotiana tabacum* in the initial cross for transferring black shank resistance factors from *N. plumbaginifolia* to *N. tabacum*. *Proceedings of Association of Southern Agricultural Workers*, 56.

Chaplin, J. F. (1959). *Interspecific hybridization, gene transfer and chromosomal substitution in Nicotiana*. PhD thesis, North Carolina State University, Raleigh.

Chaplin, J. F. (1962). Transfer of black shank resistance from *Nicotiana plumbaginifolia* to flue-cured *N. tabacum*. *Tobacco Science, 6*, 184–189.

Chaplin, J. F. (1964). Use of male sterile tobaccos in the production of hybrid seed. *Tobacco Science, 8*, 105–109.

Chaplin, J. F., & Mann, T. J. (1961). Interspecific hybridization, gene transfer and chromosomal substitution in *Nicotiana*. *North Carolina Agricultural Experiment Station Technical Bulletin, 128*.

Chaplin, J. F., & Sisson, V. A. (1984). Interspecific hybridization between *N. rustica* and *N. tabacum* for different alkaloid levels. *Bull Spec CORESTA Congress, 93*, A17. ISSN.0525-6240.

Chen, S. Y. (1971). Preliminary study on interspecific transfer of leaf curl resistance in tobacco. *Taiwan Tob Wine Monop Bur Tob Res Annu Rep*, 47–50.

Chen, C. E. Z., & Lin, H.-X. (2016). Evolution and molecular control of hybrid incompatibility in plants. *Frontiers in Plant Science, 7*, 1208. https://doi.org/10.3389/fpls.2016.01208

Chen, X., Fang, D., Gao, Y., Xiao, B., & Li, Y. (2013). Somatic hybrids between *Nicotiana stocktonii* and *N. tabacum*, as a source of black shank (*Phytophthora parasitica* var. *Nicotianae*) resistance in *N. tabacum*. *CORESTA Meet Brufa di Torgiano AP, 31*.

Chimoyo, H. M., & Pupert, E. A. (1988). *The effect of UV irradiation, toluidine blue, and environment on maternal haploid frequencies from the cross between Nicotiana tabacum and N. africana* (p. 71). Bull Spec CORESTA Congress. AP-2.

Chobanova, Y. (1977). Nicotiana *plumbaginofolia* and *N. rustica* hybrids with *N. tabacum*. *Genetics and Plant Breeding, 10*, 50–57.

Choi, S. J., & Hong, B. H. (1992). Production of sexual hybrids *Nicotiana rustica* X *N. tabacum* via *in vitro* culture of fertilized ovules. *Journal of Korean Society for Research on Tobacco Science, 14*, 3–11.

Choi, S. J., & Lee, S. C. (1991). Interspecific hybrid of *Nicotiana trigonophylla* x *N. tabacum* through in vitro culture of fertilized ovules. *Journal of Korean Society for Research on Tobacco Science, 15*, 123–129.

Choi, S. J., Lee, S. C., & Hong, B. H. (1993). Somatic hybridization between *Nicotiana rustica* and *N. tabacum* through protoplast fusion. *Journal of the Korean Society of Tobacco Science, 14*, 1–11.

Choi, S. J., Keum, W. S., Jung, S. H., Chung, Y. H., & Cho, M. C. (1998). Interspecific hybridization between *Nicotiana tabacum* and *N. repanda* by in vitro pollination and ovule culture; transfer to TMV resistance. *Bull Spec CORESTA Congr, Brighton, 130*, P16.

Christie, S. R. (1969). *Nicotiana* hybrid developed as a host for plant viruses. *Plant Disease Report, 53*, 939–941.

Christoff, M. (1928). Cytological studies in the genus *Nicotiana*. *Genetics, 13*, 233–277.

Chung, C. S., Nakjima, T., & Takeda, G. (1988). Interspecific hybridization between *Nicotiana* trigonophylla Dun. And *N. tabacum* L. through ovule culture. *Japanese Journal of Breeding, 38*, 319–326.

Chung, S. C., Yoshida, K. T., & Takeda, G. (1996). Reproductive characteristics of amphidiploids derived from an interspecific hybrid between *Nicotiana trigonophylla* Dun. and *N. tabacum* L. *Japanese Journal of Breeding, 46*, 29–33.

Chupeau, Y. (1987). Somatic hybridization : Examples of application to tobacco. *Ann Tabac 1 Sect, 2*, 51–56.

Clausen, R. E. (1928). Interspecific hybridization in *Nicotiana*. VII. The cytology of hybrids of the synthetic species, digluta, with its parents: *Glutinosa* and *tabacum*. *University of California Publications in Botany, II*, 177–211.

Clausen, R. E. (1931). Inheritance in *Nicotiana tabacum*. XI. The fluted assemblage. *American Naturalist, 65*, 316–331.

Clausen, R. E., & Cameron, D. R. (1944). Inheritance in *Nicotiana tabacum*. XVIII. Monosomic analysis. *Genetics, 29*, 447–477.

Clausen, R. E., & Cameron, D. R. (1957). Inheritance in *Nicotiana tabacum*. XXVIII. The cytogenetics of introgression. *Proceedings of the National Academy of Sciences of the United States of America, 43*, 908–913.

Clausen, R. E., & Goodspeed, T. H. (1925). Interspecific hybridization in *Nicotiana*. II. A tetraploid *glutinosa-tabacum* hybrid, an experimental verification of Winge's hypothesis. *Genetics, 10*, 278–284.

Clausen, R. E., & Goodspeed, T. H. (1926). Interspecific hybridization in *Nicotiana*. III. The monosomic Tabacum derivative, "corrugated," from the *sylvestris-Tabacum* hybrid. *University of California Publications in Botany, 11*, 61–82.

Clausen, R. E., & Lammerts, W. E. (1929). Interspecific hybridization in *Nicotiana*. *Haploid and diploid merogony. American Naturalist, 63*, 270–322.

Clausen, R. E., & Mann, M. C. (1924). Inheritance in *Nicotiana tabacum*. V. The occurrence of haploid plants in interspecific progenies. *Proceedings of the National Academy of Sciences of the United States of America, 10*, 121–124.

Clayton, E. E. (1947). A wildfire resistant tobacco. *The Journal of Heredity, 38*, 35–40.

Clayton, E. E. (1950). Male sterile tobacco. *The Journal of Heredity, 4*, 171–175.

Clayton, E. E. (1954). Identifying disease resistance suited to interspecific transfer. *The Journal of Heredity, 45*, 273–277.

Clayton, E. E. (1958). The genetics and breeding progress in tobacco during the last 50 years. *Agronomy Journal, 50*, 352–356.

Clayton, E. E. (1968). The transfer of blue mold resistance to tobacco from *Nicotiana debneyi*. Part IV. Breeding progress 1957-1967. *Tobacco Science, 12*, 112–124.

Clayton, E. E., Heggestad, H. E., Grosso, J. J., & Burk, L. G. (1967). The transfer of blue mold resistance to tobacco from *Nicotiana debneyi*. Part I. Breeding Progress 1937-1954. *Tobacco Science, 11*, 91–99.

Dang, J., Wang, J., Yang, Y., Shang, W., Guo, K., & Liang, G. (2019). Resistance of *Nicotiana tabacum* to Phytophthora parasitica var. nicotianae race 0 is enhanced by the addition of *N. plumbaginifolia* chromosome 9 with a slight effect on host genomic expression. *Crop Science, 59*, 2667–2678. https://doi.org/10.2135/cropsci2019.01.0005

Davey, M. (2017). Somatic hybridization. In B. Thomas, B. G. Murray, & D. J. B. T. Murphy (Eds.), *Encyclopedia of applied plant sciences* (2nd ed., pp. 480–486). Academic. https://doi.org/10.1016/B978-0-12-394807-6.00151-9

Davey, M. R., Blackhall, N. W., Lowe, K. C., & Power, J. B. (1996). Gametosomatic hybridization. In S. M. Jain, S. K. Sopory, & R. E. Veilleux (Eds.), *In Vitro haploid production in higher plants. Current plant science and biotechnology in agriculture* (Vol. 24). Springer. https://doi.org/10.1007/978-94-017-0477-9_14

Depta, A., & Doroszewska, T. (2019). Development and cytometric evaluation of interspecific F_1 hybrids *N. tabacum* × *N. africana*. *Polish Journal of Agronomy, 38*, 3–14.

Desprez, B., Chupeau, M. K., Vermeulen, A., Delbreil, B., Chupeau, Y., Bourgin, J.-P., Giddings, G. D., & Rees, H. (1992). A *Nicotiana* gametosomatic hybrid and its progenies. *Journal of Experimental Botany, 43*, 419–425.

DeVerna, J. W. (1984). *In vitro-facilitated wide hybridization in Nicotiana.* PhD Dissertation,. University of Kentucky.

DeVerna, J. W., Myers, J. R., & Collins, G. B. (1987). Bypassing prefertilization barriers to hybridization in *Nicotiana* using in vitro pollination and fertilization. *Theoretical and Applied Genetics, 73*, 665–671.

Dijak, M., Sproule, A., Keller, W., Pandeya, R., & Gleddie, S. (1991). Transformation of *Nicotiana tabacum, N. debneyi*, and *N. rustica* inheritance and protoplast expression of antibiotic resistance. *Plant Cell, Tissue and Organ Culture, 25*, 189–197.

Donaldson, P., Sproule, A., Bevis, E., Pandeya, R., Keller, W. A., & Gleddie, S. (1993). Non-random chloroplast segregation in *Nicotiana tabacum* + *N.Rustica* somatic hybrids selected by dual nuclear-encoded resistance. *Theoretical and Applied Genetics, 86*, 465–473.

Donaldson, P. A., Bevis, E., Pandeya, R. S., & Gleddie, S. C. (1994). Random chloroplast segregation and frequent mt DNA rearrangements in fertile somatic hybrids between *Nicotiana tabacum* L. and *N. glutinosa* L. *Theoretical and Applied Genetics, 87*, 900–908.

Donaldson, P. A., Bevis, E., Pandeya, R., & Gleddie, S. C. (1995). Rare symmetric and asymmetric *Nicotiana tabacum* (+) *N. megalosiphon* somatic hybrids recovered by selection for nuclear-encoded resistance genes and in the absence of genome inactivation. *Theoretical and Applied Genetics, 91*, 747–755.

Dorossiev, L., & Palakarcheva, M. (1990). Production of the resistant forms to aphids (*Myzus persicae* Sulzer), powdery mildew (*Erysiphe cichoracearum* DC) and the tobacco mosaic virus (*Nicotiana virus* L) by the hybridization of *N. gossei* D ($2n = 36$) × *N. tabacum* ($2n = 48$) and using in vitro methods. *Comptes rendus de l'Académie bulgare des Sciences, 43*, 79–81.

Dorossiev, L., Palakarcheva, M., Stanoeva, M., & Petkova, M. (1978). Overcoming the sterility in F_1 of interspecific hybrids of the genus *Nicotiana* using the methods of tissue culture. *Bulletin Spéc CORESTA, Symposium Sofia*, 80–81. P 03.

Dorossiev, L., Palakarcheva, M., & Yancheva, A. (1990). Application of *in vitro* methods in the development of disease resistant oriental tobacco hybrids and lines. *Genetics and Breeding, 23*, 306–315.

Doroszewska, T., & Berbeć, A. (1990). Investigations of the hybrids of *Nicotiana tabacum* L. × *Nicotiana africana* Merxm. *Bul Spéc CORESTA Symposium Kallithea, Greece:176*, P19.

Doroszewska, T., & Berbeć, A. (1996). Chromosome pairing and microsporogenesis in interspecific F_1 hybrids of *Nicotiana africana* with different cultivars of *N.Tabacum. Journal of Genetics and Breeding, 50*, 75–82.

Doroszewska, T., & Berbeć, A. (2000). Cytogenetical investigations of polyploid interspecific hybrids of *Nicotiana africana* with different cultivars of *N. tabacum. Journal of Genetics and Breeding, 54*, 77–82.

Douglas, G. C., Wetter, L. R., Keller, W. A., & Setterfield, S. (1983). Production of sexual hybrids of *Nicotiana rustica* × *Nicotiana tabacum* and *N. rustica* × *N. glutinosa* via *in vitro* culture of fertilized ovules. *Z Pflanzenzuecht, 90*, 116–129.

Dragoeva, A., Jankulova, M., Gorinova, N., Atanassov, I., & Atanassov, A. (1977). Somatic hybridization between *N. tabacum* (NR−, Km+) and *N. sanderae*. Hybrid genome constitution and expression of resistance to tomato spotted wilt virus. *Biotechnology & Biotechnological Equipment, 11*, 12–16.

Durbin, R. D., & Uchytil, T. F. (1977). Cytoplasmic inheritance of chloroplast coupling factor 1 subunits. *Biochemical Genetics, 15*, 1143–1146.

Eapen, P., Cates, J., Mundell, R., Palmer, K. E., & Fuqua, J. L. (2020). In preparation for outdoor pharming: Griffithsin can be expressed in *Nicotiana excelsiana* and retains activity after storage as silage. *Frontiers in Bioengineering and Biotechnology, 8*, 199. https://doi.org/10.3389/fbioe.2020.00199

East, E. M. (1928). The genetics of the genus *Nicotiana*. *Bibliography Genetics, 4*, 243–318.

East, E. M. (1930). The origin of the plants of maternal type which occur in connection with interspecific hybridizations. *Proceedings of the National Academy of Sciences of the United States of America, 16*, 377–380.

East, E. M. (1935). Genetic reactions in *Nicotiana.* I. Compatibility. *Genetics, 20*, 403–413.

East, E. M., & Hayes, H. K. (1912). Heterozygosis in evolution and in plant breeding. *USDA Bul, 243*, 26–32.

Eghis, S. A. (1927). Опыты по межвидовой гибридизации, в роде *Nicotiana*. Гибридизация между видами *N.tabacum* x *N.rustica* (Experiments on interspecific hybridisation in the genus *Nicotiana*. (Hybridization bwtween the species *N.tabacum* x *N.rustica*)). *Труды по прикладной ботан генет селекции Bulletin Applied Botany Genetics and Plant Breeding Leningrad, 17*, 151–190.

Eghis, S. A. (1930). Experiments on interspecific hybridization in the genus *Nicotiana.* II. The fertile hybrids between *N. tabacum* L. and *N. sylvestris* Speg. and Games. *Proceedings of USSR Congress on Genetics, Plant and Animal Breeding, 2*.

Evans, D. A., Wetter, L. R., & Gamborg, O. L. (1980). Somatic hybrid plants of *Nicotiana glauca* and *Nicotiana tabacum* obtained by protoplast fusion. *Physiologia Plantarum, 48*, 225–230.

Evans, D. A., Flick, C. E., & Jensen, R. A. (1981). Disease resistance: Incorporation into sexually incompatible somatic hybrids of the genus *Nicotiana. Science, 213*, 907–909.

Evans, D. A., Flick, C. E., & Kut, S. A. (1982). Comparison of *Nicotiana tabacum* and *Nicotiana nesophila* hybrids produced by ovule culture and protoplast fusion. *Theoretical and Applied Genetics, 62*, 193–198.

Evans, D. A., Bravo, J. E., Kut, S. A., & Flick, C. E. (1983). Genetic behaviour of somatic hybrids in the genus *Nicotiana*: *N. otophora* + *N. tabacum* and *N. sylvestris* + *N. tabacum. Theoretical and Applied Genetics, 65*, 93–101.

Fardy, A., & Hitier, H. (1945). Espèces tétraploides et hybrides interspécifiques amphidiploides et triples diploides ed *Nicotiana*, obtenus par l'action de la colchicine. *Mémorial du Serv d'Expl Industr des Tabacs et des Allumettes (SEITA) Ser B*, 1–117.

Fardy, A., & Hitier, H. (1947). Observations sur les hybrides polygenomiques de *Nicotiana* et leurs descendants. *Comptes Rendus de l'Académie des Sciences, 224*, 292–294.

Fitter, J. T., Thomas, M. R., Chen, N., & Ray, J. R. (2005). Investigation of Nicotiana *tabacum* (+) *N. suaveolens* cybrids with carpelloid stamens. *Plant Physiology, 162*, 225–235.

Fitzmaurice, W. P. (2002). Interspecific *Nicotiana* hybrids and their progeny. *US Patent, 6344597*.

Flick, C. E., & Evans, D. A. (1982). Evaluation of cytoplasmic segregation in somatic hybrids of *Nicotiana*: Tentoxin sensitivity. *The Journal of Heredity*, 264–266.

Foster, H. H. (1943). Resistance in the genus *Nicotiana* to *Phytophtora parasitica* Dastur var. *nicotianae* Tucker. *Phytopathology, 33*, 403–404.

Fuentes, I., Stegemann, S., Golczyk, H., Karcher, D., & Bock, R. (2014). Horizontal genome transfer as an asexual path to the formation of new species. *Nature, 511*, 232–235.

Furusato. (1960). Breeding of amphidiploids from a hybrid between *Nicotiana rustica* and *N. tabacum. National Institution on Genetics Annual Report, 11*, 61.

Gajos, Z. (1975). Rozmnażanie roślin międzygatunkowego mieszańca *Nicotiana tabacum* L. x *Nicotiana alata* link z kalusa wytwarzanego in vivo (Propagation of interspecific hybrid *Nicotiana tabacum* L. x *Nicotiana alata* link. From callus produced i*n vivo*). *Biul Inf Centr Lab Przem Tyton, 1-2*, 15–20. (in Polish).

Gajos, Z. (1979). Próby wykorzystania mieszańców *Nicotiana tabacum* L. x *Nicotiana otophora* Gris. W hodowli tytoniu odpornego na *Peronospora tabacina* Adam (PT-2) attempt to use hybrids of *Nicotiana tabacum* L. × *N. otophora* Gris for breeding tobacco resistant to *Peronospora tabacina* Adam (PT-2) and other diseases (in Polish). *Biul Inf Centr Lab Przem Tyton, 1-2*, 11–23.

Gajos, Z. (1981). Przeniesienie odporności na wirus brązowej plamistości pomidora (Tomato spotted wilt virus) z *Nicotiana alata* Link et Otto do tytoniu szlachetnego przez skrzyżowanie obu gatunków (Transfer of resistance to tomatto spotted wilt virus from *N. Nicotiana alata* Link

et Otto to cultivated tobacco by crossing the two species). *Biul Inf Centr Lab Przem Tyton, 1-2*, 3–24. (in Polish).
García Cruz, H., Perez-Lara, E., & Castro, M. C. (2008). Advances in obtaining interespecific hybrids (*Nicotiana megalosiphon* Heurck and Muller x *Nicotiana tabacum* L.) resistant to blue mold *Peronospora hyoscyami* sp. starting from seedling stage (In Spanish). *Cuba Tabaco, 9*(1), 37–41.
Gentscheff, G. (1931). Investigations on interspecific hybridisation in the genus *Nicotiana*. *Zemedlska Misl, 2*, 91–111. (in Bulgarian).
Gerstel, D. U. (1960). Segregation in new allopolyploids of *Nicotiana*. Comparison of 6x (*N. tabacum* × *N. tomentosiformis*) and 6x (*N. tabacum* × *N. otophora*). *Genetics, 45*, 1713–1734.
Gerstel, D. U., & Burns, J. A. (1983). Anther restoration in male sterile tobacco by nucleolus organizing chromosomes: A restorer from *Nicotiana bigelovii*. *Tobacco Science, 27*, 151–153.
Gerstel, D. U., & Wernsman, E. A. (1979). Maternal haploids of *Nicotiana tabacum* from seed. *Science, 206*, 585.
Gerstel, D. U., Burns, J. A., & Burk, L. G. (1979). Interspecific hybridizations with an African tobacco, *Nicotiana africana* Merxm. *The Journal of Heredity, 70*, 342–344.
Giddings, G. D., & Rees, H. (1992). A *Nicotiana* gametosomatic hybrid and its progenies. *Journal of Experimental Botany, 43*, 419–425.
Gillham, F. E. M., Wark, D. C., & Harrigan, E. K. S. (1977). Disease resistant flue-cured tobacco breeding lines for North Queensland. I. Resistance to blue mould, Peronospora tabacina. *Australian Journal of Experimental Agriculture and Animal Husbandry, 17*, 652–558.
Gisquet, P., Dufrenoy, J., & Dusseau, A. (1940). Hybrides interspecifiques de *Nicotiana tabacum* var. Purpurea × *N. petunoides* var. Sylvestris. *Science Genetics (Torino), 2*, 67–78.
Gleba, Y. Y., Kolesnik, N. N., Meskene, I. W., Cherep, N. N., & Parokonny, A. S. (1984). Transmission genetics of somatic hybridization process in Nicotiana. 1. Hybrids and cybrids among the regenerants from cloned protoplast fusion products. *Theoretical and Applied Genetics, 69*, 121–128.
Goldschmidt, E. E. (2014). Plant grafting: New mechanisms, evolutionary implications. *Frontiers in Plant Science, 5*, 727. https://doi.org/10.3389/fpls.2014.00727
Goodspeed, T. H. (1915). Parthenogenesis, parthenocarpy and phenospermy in *Nicotiana*. *University of California publications in Botany, 5*, 249–272.
Goodspeed, T. H. (1945). Cytotaxonomy of *Nicotiana*. *The Botanical Review, 11*, 533–592.
Goodspeed, T. H. (1954). *The Genus Nicotiana*. Chronica Botanica Co.
Goodspeed, T., & Bradley, M. (1942). Amphidiploidy. *The Botanical Review, 8*, 272–316.
Goodspeed, T. H., & Clausen, R. E. (1917). Mendelian factor differences versus reaction system contrasts in heredity. *The American Naturalist, 51*(31–46), 92–101.
Goodspeed, T. H., & Clausen, R. E. (1928). Interspecific hybridization in *Nicotiana*. VIII. The *sylvestris-tomentosa-tabacum* triangle and its bearing on the origin of tobacco. *University of California Publications in Botany, 11*, 127–140.
Goodspeed, T. H., & Thompson, M. C. (1959). *Cytotaxonomy of Nicotiana II*. University of California Botanical Garden.
Gopinath, D. M., Krishnamurthy, K. V., & Krishnamurthy, A. S. (1970). A note on the cytogenetic status of *Nicotiana amplexicaulis*. *BurbidgeCurrent Science, 39*, 410–411.
Grebenkin, A. P. (1968). Study of the reaction of certain wild *Nicotiana* species during the crossing with unisexual forms of tobacco. *Genetika, 4*, 5–10.
Grebenkin, A. P. (1970). The role of cytoplasm in interspecific crosses of tobacco. *CORESTA Intl Sci Tob Congress Bull Spec, 12*, A104.
Greenleaf, W. H. (1938). Induction of polyploidy in *Nicotiana*. *The Journal of Heredity, 29*, 451–464.
Greenleaf, W. H. (1941). Sterile amphidiploids: Their possible relation to the origin of *Nicotiana tabacum*. *The American Naturalist, 75*, 394–399.

Gurdon, C., Svab, Z., Feng, Y., Kumar, D., & Maliga, P. (2016). Cell-to-cell movement of mitochondria in plants. *PNAS, 113*, 3395–3400.

Hagimori, M., Matsui, M., Matsuzaki, T., Shinozaki, Y., Shinoda, T., & Harada, H. (1993). Production of somatic hybrids between *N. benthamiana* and *N. tabacum* and their resistance to aphids. *Plant Science, 91*, 213–222.

Hamill, J. D., Pental, D., & Cocking, E. C. (1984). Analysis of fertility in somatic hybrids of *N. rustica* and *N. tabacum* and progeny over two sexual generations. *Theoretical and Applied Genetics, 71*, 486–490.

Hamill, J. D., Watts, J. W., & King, J. M. (1987). Somatic hybridization between *N.Tabacum* and *N. Plumbaginifolia* by electrofusion of mesophyl protoplasts. *Journal of Plant Physiology, 129*, 111–118.

Hancock, W. G., Kuraparthy, V., Kernodie, S. P., & Lewis, R. S. (2015). Identification of maternal haploids of *Nicotiana tabacum* aided by transgenic expression of green fluorescent protein: Evidence for chromosome elimination in the N-*tabacum* × N-*africana* interspecific cross. *Molecular Breeding, 35*, 179.

He, H., Iizuka, T., Maekawa, M., et al. (2019). (2019) *Nicotiana suaveolens* accessions with different ploidy levels exhibit different reproductive isolation mechanisms in interspecific crosses with *Nicotiana tabacum. Plant Research, 132*, 461–471. https://doi.org/10.1007/s10265-019-01114-w

He H, Yokoi S, Tezuka T (2020) A high maternal genome excess causes severe seed abortion leading to ovary abscission in *Nicotiana* interploidy-interspecific crosses. Plant Dir:1–11.

He, H., Sadahisa, K., Yokoi, S., & Tezuka, T. (2022). Parental genome imbalance causes hybrid seed lethality as well as ovary abscission in interspecific and interploidy crosses in *Nicotiana.* Front plant Sci:899206, *13*. https://doi.org/10.3389/fpls.2022.899206

He, H., Shiragaki, K., & Tezuka, T. (2023). Understanding and overcoming hybrid lethality in seed and seedling stages as barriers to hybridization and gene flow. *Frontiers in Plant Science, 14*, 1219417. https://doi.org/10.3389/fpls.2023.1219417

Holmes, F. O. (1937a). Genes affecting response of *Nicotiana tabacum* hybrids to tobacco mosaic virus. *Science, 85*, 104–105.

Holmes, F. O. (1937b). Hereditary factors affecting tobacco-mosaic disease in solanaceous plants. *Phytopathology, 27*, 131–132.

Holmes, F. O. (1938). Inheritance of resistance to tobacco-mosaic disease in tobacco. *Phytopathology, 28*, 553–561.

Horlow, C., Goujaud, J., Lepingle, A., Missionier, C., & Bourgin, J. P. (1990). Transmission of paternal chloroplasts in tobacco (*Nicotiana tabacum*). *Plant Cell Reports, 9*, 249–252.

Horlow, C., Defrance, M. C., Pollien, J. M., Goujaud, J., Delon, R., & Pelletier, G. (1993). Transfer of cytoplasmic male sterility by spontaneous androgenesis in tobacco. *Euphytica, 66*, 45–53.

Horn, M. E., Kameya, T., Brotherton, J. E., & Widholm, J. M. (1983). The use of amino-acid analog resistance and plant regeneration ability to select somatic hybrids between *Nicotiana tabacum* and *N. glutinosa*. *Molecular & General Genetics, 192*, 235–240.

Hranov, A. (1970). Some problems on crossability between *Nicotiana tabacum* L. and *N. megalosiphon* Heurk et Muell (in Bulgarian). *Bulgarian Tobacco, 10*, 45–48.

Hu, C. (1956). Cytogenetic studies in *Nicotiana*, XII. Reduction divisions in hybrids between *N. langsdorffii* and three other species. *Japanese Journal of Breeding, 6*, 117–121.

Huesing, J., Jones, D., DeVema, J., Myers, J., Collins, G., Severson, R., & Sisson, V. (1989). Biochemical investigations of antiobiosis material in leaf exudate of wild *Nicotiana* species and interspecific hybrids. *Journal of Chemical Ecology, 15*, 1203–1217.

Iizuka, T., Kuboyama, T., Marubashi, W., Oda, M., & Tezuka, T. (2012). *Nicotiana debneyi* has a single dominant gene causing hybrid lethality in crosses with *N. tabacum. Euphytica, 186*, 321–328.

Ilcheva, V., & San, I. H. (1997). Hybridation somatique chez le gendre *Nicotiana*—revue bibliographique. *Ann Tabac, Sect, 2*(29), 9–25.

Ilcheva, V., San, L. H., & Dorlhac, F. (1997). Morphological and cytological characteristics of somatic hybrids between *N. tabacum* and *N. megalosiphon*. *CORESTA Agro/Phyto Meet, Montreux, PT13*.

Ilcheva, V., San, L. H., Dimitrov, B., & Zagorska, N. (2000). Morphological and cytological characteristics of somatic hybrids of *Nicotiana tabacum*L. + *N. megalosiphon* Heurk. et Muell. *In Vitro Cellular & Developmental Biology. Plant, 36*, 69–73.

Ilcheva, V., San, L. H., Zagorska, N., & Dimitrov, B. (2001). Production of male sterile interspecific somatic hybrids between transgenic *N. tabacum* (bar) and *N. rotundifolia* (npt ii) and their identification by AFLP analysis. *In Vitro Cellular & Developmental Biology. Plant, 37*, 496–502.

Inoue, E., Marubashi, W., & Niwa, M. (1994). Simple method for overcoming the lethality observed in the hybrid between *Nicotiana suaveolens* and *N. tabacum*. *Breeding Science, 44*, 333–336.

Inoue, E., Marubashi, W., & Niwa, M. (1996). Genomic factors controlling the lethality exhibited in the hybrid between *Nicotiana suaveolens* Lehm. and *N. tabacum* L. *Theoretical and Applied Genetics, 93*, 341–347.

Inoue, E., Marubashi, W., & Niwa, M. (1997). Improvement of the method for overcoming the hybrid lethality between *Nicotiana suaveolens* and *N. tabacum* by culture of F_1 seeds in liquid media containing cytokinins. *Japanese Journal of Breeding, 47*, 211–216.

Ivancheva-Gabrovska, T., & Manolov, A. (1982). Investigations on the resistance of interspecific hybrids *N. tabacum* x *N. alata* and *N. tabacum* x *N. sanderae* to tomatto spotted wilt virus. *Reports of the 2nd National Symposium of Plant Immunity, Plovdiv, 2*, 65–78.

Iwai, S., Kishi, C., Nakata, K., Kawashima, N., & Kubo, S. (1984). Production of interspecific hybrids by ovule culture. *Japanese Journal of Breeding, 34*(suppl 1), 60–61.

Iwai, S., Kishi, C., Nakata, K., & Kubo, S. (1985). Production of a hybrid of *N. repanda* × *N. tabacum* by ovule culture. *Plant Science, 41*, 175–178.

Iwai, S., Kishi, C., Nakata, K., & Kawashima, N. (1986). Production of *Nicotiana tabacum* x *N. acuminata* hybrid by ovule culture. *Plant Cell Reports, 5*, 403–404. [Tob.Abstr. 31:639 (1987)].

Izard, C., & Hitier, H. (1955). Observations sur un hybride complexe susceptible de produire des plantes a sterilite male. *Ann Du Tabac Seita, Sect, 2*(2), 1–13.

Kandra, G. (1984). Investigations of the interspecific hybridizations carried out in *Nicotiana* genus. *Dohanyipar*, 128–130. ISSN. 0865-1396.

Kasza, A., & Kandra, G. A. (1990). Protoplast fusion transferred cytoplasmic male sterility and its utilization in tobacco breeding. *CORESTA Inf Bul CORESTA Symp Hellas, 140*(A31).

Katsuyama, Y., Doi, M., Shioya, S., et al. (2021). The role of chaperone complex HSP90-SGT1-RAR1 as the associated machinery for hybrid inviability between Nicotiana gossei Domin and *N.* tabacum L. *Gene, 776*, 145443. https://doi.org/10.1016/j.gene.2021.145443

Kaul, M. H. L. (1988). *Male sterility in higher plants* (p. 10). Monogr Theor Appl Genet Springer Verlag.

Kawaguchi, K., Ohya, Y., Maekawa, M., Iizuka, T., Hasegawa, A., et al. (2021). Two *Nicotiana occidentalis* accessions enable gene identification for type II hybrid lethality by the cross to *N. sylvestris*. *Scientific Reports, 11*, 17093. https://doi.org/10.1038/s41598-021-96482-6

Kehr, A. E., & Smith, H. H. (1952). Multiple genome relationships in *Nicotiana*. *Cornell University Agricultural Experiment Station Memoir, 311*, 3–19.

Keum, W. S., Chung, Y. H., Choi, S. J., Myung, J. C., & Lee, S. C. (1991). Inheritance of resistance to potato virus Y—Necrosis strain of Nicotiana *africana*. *Journal of Korean Society for Research on Tobacco Science, 13*, 48–51.

Keum, W. S., Chung, Y. H., Choi, S. J., Myung, J. C., & Lee, S. C. (1994). Transfer of potato virus Y (necrotic strain) resistance from *Nicotiana africana* to *Nicotiana tabacum*. *Journal of Korean Society for Research on Tobacco Science, 16*, 102–107.

Kim, J. C., & Choi, S. J. (1991). Endonuclease restriction patterns of chloroplast DNA in somatic hybrids obtained by protoplast fusion of *Nicotiana tabacum* and *Nicotiana glutinosa*. *Korean Journal of Botany, 34*, 37–44.

Kincaid, R. R. (1949). Three interspecific hybrids of tobacco. *Phytopathology, 39*, 284–287.

Kitamura, S., Inoue, E., Ohmido, N., Fukui, K., & Tanaka, A. (2003). Chromosomal rearrangements in interspecific hybrids between *Nicotiana gossei* Domin and *N. tabacum* L., obtained by crossing with pollen exposed to helium ion beams or gamma-rays. *Nuclear Instruments and Methods in Physics Research Section B: Beam Interactions with Materials and Atoms, 206*, 548–552.

Kobus, I. (1971). Investigations in the polyploids of *Nicotiana tabacum* L. and their hybrids with wild tobacco species. *Genetica Polonica, 12*, 324–328.

Kostoff, D. (1930). Tumours and other malformations on certain *Nicotiana* hybrids. *Zbl Bakt II Abt., 81*, 244–260.

Kostoff, D. (1932). Pollen abortion in species hybrids. *Cytologia, 5*, 337–339.

Kostoff, D. (1935). Heritable tumors in plants experimentally produced. *Genetica, 17*, 367–376.

Kostoff, D. (1936). The origin of cultivated tobacco. *Current Science, 4*, 872.

Kostoff, D. (1937). Studies on polyploid hybrid plants (in Russian). XVI. *Nicotiana rustica* x *Nicotiana tabacum* amphidiploid. *CR Acadamic Science URSS, 11*, 453–455.

Kostoff, D. (1938a). Studies on polyploid plants XVIII. Cytogenetic studies on *Nicotiana sylvestris* x N. *tomentosiformis* hybrids and amphidiploids and their bearing on the problem of the origin of *N. tabacum. CR Acadamic Science URSS, 18*, 459–462.

Kostoff, D. (1938b). Studies on polyploid plants. XXI. Cytogenetic behaviour of the allopolyploid hybrids *Nicotiana glauca* Grah. × *Nicotiana langsdorffii* and their evolutionary significance. *Journal of Genetics, 37*, 129–209.

Kostoff, D. (1939a). Abnormal mitosis in tobacco plants forming hereditary tumours. *Nature, 144*, 599.

Kostoff, D. (1939b). Lethality of gametes conditioned by exchange of segments between partially homologous chromosomes in a *Nicotiana* species hybrid. *Current Science, 8*, 260.

Kostoff, D. (1943). *Cytogenetics of the genus Nicotiana*. State Printing House.

Kramer, M. G., & Reed, S. M. (1988). An evaluation of maternal nullihaploidy for the production of nullisomics of *Nicotiana tabacum* L. I. An interspecific hybridization approach. *Journal of Heredity, 79*, 24–27.

Krishnamurthy, K. V., & Gopinath, D. M. (1969). Enlargement of the genus *Nicotiana* I. Establishment of the stable amphiploids. *Japanese Journal of Genetics, 44*, 139–147.

Krishnamurthy, K. V., Murty, G. S., & Appa Rao, K. (1960). Cytogenetics of the trispecific hybrid *Nicotiana tabacum* × (*N. glutinosa* × *N. trigonophylla*) and its reciprocal. *Euphytica, 9*, 111–121.

Krügel, T. (2010). *Nicotiana* × *mierata* und *N.* × *obtusiata*—Zwei neue synthetische Hybriden. *Haussknechtia, 12*, 23.

Krusteva, D., Nikova, V., Vladova, R., Dimitrov, B., Petkova, A., & Iancheva, A. (2003). *Nicotiana benthamiana DomiN. as a source of aphid resistance in N. tabacum L.* CORESTA Meeting, Agro-Phyto Groups, 2003, Bucharest APOST11.

Kubo, T. (1985). Studies on hybrid breeding by means of cytoplasmic male sterility in tobacco. *Bulletin of the Iwata Tobacco Experiment Station, 17*, 69–138.

Kuboyama, T., Chung, C. S., & Takeda, G. (1994). The diversity of interspecific pollen-pistil incongruity in *Nicotiana*. *Sexual Plant Reproduction, 7*, 250–258.

Kumashiro, T., & Kubo, T. (1986). Cytoplasm transfer of *N. debneyi* to *N. tabacum* by protoplast fusion. *Japanese Journal of Breeding, 36*, 39–48.

Kumashiro, T., & Oinuma, T. (1985). Comparison of genetic variability among anther derived and ovule derived doubled haploid lines of tobacco. *Japanese Journal of Breeding, 35*, 301–310.

Kumashiro, T., Asahi, T., & Komari, T. (1988). A new source of cytoplasmic male sterile tobacco obtained by fusion between *N. tabacum* and X-irradiated *N. africana* protoplasts. *Plant Science, 55*, 247–254. (CORESTA Inf Bul 1989:79).

Larkina, N. I. (1980). Overcoming incompatibility between *Nicotiana* species by means of *in vitro* pollination. *Plant Breeding Abstract, 50*(8067), 690.

Larkina, N. I. (1983). Создание путем межвидовой гибридизации нового исходного материала для селекции табака (Development of starting material for tobacco improvement by using interspecific hybridization). DSc thesis (in Russian). WITIM.

Larkina, N. I. (2015). *Scientific principles of interspecific hybridization using maternal Nicotiana tabacum*. Monograph. All-Russian Institute of Tobacco and Makhorka. Krasnodar ISBN 978-5-93491-681-8.

Larkina, N. I. (2017). Creation of amphidiploids—Fertile interspecific *Nicotiana* hybrids. *Sciences of Europe. Biological Sciences, 2*, 95–105.

Laskowska, D., & Berbeć, A. (2005). Cytology and fertility of viable hybrids of *Nicotiana tabacum* L. cv. TB-566 with *N. alata* Link et Otto. *Journal of Applied Genetics, 46*, 11–18.

Laskowska, D., & Berbeć, A. (2012). Production and characterization of amphihaploid hybrids between *Nicotiana wuttkei* Clarkson et Symon and *N. tabacum*. *Euphytica, 183*, 75–82.

Laskowska, D., Berbeć, A., Van Laere, K., Kirov, I., Czubacka, A., & Trojak-Goluch, A. (2015). Cytology and fertility of amphidiploid hybrids between *Nicotiana wuttkei* Clarkson et Symon and *N. tabacum* L. *Euphytica, 206*, 597–608. https://doi.org/10.1007/s10681-015-1459-3

Lee, C. B., Page, L. E., McClure, B. A., & Holtsford, T. P. (2008). Post-pollination hybridization barriers in *Nicotiana* section *Alatae*. *Sexual Plant Reproduction, 21*, 183–195.

Legg, P. D., & Mann, T. J. (1961). A study of introgression of *N. rustica* germplasm into flue-cured varieties of *N.* tabacum. *Tobacco Science, 5*, 136–139.

Lehmann, H. (1936). On interspecific cross sterility in the genus Nicotiana (in German). *Plant Breeding Abstract, 7*. abst 1023.

Lewis, R. S. (2011). Nicotiana. In C. Kole (Ed.), *Wild crop relatives: Genomic and breeding resources, plantation and ornamental crops*. Springer. https://doi.org/10.1007/978-3-642-21201-7_10

Lewis, R. S., Milla, S. R., & Levin, J. S. (2005). Molecular and genetic characterization of *Nicotiana glutinosa* L. chromosome segments in tobacco mosaic virus-resistant tobacco accessions. *Crop Science, 45*, 2355–2362.

Liao, J., Dai, J., Yang, S., Zhou, X., Ren, L., Chen, Z., He, H., & Chen, S. (2017). Interspecific cross-hybrids of *Nicotiana tabacum* L. cv. (gla.) S 'K326' with *Nicotiana alata*. *Plant Breeding, 136*, 427–435. https://doi.org/10.1111/pbr.12474

Lim, K. Y., Souckova-Skalicka, K., Sarasan, V., Clarkson, J. J., Chase, M. W., Kovarik, A., & Leitch, A. R. (2006). A genetic appraisal of a new synthetic *Nicotiana tabacum* (Solanaceae) and the Kostoff synthetic tobacco. *American Journal of Botany, 93*, 875–883.

Ling, H.-Y., Edwards, A. M., Gantier, M. P., DeBoer, K., Neale, A. D., Hamil, J. D., & Walmsley, A. M. (2012). An interspecific *Nicotiana* hybrid as a useful and cost-effective platform for production of animal vaccines. *PLoS One, 7*(4), e35688. https://doi.org/10.1371/journal.ponc.0035688

Liu, H., & Marubashi, W. (2014). Species origin of genomic factors in *Nicotiana nudicaulis* Watson controlling hybrid lethality in interspecific hybrids between *N. nudicaulis* Watson and *N. tabacum* L. *PLoS One, 9*, e97004. https://doi.org/10.1371/journal.pone.0097004

Liu, H., Ito, Y., Muraida, N., Hayakawa, Y., Itoyama, K., Ohsato, S., & Marubashi, W. (2017). Development of a method to produce chromosome lacking lines (CLLs) in Nicotiana tabacum L. "red Russian". *American Journal of Plant Sciences, 8*, 2923–2943.

Liu-Bao, X.-H., He-Mengyuan, X.-M., & Tao-Wenjing, H.-S. (1995). Interspecific somatic hybridization in *Nicotiana*: I. Transfer of partial genome from *N. undulata* into *N. tabacum* by protoplast fusion. *Acta Genetica Sinica, 22*, 463–469.

Lloyd, R. (1975). Tissue culture as a means of circumventing lethality in an interspecific *Nicotiana* hybrid. *Tobacco Science, 19*, 4–6.

Lu, P. Z. C., & Yang, H. Y. (1996). Pollen-mesophyll protoplast fusion and hybrid plant regeneration in *Nicotiana*. *Acta Botanica Sinica, 38*, 342–346.

Ma, J. M. Y. (2017). *The fine mapping of two black shank resistance loci and identification of a hybrid lethality gene in tobacco*. PhD thesis, North Carolina State University.

Ma, J., Hancock, W. G., Nifong, J. M., Kernodle, S. P., & Lewis, R. S. (2020). Identification and editing of a hybrid lethality gene expands the range of interspecific hybridization potential in *Nicotiana*. *Theoretical and Applied Genetics, 133*, 2915–2925. https://doi.org/10.1007/s00122-020-03641-w

Malecka, J. (1977). Cyto-embryological studies in *Nicotiana* hybrids. *Acta Biologica Cracoviensia Series Botanica, 20*, 89–101.

Maliga, P., Kiss, R., Nagy, A., & Lazar, G. (1978). Genetic instability in somatic hybrids of *N. tabacum* and *N. knightiana*. *Molecular & General Genetics, 163*, 145–151.

Malinowski, E. (1916). On the occurrence of new forms in the progeny of the hybrids *Nicotiana atropurpurea* × *Nicotiana silvestris* (in Polish). *Compt Rend Soc Sci Varsovie, 9*, 827–864.

Malloch, W. S., & Malloch, F. W. (1924). Species crosses in *Nicotiana* with special reference to *N. longiflora* × *N. tabacum*, *N. longiflora* × *N. sanderae*, *N. tabacum* × *N. glauca*. *Genetics, 9*, 261–291.

Manabe, T., Marubashi, W., & Onozawa, Y. (1989). Temperature-dependent conditional lethality in interspecific hybrids between *Nicotiana suaveolens* Lehm. and *N. tabacum* L. *Proceedings of 6th International Congress SABRAO*, 459–462.

Manolov, A., Leshcheva, D., & Baldzhiyeva, E. (1978). The use of interspecific hybridization for the development of initial material resistant to diseases. *First National Symposium of Plant Immunity, 2*, 36–45. (in Bulgarian).

Marubashi, W., & Nakajima, T. (1985). Overcoming cross-incompatibility between *Nicotiana tabacum* L. and *Nicotiana rustica* L. by test-tube pollination and ovule culture. *Japanese Journal of Breeding, 35*, 429–437.

Marubashi, W., & Onosato, K. (2002). Q chromosome controls the lethality of interspecific hybrids between *Nicotiana tabacum* and *Nicotiana suaveolens*. *Breeding Science, 52*, 137–142.

Marubashi, W., Nakamura, N., & Onozawa, Y. (1988). Production of interspecific seedlings between *Nicotiana tabacum L.* and *N. suaveolens* Lehm. via test-tube pollination and exhibition of seedling lethality. *Japanese Journal of Breeding, 37*(Suppl i), 116–117.

Mayr E (1986) Joseph Gottlieb Koelreuter's contribution to biology. Osiris 2: 135-176, University of Chicago Press.

McCarthy, E. W., Arnold, S. E. J., Chittka, L., et al. (2015). The effect of polyploidy and hybridization on the evolution of flower colour in Nicotiana (Solanaceae). *Annals of Botany, 115*, 1117–1131.

McClure, B. (1996). *The relationship between interspecific cross compatibility and self-incompatibility in Nicotiana*. 14th Intl Congr Sex Plant Reprod, Lorne, Australia.

McClure, B., Haring, V., Eber, P., Anderson, M., Simpson, R., Sakiyama, F., & Clarke, A. (1989). Style self-incompatibility gene products of *Nicotiana alata* are ribonucleases. *Nature, 342*, 955–977.

McCray, F. A. (1932). Compatibility of certain *Nicotiana* species. *Genetics, 17*, 621–636.

McCray, F. A. (1933). Embryo development in *Nicotiana* species hybrids. *Genetics, 18*, 95–110.

Medgyesy, P., Menczel, L., & Maliga, P. (1980). The use of cytoplasmic streptomycin resistance: chloroplast transfer from Nicotiana tabacum into *N. sylvestris* and isolation of their somatic hybrids. *Molecular & General Genetics, 179*, 693–698.

Medgyesy, P., Fejes, E., & Maliga, P. (1985). Interspecific chloroplast recombination in a *Nicotiana* somatic hybrid. *Proceedings of the National Academy of Sciences of the United States of America, 82*, 6990–6964.

Menczel, L., Nagy, F., Kiss, Z., & Maliga, P. (1981). Streptomycin resistant and sensitive somatic hybrids of *N. tabacum* + *N. knightiana*. Correlation of resistance to *N. tabacum* plastids. *Theoretical and Applied Genetics, 59*, 191–195.

Menczel, L., Polsby, L. S., Steinback, K. E., & Maliga, P. (1986). Fusion-mediated transfer of triazine-resistant chloroplasts: Characterization of *Nicotiana tabacum* cybrid plants. *Molecular & General Genetics, 205*, 201–205.

Mihaylova-Kroumova, A. B., & Wagner, J. G. (2016). Trichome extracts from *Nicotiana* hybrids – A resource for testing disease/insect resistance. CORESTA Meeting, Agronomy/Phytopathology, 2017, Santa Cruz do Sul, AP 50.

Mihaylova-Kroumova, A. B., Artiouchine, I., Korenkov, V. D., & Wagner, J. G. (2020). Patterns of inheritance of acylsugar acyl groups in selected interspecifc hybrids of genus *Nicotiana*. *Journal of Plant Research, 133*, 509–523. https://doi.org/10.1007/s10265-020-01188-x

Mino, M., Maekawa, K., Ogawa, K., Yamagishi, H., & Inoue, M. (2002). Cell death processes during expression of hybrid lethality in interspecific F1 hybrid between *Nicotiana gossei* Domin and *Nicotiana tabacum*. *Plant Physiology, 130*, 1776–1787.

Mino, M., Tezuka, T., & Shomura, S. (2022). The hybrid lethality of interspecific F1 hybrids of Nicotiana: A clue to understanding hybrid inviability—A major obstacle to wide hybridization and introgression breeding of plants. *Molecular Breeding, 42*, 10. https://doi.org/10.1007/s11032-022-01279-8

Moav, R. (1958). Inheritance in Nicotiana tabacum XXIX: Relationship of residual-chromosome homology to interspecific gene transfer. *The American Naturalist, 92*, 267–278.

Moav, R., & Cameron, D. R. (1960). Genetic instability in *Nicotiana* hybrids. I. The expression of instability in *N. tabacum* × *N. plumbaginifolia*. *American Journal of Botany, 47*, 87–93.

Moav, R., & Cameron, D. R. (1961). Chromosome complement dosage in relation to seed development of species hybrids in *Nicotiana*. *Botanical Gazette, 123*, 70–77.

Modilevsky, J. (1936). Cytogenetical investigation of the genus *Nicotiana* I. Cytology and embryology of the amphidiploid *Nicotiana ditagla* (in Russian). *Journall Institute of Botany, National Academy of Sciences of Ukraine, 7*, 21–39.

Modilevsky, J. (1939). Cytogenetic investigations of the genus Nicotiana. VII. Crossing amphidiploid *N. disualovii* with some species of genus *Nicotiana* (in Russian). *Journall Institute of Botany, National Academy of Sciences of Ukraine, 21–22*(29–30), 107–137.

Morgan, O. D. (1964). Experimental infection of *Nicotiana* species and interspecific crosses with *Cercospora nicotianae* and *Cercospora kikuchii*. *Plant Disease Report, 48*, 693–695.

Mueller-Gensert, E., & Schieder, O. (1987). Interspecific t-DNA transfer through plant protoplast fusion. *Molecular & General Genetics, 208*, 235–241.

Mukhopadhyay, A., Pradhan, A. K., & Pental, D. (1991). Mitochondrial-DNA patterns are similar in gametosomatic and somatic hybrids of 2 *Nicotiana* species. *Plant Cell Reports, 10*, 522–524.

Muraida, N., & Marubashi, W. (2015). Characterization of hybrid seedlings from crosses of *Nicotiana stocktonii* Brandegee × *N. tabacum* L. and *N. stocktonii* × progenitors of *N. tabacum*. *Plant Biotechnology, 32*, 139–147.

Murfett, J., Strabala, T. J., Zurek, D. M., Mou, B., Beecher, B., & McClure, B. A. (1996). S RNase and interspecific pollen rejection in the genus *Nicotiana*: Multiple pollen-rejection pathways contribute to unilateral incompatibility between self-incompatible and self-compatible species. *Plant Cell, 8*, 943–958. https://doi.org/10.1105/tpc.8.6.943

Murthy, T. G. K., & Subbarao, I. V. (2004). Some new interspecific hybrids in the genus *Nicotiana*—Characterization and utilization. *Tobacco Research, 30*, 33–41.

Murthy, T. G. K., Sredhaar, U., & Siva Raju, K. (2014). *Incorporation of aphid resistance from N. gossei. N. repanda, N x. umbratica-nesophila and N × benthamiana–repanda* (pp. 17–20). CTRT Annual Report 2013–2014.

Näf, U. (1958). Studies on tumour formation in *Nicotiana* hybrids I. The classification of parents into two etiologically significant groups. *Growth, 22*, 167–180.

Nagao, T. (1978). Somatic hybridization by fusion of protoplasts. 1. The combination of *Nicotiana tabacum* and *Nicotiana rustica*. *Japanese Journal of Crop Science, 47*, 491–498.

Nagao, T. (1979). Somatic hybridization by fusion of protoplasts. II. Combinations of *Nicotiana tabacum* with *N. glutinosa* and of *N. tabacum* with *N. alata*. *Japanese Journal of Crop Science, 48*, 385–392.

Nagao, T. (1982). Somatic hybridization by fusion of protoplasts III. Somatic hybrids of sexually incompatible combinations *Nicotiana tabacum* + *N. repanda* and *N. tabacum* + Salpiglosis sinuata. *Japanese Journal of Crop Science, 51*, 35–42.

Nakata, K., Nagashima, H., Yamashita, H., Inaba, N., Shinozaki, Y., Kanekatsu, M., Marubashi, W., & Yamada, T. (2021). Analysis of the possible cytogenetic mechanism for overcoming hybrid lethality in an interspecifc cross between *Nicotiana suaveolens* and *Nicotiana tabacum*. *Scientific Reports, 11*, 7812. https://doi.org/10.1038/s41598-021-87242-7

Naton, B., Ecke, M., & Hampp, R. (1992). Production ef fertile hybrids by electrofusion of vacuolated and evacuolated tobacco mesophyll protoplasts. *Plant Science, 85*, 197–208.

Naumenko, S. A. (2010). Роль апомиксиса в селекции табака сортотипов Вирджиния и Берлей (Role of apomixis in the improvement of Virginia and Burley tobacco market types). In S. A. Naumenko, K. I. Ivanicky, & B. A. Vinogradov (Eds.), *Sb Nauchn Tr WNIITTI* (Vol. 179, pp. 180–185). Krasn odar.

Naumenko, S. A. (2012). *Особенности селекции сортов табака сортотипов Вирджиния и Берлей в России (Particularities of the development of flue-cured and Burley tobacco cultivars in Russia)*. DSc thesis, Russian Timiryazev State Agrarian University.

Nielsen, M. T., & Collins, G. B. (1989). Variation among androgenic and gynogenic doubled haploids of tobacco (*Nicotiana tabacum*). *Euphytica, 43*, 263–267.

Nifong, J. M. (2008). *Assessment of the diversity and disease resistance properties within a collection of Nicotiana rustica*. PhD thesis, NCSU, Raleigh, North Carolina.

Nikova, V. M. (1986). *Nicotiana excelsior* black as a source of cytoplasmic male sterility in *Nicotiana tabacum* L. *Comptes rendus de l'Académie bulgare des Sciences, 39*, 105–107.

Nikova, V., & Vladova, R. (2002). Wild *Nicotiana* species as a source of cytoplasmic male sterility in *Nicotiana tabacum*. *Beitrage zur Tabakforschung Internationalt/ Contributions to Tobacco & Nicotine Research, 20*, 301–311.

Nikova, V. M., & Zagorska, N. A. (1990). Overcoming hybrid incompatibility between *Nicotiana africana* Merxm and *Nicotiana tabacum* and development of cytoplasmically male sterile tobacco forms. *Plant Cell, Tissue and Organ Culture, 32*, 71–75.

Nikova, V., Petkova, M., & Zagorska, N. (1988). Overcoming hybrid incompatibility of *Nicotiana africana* Merxm. with *N. tabacum* L. by in vitro technique. *Genetics Selection, 21*, 283–289.

Nikova, V. M., Zagorska, N. A., & Pundeva, R. S. (1991). Development of four tobacco male sterile sources using *in vitro* techniques. *Plant Cell, Tissue and Organ Culture, 27*, 289–295.

Nikova, V., Vladova, R., Pundeva, R., & Shabanov, D. (1997). Cytoplasmic male sterility in *Nicotiana tabacum* L. obtained through interspecific hybridization. *Euphytica, 94*, 375–378.

Nikova, V., Palakarcheva, M., Pundeva, R., & Krusteva, D. (1998a). Somaclonal variation in cultured in vitro tobacco plants from the hybrid *Nicotiana gossei* Domin. x *N. tabacum* L. *Israel Journal of Plant Sciences, 46*, 35–40.

Nikova, V., Vladova, R., Pundeva, R., Petkova, A., & Vladovska, A. (1998b). Study of *Nicotiana ingulba* Black. and its crossability with *N. tabacum* L. *Bull Spec. CORESTA Congress Brighton*, 150, PPOST 12 (b).

Nikova, V., Pundeva, R., & Petkova, A. (1999). *Nicotiana tabacum* L. as a source of cytoplasmic male sterility in interspecific cross with *N. alata* Link & Otto. *Euphytica, 107*, 9–12.

Nikova, V., Pundeva, R., Vladova, R., & Petkova, A. (2001). A new tobacco cytoplasmic male sterile source from the hybrid combination *Nicotiana longiflora* Gav. and *N. tabacum* L. using in vitro techniques. *Israel Journal of Plant Sciences, 49*, 9–13.

Nikova, V., Krusteva, D., Dimitrov, B., Pandeva, R., & Vladova, R. (2003). *Investigation of the interspecific hybrid Nicotiana tabacum L. × N. sanderae Hort. obtained via in vitro techniques*. CORESTA Meeting, Agro-Phyto Groups, Bucharest, APOST12.

Nikova, V., Pundeva, R., Vladova, R., & Petkova, A. (2004). Application of tissue cultures to overcome the complete sterility of Nicotiana plumbaginifolia Viviani ´ *N.* tabacum L. F_1 hybrid. *Israel Journal of Plant Sciences, 52*, 45–49.

Nikova, V., Iancheva, A. M., Vladova, R., Pandeva, R., & Petkova, A. (2006). *Genetic improvement of Nicotiana tabacum by applying remote hybridization and biotechnological methods*. CORESTA Congress, Paris, Agro-Phyto Groups, APOST 07.

Nitsch, J. P., & Nitsch, C. (1969). Haploid plants from pollen grains. *Science, 163*, 85–87.

Ohashi, Y. (1985). Breeding studies of wild species related to tobacco, with special reference to disease resistance. *Bulletin of the Iwata Tobacco Experiment Station, 17*, 1–62.

Ohta, Y. (1991). Graft-transformation, the mechanism for graft-induced genetic changes in higher plants. *Euphytica, 55*, 91–99. https://doi.org/10.1007/BF00022565

Olmo, H. P. (1935). Genetical studies of monosomic types of *Nicotiana tabacum*. *Genetics, 20*, 286–300.

Pal, B. P., & Nath, P. (1936). A note on the sterile hybrid between *N. tabacum* and *N. plumbaginifolia*. *Indian Journal of Agricultural Sciences, 6*, 828–832.

Palakarcheva, M. (1974). New amphidiploids *Nicotiana goodspeedii* × *N. tabacum* (2n=88), important breeding source for disease resistance (in Russian). *Comptes rendus de l'Académie bulgare des Sciences, 7*, 67–71.

Palakarcheva, M. (1975). Interspecific hybridization in tobacco (in Bulgarian). *Bulg Tiutun, 11*, 30–34.

Palakarcheva, M. (1978). Immunogenetic investigations in tobacco. *Bul Spec CORESTA Symposium Sophia, 86*, P11.

Palakarcheva, M. T. (1992). Utilization of the genetic potential in wild Nicotiana species. *Bull Spec CORESTA Congress, Jerez de la Frontera, 163*, PPOST1. ISSN.0525-6240.

Palakarcheva, M., & Bailov, D. (1976). Disease resistant hybrids and varieties of tobacco obtained from interspecific hybridization. *Rasteniev Nauki, 13*, 35–42. (Review of Plant Pathology 56: 3705).

Palakarcheva, M. T., & Dorossiev, L. (1992). *Results of hybridization between the species Nicotiana gossei D. (2n=36) and Nicotiana tabacum L. (2n=48)*. Bull Spec CORESTA Congress, Jerez de la Frontera 163, PPOST1.

Palakarcheva, M., Edreva, A., Cholakova, N., & Noveva, S. (1978). Morphological, cytological, and biochemical studies of the amphidiploid *Nicotiana goodspeedii* W. × *Nicotiana tabacum* L. (2n=88). *Z Pflanzenzuechtg, 80*, 49–63.

Palakarcheva, M., Staneva, M., & Tsanova, E. (1995). Hybridization between *Nicotiana gossei* Domin and *N. tabacum* L. for development of Oriental tobacco lines resistant to tobacco aphids and diseases. *Tobacco Science, 39*, 38–42.

Pandey, K. K. (1974). Overcoming interspecific pollen incompatibility through the use of ionising radiation. *Heredity, 33*, 279–284.

Pandey, K. K. (1975). Sexual transfer of specific genes without gametic fusion. *Nature, 256*, 310–313.

Pandey, K. K. (1977). Genetic control of interspecific incompatibility between *Nicotiana alata* and *N. langsdorffii*: Correction of Yakahashi's observations. *Japanese Journal of Genetics, 52*, 431–433.

Pandey, K. K., & Phung, M. (1982). "Hertwig effect" in plants: Induced parthenogenesis through the use of irradiated pollen. *Theoretical and Applied Genetics, 62*, 295–300.

Pandeya, R. S., & White, W. H. (1981). Nordel, a new bright tobacco cultivar. *Canadian Journal of Plant Science, 61*, 795–797.

Pandeya, R. S., & White, W. H. (1984). Delgold: A new flue-cured tobacco. *Canadian Journal of Plant Science, 64*, 233–236.

Papadopoulou, A. D., Roupakias, D. G., Lazaridou, T. B., & Manromatis, A. G. (1997). Effect of nutrient medium, age of ovules and female parent on the in vitro embryo rescue of the interspecific hybrid *Nicotiana tabacum* × *Nicotiana alata*. *Geotechnic Scientific Issues., 8*, 4–12.

Parr, J. C., & Thurston, R. (1968). Toxicity of *Nicotiana* and Petunia species to the larvae of the tobacco hornworm. *Journal of Economic Entomology, 61*, 1525–1531.

Patel, M. S. (1960). *Indian tobacco*. A monograph. Indian Central Tobacco Committee.

Patrascu, M., Paunescu, D., & Ciuperca, A. (1999). *Transfer of the resistance to TSWV from Nicotiana alata to some tobacco cultivars, using the in vitro culture technique*. CORESTA Meet Agro-Phyto Groups Suzhou, China, POST8.

Pearse, I. S., Krugel, T., & Baldwin, I. T. (2006). Innovation in anti-herbivore defense systems during neopolypoloidy—The functional consequences of instantaneous speciation. *The Plant Journal, 47*, 196–210.

Pental, D., Mukhopadhyay, A., Grover, A., & Pradhan, A. K. (1988). A selection method for the synthesis of triploid hybrids by fusion of microspore protoplasts (n) with somatic cell protoplasts (2n). *Theoretical and Applied Genetics, 76*, 237–243.

Pental, D., Pradhan, A. K., & Mukopadhyay, A. (1989). Transmission of organelles in triploid hybrids produced by gametosomatic fusions of two *Nicotiana* species. *Theoretical and Applied Genetics, 78*, 547–552.

Ping, L., Hang, Z., & Hong-yuan, Y. (1996). Pollen-mesophyll protoplast fusion and hybrid plant regeneration in *Nicotiana* L. *Acta Botanica Sinica, 38*, 342–346.

Pirrie, A., & Power, J. B. (1986). The production of fertile, triploid somatic hybrid plants (*Nicotiana glutinosa*) (n)+*N. tabacum* (2N) via gametic: Somatic protoplast fusion. *Theoretical and Applied Genetics, 72*, 48–52.

Pittarelli, G. W., & Sisson, V. A. (1989). Registration of cytoplasmic male sterile tobacco germplasm Bel MS-2. *Crop Science, 29*, 836–837.

Pittarelli, G. W., & Stavely, J. R. (1975). Direct hybridization of *Nicotiana repanda* × *N. tabacum*. *The Journal of Heredity, 66*, 281–284.

Pontes, O. F. S., Pulcinelli, C. E., Lopes, F. C. C., Latado, R. R., Neto, A. T., & Camargo, N. (2005). *Obtaining Nicotiana repanda and Nicotiana tabacum fertile hybrids using gamete cell irradiation and immature embryo recovery*. CORESTA Meeting, Agro-Phyto Groups, Santa Cruz do Sul, APOST27.

Powell, J. R. (1979). Ovens 62. *Journal of the Australian Institute of Agricultural Science, 46*, 144.

Prasad, G. S. R., Subhashini, U., & Venkateswarlu, T. (1985). A new technique for mitotic chromosomes from *in vitro* shoot tips of *Nicotiana* hybrid. *Tobacco Research, 11*, 182–183.

Protassenya, G. D. (1935). Development of polyploid plants by means of regeneration (in Russian). *Trudy Gosudarstvennogo Nikitskogo Botaničeskogo Sada, 19*, 1–19.

Raicu, P., Badea, E., Stoian, V., & Gregorian, L. (1978). Variation of chromosome number in tobacco plants regenerated from haploid and diploid callus. *Bul Spec CORESTA Symposium Sofia, 79*, P01.

Ramavarma, K. T., Apparao, K., & Narrayanan, A. I. (1977). Cytogenetics of a trispecific hybrid (*N. benthamiana* × *N. glutinosa*) × *N. tabacum* and its alloploids in genus *Nicotiana*. *Cytologia, 42*, 703–709.

Reed, S., & Collins, G. B. (1978). Interspecific hybrids in *Nicotiana* through in vitro culture of fertilized ovules. *The Journal of Heredity, 69*, 311–315.

Reed, S. M., & Collins, G. B. (1980a). Histological evaluation of seed failure in three *Nicotiana* interspecific hybrids. *Tobacco Science, 24*, 154–156.

Reed, S., & Collins, G. B. (1980b). Chromosome pairing relationship and black shank resistance in three *Nicotiana* interspecific hybrids. *The Journal of Heredity, 71*, 423–426.

Rice, H., Mundell, R. E., Millwood, R. J., Chambers, O. D., Stewart, C. N., & Davies, H. M. (2013). Assessing the bioconfinement potential of a *Nicotiana* hybrid platform for use in plant molecular farming applications. *BMC Biotechnology, 13*, 63.

Rybin, V. A. (1927). Polyploid hybrids of *Nicotiana tabacum* L x *Nicotiana rustica* L. *Bulletin of Applied Botany and Genetics and Plant-Breeding, 17*, 191–240.

Rybin, V. A. (1929). Uber einen allotetraploiden bastard von *Nicotiana tabacum* × *Nicotiana sylvestris*. *Berichte. Deutsche Botanische Gesellschaft, 37*, 385–394.

Sachs-Skalińska, M. (1917). Beitrag zur Cytologie des sterilen Bastardes *Nicotiana atropurpurea* × *Nicotiana silvestris*. *Compt Rend Soc Sci Varsovie, 10*(6).

Sand, S. A. (1968). Genetic modification of cytoplasmic male sterility in tobacco. *The Journal of Heredity, 59*, 175–177.

Sarala, K., Prabhakara Rao, K., & Sheshu Madhav, M. (2023). Interspecific hybridization. Division of Crop Improvement. *ICAR -Central Tobacco Research Institute Technical Bulletin, 052023*, 11–14.

Sarana, M. O. (1934). Амфидиплоидный побег *N.glauca* X *N.tabacum* (Amphidiploid combination of F_1 *glauca-tabacum*). *Sb Nauch Issled Rabot WITIM, 10*, 221–224.

Sarychev, Y. F. (1987). A new method of inducing diploid apomixis in *Nicotiana tabacum* L. *Soviet Genetics, 22*, 871–875. (Tob Abstr 31:887 Abstr 1577).

Semenova, I. V., & Ivanova, T. Z. (1973). Influence of cultivation conditions in obtaining plants of the interspecific hybrid *Nicotiana occidentalis* × Ostrolist 1519 cultivated in vitro (in Russian). *Sborn Nauchno Issled Rabot WITIM, 158*, 38–42.

Shinkareva, I. K. (1979). Meiosis in hybrids of tobacco with the wild species *Nicotiana suaveolens* Lehm. *Soviet Genetics, 15*, 588–592. (Genetika 5:887-889).

Shintaku, Y., Yamamoto, K., & Nakajima, T. (1985). Overcoming hybrid inviability in interspecific cross between *N. repanda* and *N. tabacum* L. *Japanese Journal of Breeding, 35*, 76–79.

Shintaku, Y., Yamamoto, K., & Nakajima, T. (1986). Interspecific hybridization between *Nicotiana repanda* Willd. and *Nicotiana tabacum* L. through *in vitro* culture of irradiated ovules. *Japanese Journal of Breeding, 36*, 420–423.

Shiragaki, K., Nakamura, R., Nomura, S., He, H., Yamada, T., Marubashi, W., Oda, M., & Tezuka, T. (2020). Phenylalanine ammonia-lyase and phenolic compounds are related to hybrid lethality in the cross *Nicotiana suaveolens* × *N. tabacum*. *Plant Biotechnol (Tokyo), 37*, 327–333. https://doi.org/10.5511/plantbiotechnology

Shizukuda, N., & Nakajima, T. (1982). Production of interspecific hybrids between *Nicotiana rustica* L. and *N. tabacum* L. through ovule culture. *Japanese Journal of Breeding, 32*, 371–377.

Shizukuda, N., Yamamoto, K., & Nakajima, T. (1983). Sexual transfer of an incomplete chromosome complement from *Nicotiana tabacum* L. to *N. rustica* L. *Japanese Journal of Breeding, 33*, 15–22.

Sievert, R. C. (1972a). Resistance to anthracnose in the genus *Nicotiana*. *Tobacco Science, 16*, 32–34.

Sievert, R. C. (1972b). Sources of resistance to potato virus Y in the genus *Nicotiana*. *Tobacco Science, 106*, 92–94.

Skucińska, B., Miszke, W., & Kruczkowska, H. (1977). Studies on the use of interspecific hybrids in tobacco breeding. Obtaining of fertile hybrids by propagation in vitro. *Acta Biologica Cracoviensia, 20*, 81–88.

Slusarkiewicz-Jarzina, A., & Zenkteler, M. (1983). Development of hybrid plants from ovules of *Nicotiana tabacum* pollinated in vitro with pollen grains of *Nicotiana knightiana*. *Experimentia, 39*, 1399–1400.

Smith, H. H. (1939). The induction of polyploidy in *Nicotiana* species and species hybrids by treatment with colchicine. *The Journal of Heredity, 30*, 291–306.

Smith, H. H. (1941). Polyploidy in *Nicotiana* (discussion). *The American Naturalist, 75*, 307–309.

Smith, II. II. (1968). Recent cytogenetic studies in the genus *Nicotiana*. *Advances in Genetics, 14*, 1–43.

Smith, H. H. (1979). The genus as the genetic resource. (in) Durbin RD (ed) *Nicotiana*. Procedures for experimental use. *USDA Department of Agriculture Technical Bulletin, 1586*, 1–16.

Smith, H. H., & Abashian, D. V. (1963). Chromatographic investigations on the alkaloid content of *Nicotiana* species and interspecific combinations. *American Journal of Botany, 50*, 435–447.

Smith, H. H., Hamill, D. E., Weaver, E. A., & Thompson, K. H. (1970). Multiple molecular forms of peroxidase and esterase among *Nicotiana* species and amphidiploids. *The Journal of Heredity, 61*, 203–212.

Sproule, A., Donaldson, P., Dijak, M., Bevis, E., Pandeya, R., Keller, W. A., & Gleddie, S. (1991). Fertile somatic hybrids between transgenic *Nicotiana tabacum* and transgenic *N. debneyi* selected by dual antibiotic resistance. *Theoretical and Applied Genetics, 82*, 450–456.

Stanoeva I, Petkova M (1978) Overcoming the sterility in F_1 of interspecific hybrids of the genus *Nicotiana* using the method of tissue cultures. CORESTA symposium Bul Sofia:80.

Stavely, J. R. (1979). Disease resistance. (in) Durbin RD ed. *Nicotiana*. Procedures for experimental use. *USDA Department of Agriculture Technical Bulletin, 1586*, 111–120.

Stebbins, G. L. (1950). *Variation and evolution in plants*. Columbia University Press.

Stegemann, S., Keuthe, M., Greiner, S., & Bock, R. (2012). Horizontal transfer of chloroplast genomes between plant species. *PNAS, 109*, 2434–2438. https://doi.org/10.1073/pnas.1114076109

Stoyanova, M. (1978). *Results from the hybridization of the species Nicotiana tabacum and N. alata Link*. Bul Spec CORESTA Symposium Sophia: 86 P12.

Stoyanova, M. P. (1979). Resultats de l'hybridation entre les especies *Nicotiana tabacum* L. Et *Nicotiana alata* Link. *Comptes rendus de l'Académie bulgare des Sciences, 32*, 683–686.

Stoyanova, M., & Konotop, A. (1975). A study on the resistance to Lycopersicon virus 3 Smith in progenies of *Nicotiana tabacum* × *N. glauca* Grah. interspecies hybridization. *Doklady Akademii Selskokhozaistv Akademii Nauk im Georgija Dimitrova, 8*(13), 43–44. (Tob Abstr 19:3469).

Subhashini, U., Venkateswarlu, T., & Anjani, K. (1986). Embryo rescue in *Nicotiana* hybrids by in vitro culture. *Plant Science, 43*, 219–222.

Sun, Y., Xue, Q., Zhang, X., Ding, C., Zhang, L., Gong, M., Bu, G., & Wang, W. (2007). Morphological, cytological and molecular characterization of a novel symmetric somatic hybrid between *N. tabacum* and *N. glauca*. *Plant Biosystems, 141*, 129–133.

Svab, Z., & Maliga, P. (2007). Exceptional transmission of plastids and mitochondria from the transplastomic pollen parent and its impact on transgene containment. *Proceedings of the National Academy of Sciences, 104*, 7003–7008.

Swaminathan, M. S., & Murthy, B. R. (1957). One-way incompatibility in some species crosses in the genus *Nicotiana*. *Indian Journal Of Genetics And Plant Breeding, 17*, 23–26.

Szilagyi, L. (1975). Elimination of chromosomes in an alloploid hybrid of *Nicotiana tabacum* × *N. glauca*. *Acta Botanica Academiae Scientiarum Hungaricae, 21*, 433–441.

Takahashi, H. (1973). Anther culture with a sterile interspecific hybrid between *Nicotiana tabacum* and *N. sylvestris*. *Japanese Journal of Breeding, 23*, 250–254.

Takanashi, H., & Marubashi, W. (2017). Tumorigenesis inheritance from the putative progenitor species of Nicotiana rustica. *Plant Biotechnology, 34*, 79–87.

Takenaka, Y. (1953). Cytogenetic studies on the genus *Nicotiana*. V. Reduction divisions in hybrids between *N. tabacum* and three other species. *National Institute of Genetics (Japanese) Annual Report, 4*, 33–34.

Takenaka, Y. (1954). Cytogenetic studies on the genus *Nicotiana*. VI. Reduction divisions in hybrids between *N. tabacum* and three other species. *National Institute of Genetics (Japanese) Annual Report, 5*, 62–64.

Takenaka, Y. (1956a). Cytogenetic studies in *Nicotiana*. VI. On three hybrids between *N. tabacum* and other species. *Bot Ag Tokyo, 68*, 358–362. (J Jpn Bot 68:358-362).

Takenaka, Y. (1956b). Cytogenetic studies in *Nicotiana*. XIV. Reduction divisions in five interspecific hybrids. *Japanese Journal of Genetics, 31*, 155–161.

Takenaka, Y. (1956c). Cytogenetic studies in *Nicotiana*. IX. Reduction divisions in hybrids between *N.* tabacum and two other species. *Japanese Journal of Genetics, 31*, 104–108.

Takenaka, Y. (1957). Cytogenetic studies on the genus *Nicotiana. X. National Institution of Genetics Annual Report, 8*, 43–44.

Takenaka, Y. (1958). Cytogenetic studies of the genus Nicotiana XI. *National Institute of Genetics Annual Report, 9*, 50–51.

Takenaka, Y. (1960). Cytogenetic studies of the genus Nicotiana XIII. *National Institute of Genetics Annual Report, 11*, 63–64.

Takenaka, Y. (1962a). Cytogenetic studies in *Nicotiana* XVI: Reduction divisions in six interspecific hybrids between *N. tabacum* and other six species. *Shokubutsugaku Zasshi*, 75237–75241.

Takenaka, Y. (1962b). Cytogenetic studies in Nicotiana XVIII. Reduction divisions in three interspecific hybrids between autotetraploid *N. Tabacum* and three other species. *Japanese Journal of Plant Breeding, 12*, 275–277.

Takenaka, Y. (1962c). Cytogenetic studies in Nicotiana. XVII. Reduction divisions in five interspecific hybrids. *Japanese Journal of Genetics, 37*, 343–347.

Takenaka, Y. (1963). Cytogenetic studies in *Nicotiana.* XV. Reduction divisions in three interspecific hybrids and one amphidiploid. *Japanese Journal of Genetics, 38*, 135–140.

Takenaka, Y., & Yoneda, Y. (1964). Tumor formation in interspecific hybrids of *Nicotiana,* especially in relation to injury. *National Institute of Genetics Annual Report, 15*, 87.

Takenaka, Y., Hu, C. H., & Shimoyama, K. (1955). Cytogenetic studies on the genus Nicotiana, VII. *National Institution of Genetics Annual Reports*, 58–61.

Takenaka, Y., Amano, Y., & Tateoka, T. (1956). Cytogenetic studies on the Genus Nicotiana. IX. *National Institute of Genetics Annual Report, 7*, 42–44.

Tanaka, M. (1961). The effect of irradiated pollen grains on species crosses of *Nicotiana. Hatano Tobacco Experiment Stationa Bulletin, 51*, 1–38.

Tanaka, M., & Kurihara, T. (1968). Induction of haploid tobacco by means of X-irradiated pollen grains. *Japanese Journal of Breeding, 18*, 107–109.

Ternovsky, M. F. (1934). Die Fragen der Immunitat bei Vertretern der Gattung *Nicotiana. Der Zuchter, 6*, 140–144.

Ternovsky, M. F. (1936a). Uber einen allotetraploiden bastard von *Nicotiana Tabacum* × *Nicotiana sylvestris. Berichte. Deutsche Botanische Gesellschaft, 37*, 385–394.

Ternovsky, M. F. (1936b). Poliploidy i haploidy pri mezduvidovoj hibridizacji w rodie Nicotiana (Polyploids and haploids in *Nicotiana* interspecific hybridization) (in Russian). *Sb Nauch Issled Rabot WITIM, 132*, 59–108.

Ternovsky, M. F. (1939). Amphidiploid *Nicotiana tabacum* L. x *N.sylvestris* Speg. et Comes. *Sb Nauch Issled Rabot WITIM, 139*, 141–144.

Ternovsky, M. F. (1962). Summary and prospects of interspecific hybridization in the genus *Nicotiana.* In *Wide hybridization of plants (Otdalennaya gibridizatsiya rastenii)* (pp. 331–345). Proceedings of the Conference on Wide Hybridization of Plants and Animals; collection of reports. Jerusalem.

Ternovsky, M. F., & Khudina, J. P. (1938). Response of *N. glutinosa* × *N. tabacum* hybrids to ordinary tobacco mosaic. *Sb Nauch Issled Rabot WITIM, 135*, 69–70.

Ternovsky, M. F., & Larkina, N. I. (1978a). Hybridization of native tobacco cultivars with the wild species *Nicotiana rosulata. Soviet Genetics, 14*, 1038–1045.

Ternovsky, M. F., & Larkina, N. I. (1978b). Remote hybridization as a method of creating resistance to *Peronospora tabacina. Soviet Genetics, 14*, 738–745.

Ternovsky, M. F., Butenko, R. G., & Moiseeva, M. E. (1972). The use of tissue culture to overcome the barrier of incompatibility between species and sterility of interspecific hybrids. *Soviet Genetics, 8*, 27–33.

Ternovsky, M. F., Moiseyeva, M. E., & Grebenkin, A. (1973). Experimentally produced new type of male sterility in interspecies *Nicotiana* hybrids. *Soviet Genetics, 9*, 693–702.

Ternovsky, M. F., Shinkareva, I. K., & Larkina, N. I. (1976). Production of interspecific tobacco hybrids by the pollination of ovules *in vitro. Soviet Genetics, 12*, 1209–1213.

Tezuka, T. (2012). Hybrid lethality in the genus *Nicotiana.* In J. K. Mworia (Ed.), *Botany* (pp. 191–210). Rijeka InTech.

Tezuka, T., & Marubashi, W. (2004). Apoptotic cell death observed during the expression of hybrid lethality in interspecific hybrids between *Nicotiana tabacum* and *N. Suaveolens. Breeding Science, 54*, 59–66.

Tezuka, T., & Marubashi, W. (2006a). Hybrid lethality in interspecific hybrids between *Nicotiana tabacum* and *N.* suaveolens: evidence that the Q chromosome causes hybrid lethality based on Q-chromosome-specfic DNA markers. *Theoretical and Applied Genetics, 112*, 1172–1178.

Tezuka, T., & Marubashi, W. (2006b). Genomic factors lead to programmed cell death during hybrid lethality in interspecific hybrids between *Nicotiana tabacum* and *N. debneyi. SABRAO Journal of Breeding and Genetics, 38*, 69–81.

Tezuka, T., & Marubashi, W. (2012). Genes in S and T subgenomes are responsible for hybrid lethality in interspecific hybrids between *Nicotiana tabacum* and *Nicotiana occidentalis. Online Plos One, 7*, e36204. https://doi.org/10.1371/journal.pone.0036204

Tezuka, T., Kuboyama, T., Matsuda, T., & Marubashi, W. (2007). Possible involvement of genes on the Q chromosome of *Nicotiana tabacum* in expression of hybrid lethality and programmed cell death during interspecific hybridization to *Nicotiana debneyi*. *Planta, 226*, 753–764.

Tezuka, T., Kuboyama, T., Matsuda, T., & Marubashi, W. (2010). Seven of eight species in *Nicotiana* section Suaveolentes have common factors leading to hybrid lethality in crosses with *Nicotiana tabacum*. *Annals of Botany, 106*, 267–276.

Tezuka, T., Matsuo, C., Iizuka, T., Oda, M., & Marubashi, W. (2012). Identification of *Nicotiana tabacum* linkage group corresponding to the Q chromosome gene(s) involved in hybrid lethality. *PLoS One, 7*, e37822. https://doi.org/10.1371/journal.pone.0037822

Trojak-Goluch, A., & Berbeć, A. (2002). *Potential of Nicotiana glauca as a source of tolerance to black root rot caused by Chalara elegans in tobacco improvement*. CORESTA Congress, New Orleans, Agro-Phyto Groups, APPOST20.

Trojak-Goluch, A., & Berbeć, A. (2007). Meiosis and fertility in interspecific hybrids of *Nicotiana tabacum* L. × *N. glauca* Grah. and their derivatives. *Plant Breeding, 126*, 201–206.

Tsikov, D. (1966). Effectiveness of multiple pollinations in hybridizing *Nicotiana tabacum* with *N. glutinosa*. *Rastenev Nauk, 3*, 95–102.

Tsikov, D., & Tsikova, E. (1986). Male sterility in tobacco. VI. Cytoplasm of *N. gossei* (in Bulgarian). *Genetics Selection Evolution, 19*, 121–131. [CORESTA Inf. Bul. 1986(2):51–52].

Uchimiya, H. (1982). Somatic hybridization between male-sterile *Nicotiana tabacum* and *N. glutinosa* through protoplast fusion. *Theoretical and Applied Genetics, 61*, 69–72.

Ueno, N., Kashiwagi, M., Kanekatsu, M., et al. (2019). Accumulation of protein aggregates induces autolytic programmed cell death in hybrid tobacco cells expressing hybrid lethality. *Scientifc Reports, 9*, 1–10. https://doi.org/10.1038/s41598-019-46619-5

Valleau, W. D. (1952). Breeding tobacco for disease resistance. *Economic Botany, 6*, 69–102.

Venkateswarlu, T., Nagarajan, K., & Subhashini, U. (1998). Root-knot nematode resistance through interspecific hybridization. *Tobacco Symposium, 7*, B20.

Vontimitta, V., & Lewis, R. S. (2012). Mapping of quantitative trait loci affecting resistance to *Phytophthora nicotianae* in tobacco (*Nicotiana tabacum* L.) line Beinhart-1000. *Molecular Breeding, 29*, 89–98.

Wark, D. C. (1970). Development of flue cured tobacco cultivars resistant to common strain of blue mold. *Tobacco Science, 14*, 147–150.

Warmke, H. E., & Blakeslee, A. F. (1939). Induction of tetraploidy in *Nicotiana sanderae* and in the sterile hybrid *N. tabacum* × *N. glutinosa*. *Genetics, 24*, 109–110.

Watanabe, H., & Marubashi, W. (2004). Temperature-dependent programmed cell death detected in hybrids between *Nicotiana langsdorffii* and *N. tabacum* expressing lethality. *Plant Biotechnology, 21*, 151–154.

Webber, J. M. (1930). Interspecific hybridization in *Nicotiana*. XI. The cytology of a sesquidiploid hybrid between *tabacum* and *sylvestris*. *University of California Publications in Botany, 11*, 319–354.

Wellington, R. (1913). Studies of natural and artificial parthenogenesis in the genus *Nicotiana*. *The American Naturalist, 47*, 279–306.

Wernsman, E. A. (1992). Varied role for the haploid sporophyte in plant improvement. In H. T. Stalker & J. P. Murphy (Eds.), *Plant Breeding in the 1990's (Proceedings of the Symposium in Plant Breeding in the 1990s)* (pp. 461–484). CAB Int Wallingford.

Whitaker, T. W. (1934). The occurrence of tumors on certain *Nicotiana* hybrids. *Journal of the Arnold Arboretum, 15*, 144–153.

Wichert-Kobus, I. (1967). Attempts to obtain artificial polyploids of tobacco (in polish). *Pamiętnik Puławski, 28*, 79–98.

Wichert-Kobus, I. (1971). Investigations on the polyploids of *Nicotiana tabacum* L. and their hybrids with wild tobacco species (in Polish). *Pamiętnik Puławski, 43*, 33–98.

Wong, Y. N. (1975). Genome analysis of tobacco interspecific hybrids. *Taiwan Tob Wine Monop Bur Tob Res Inst Res Rep, 2*, 22–31.

Yamada, T., Marubashi, W., & Niwa, M. (1999). Detection of four lethality types in interspecific crosses among Nicotiana species through the use of three rescue methods for lethality. *Breeding Science, 49*, 203–210.

Yang, S. J. (1960). Numerical chromosome instability in *Nicotiana* hybrids. I. Interplant variation among offspring of amphiploids. *Genetics, 50*, 745–756.

Ye, S., Jiang, L., An, L., & Yu, L. (1993). Interspecies test-tube fertilization in Nicotiana tabacum x N. repanda. *Acta Botanica Boreali-Occidentalia Sinica*, 13.

Zagorska, N. A., & Palakarcheva, M. (1978). Overcoming the sterility in the interspecific hybrid *N. goodspeedii* × *N. tabacum* by the use of in vitro cultures. *Comptes rendus de l'Académie bulgare des Sciences, 29*(11), 1677–1682.

Zaitlin, D., & Mundell, R. (2006). *Nicotiana hybrids and plant varieties for use in production of pharmaceutics*. US Patent 20060236433 A1.

Zelcer, A., Aviv, D., & Galun, E. (1978). Interspecific transfer of male sterility by fusion between protoplasts of normal *N.* sylvestris and X-ray irradiated protoplasts of male sterile *N. tabacum*. *Zeitschrift für Pflanzenphysiologie, 90*, 397–407.

Zenkteler, M., & Melchers, G. (1978). In vitro hybridization by sexual methods and by fusion of somatic protoplasts: Experiments with *Nicotiana tabacum* × *Petunia hybrida, N. tabacum* × *Hyoscyamus niger, H. niger* × *P. hybrida, Melandrium album* × *P. hybrida. Theoretical and Applied Genetics, 52*, 81–90.

Zheng, Y., Liu, Z., Sun, Y., Liu, G., Yang, A., & Li, F. (2018). Characterization of genes specific to sua-CMS in *Nicotiana tabacum*. *Plant Cell Reports, 37*, 1245–1255. https://doi.org/10.1007/s00299-018-2309-2

Zhou, X., & Liu, Y. (2015). Hybridization by grafting: A new perspective? *Horticultural Science, 50*, 520–521.

Zhou, W. M., Yoshida, K., Shintaku, Y., & Takeda, G. (1991). The use of IAA to ovecome interspecific hybrid inviability in reciprocal crosses between *N. tabacum* L. and *N. repanda*. *Theoretical and Applied Genetics, 82*, 657–661.

Zhukov, N. I. (1939). Sesquidiploids *Nicotiana rustica* x *N. tabacum* and the analysis of nicotine, anabasine and citric acid contents in their crosses with *N. glauca*. Sbornik issl-nauch rabot WITIM, Krasnodar (139), 208–233 (in Russian).

Chapter 4
Experimental Introgression from *Nicotiana* Species to Cultivated Tobacco

4.1 Genetic Resistance to Diseases and Pests Among *Nicotiana* Species

Theoretically, a number of breeding goals and strategies involving the use of different *Nicotiana* species can be envisaged. In actual breeding practice, however, it is only breeding for resistance and the deployment of cytoplasmic male sterility, the latter discussed in Chap. 5, that has unequivocally profited from resorting to tobacco's wild relatives. Sources of resistance present within the cultivated species are for the most part polygenic, and their incorporation in the cultivars to be improved frequently entails undesirable effects related to linkage, dominance, epistasis, etc., which may make the recovery of the desired phenotype very troublesome and difficult. As a result, the trade-offs may effectively outweigh the added value of the newly incorporated trait. Experience has shown that the resistance factors present in wild *Nicotiana* species are often monogenic and dominant, which makes them attractive from the perspective of the interspecific breeder.

Of the two categories, pests and diseases, resistance to pests among the species of *Nicotiana* is far less widespread and is limited to only a few insects. Resistance to the aphid *Myzus persicae* was reported in *N. repanda* (Thurston, 1961; Murthy et al., 2014), *N. benthamiana* (Krusteva et al., 2003b) *N. gossei* (Thurston, 1961; Burk & Dean, 1975; Krusteva et al., 2003a) *N. goodspeedii* (Palakarcheva & Bailov, 1976); resistance to the tobacco hornworm *Manduca sexta* was found in *N. nesophila, N. repanda, N. stocktonii, N. benthamiana, N. gossei* (Parr & Thurston, 1968) and in *N. attenuata* (Baldwin, 2001). Resistance to the leaf-eating caterpillar *Spodoptera litura* was reported in *N. benthamiana* (Chari & Patel, 1972; Ramavarma et al., 1980, 1991) and *N. gossei* (Rao et al., 1980). Resistance to the whitefly *Trialeurodes vaporariorum* was studied by Neal et al. (1987). The highest ovipositional nonpreference by whiteflies was found in *N. gossei* and *N. benavidesii*. No whitefly nymphs survived on *N. fragrans*.

Resistance to 17 major diseases of tobacco was dispersed among 68 species of *Nicotiana* (Table 4.1). No resistance to any of these diseases was reported among the

A. Berbeć, *A Century of Interspecific Hybridization and Introgression in Tobacco*,
https://doi.org/10.1007/978-3-031-54964-9_4

Table 4.1 Resistance to major tobacco diseases among *Nicotiana* species (adapted after Burk & Heggestad, 1966, Lucas, 1975, Stavely, 1979)

	Column number																
	1	2	3	4	5	6	7	8	9	10	11	12	13	14	15	16	17
Species	TMV	PVY	TEV	*TSWV*	*TRV*	TSV	WF	ALS	BSh	BM	PM	BRR	TBS	FE	RK	BrRR	TCN
N. alata	1/1			4/17			X^0, 1/2	X	9/22		X, 11/29	X, 12/32, 12/33, 12/34, 11/32					17/50
N. bonariensis	1/2				1/7		X^0	1/2			1/2	12/37	X, 2/11				
N. forgetiana	1/1											12/37	1/2				17/51
N. langsdorffii	X, 1/2				1/7		X^0, 1/2			X^3, 1/2, 10/25	X		2/11				
N. longiflora							X^0	1/2	$X^{0,1}$	X^{ns}	X	X 12/35, 1/2	X		X^{ij}, 15/44, 15/45		17/50
N. plumbaginifolia							X^0		X^0, 9/23	X^{ns}	X	X, 12/36			X^i, 15/45, 15/46		X^{Gts}
N. sanderae	X, 1/3			4/18							X			14/41			
N. tabacum	X	2/8	X						X		X	X	X		X		
N. acaulis							X^0	1/2					2/11				17/50
N. glauca	X/t 1/4	X	X		1/7	6/19	X^0				X	X, 12/38			15/45, 15/46	X	
N. noctiflora		1/4									X	12/37	X	X			X^{Gtt}
N. paa																	
N. petunioides																	17/50
N. benavidesii	X	X									1/2						
N. cordifolia		2/9	1/7									12/37					17/52
N. cutleri																	
N. knightiana		X	X							X^3	X			X, 14/42, 1/2	X^i, 15/45, 15/47		
N. paniculata	1/1		X							X^3	X	X, 12/35, 12/39			X^i		X^{Gts}

N. raimondii		X	X			6/19					X, [1/2]	X, [12/35]					
N. solanifolia		2/9									1/2	12/37			1/7		
N. acuminata	X	1/2			1/7					X^{3}	X	X, [12/35]	13/40				X^{Gtt}
N. attenuata							X^{0}	1/2		X^{3}	X	X, [12/35]					
N. longibracteata																	
N. miersii	X	X, [2/10], [1/2]										12/39					17/52
N. pauciflora											X, [1/2]						
N. spegazzinii													2/11				
N. clevelandii		1/2			1/7	6/19				X^{3}							
N. quadrivalvis							X^{0}	1/2		X^{3}	X						
N. nesophila	X, [1/4]	1/2							$X^{1,3}$		1/2	X, [12/35], [1/2]	1/2				
N. nudicaulis							$X^{0,1}$		X^{0}		X	X			X^{i}, [15/46], [15/45]		
N. repanda	X, [1/5]	1/2			1/7		$X^{0,1}$		$X^{0,1}$		X	X, [1/2]	X	14/43	X^{ij}, [15/44]		
N. stocktonii	[1/6], [1/2]								$X^{0,1}$			X [12/35]	X	X			
N. rustica	X, *s*, [1/2]	X^{t}					$X^{0,1}$	8/22	9/24		X	1/2, 12/37		14/41			
N. africana		[2/11], [2/12]									1/2						
N. amplexicaulis		1/7								$X^{1,2}$	1/2			X^{R}			
N. benthamiana	X^{t}, [1/2], [1/5]										X	X [12/35]					
N. cavicola							X^{0}	1/2		$X^{1,2,3}$		12/39					
N. debneyi		1/2			1/7		X^{0}		9/24a	$X^{1,2,3}$	X	X	X	X^{R}			
*N. eastii**										$X^{1,2,3}$							
N. excelsior										X^{ns}	X	12/39					
N. exigua										$X^{1,2,3}$, [10/26]	X	X [12/38], [12/39]			[1/7], [15/45]		
N. faucicola																	
N. fragrans							X^{0}	1/2			X						

(continued)

Table 4.1 (continued)

	Column number																
	1	2	3	4	5	6	7	8	9	10	11	12	13	14	15	16	17
Species	TMV	PVY	TEV	*TSWV*	*TRV*	TSV	WF	ALS	BSh	BM	PM	BRR	TBS	FE	RK	BrRR	TCN
N. goodspeedii	X, [1/1]						7/20			1/7, 10/27	X		1/2				
N. gossei	1/1, 1/7, 1/5					6/19				1/7, 1/2	X	X [12/35]					
N. hesperis										X^{ns}	1/2		1/2				
N. ingulba										$X^{1,3}$	1/2						
N. maritima	1/2, 1/5,1/7					6/19			9/24a	X^{ns}		X [12/35]					
N. megalosiphon										$X^{1,2,3}$	X	X, [12/35]			X^{ij}, [15/48]		
N. occidentalis								X		$X^{1,2,3}$	X	12/37	1/2				
N. rosulata										$X^{1,2,3}$							
N. rotundifolia										$X^{1,2,3}$	1/2						
N. simulans										$X^{1,2,3}$	1/2						
N. suaveolens	X	1/2					X^0	1/2	9/24a	$X^{1,2,3}$	X		1/2				
N. truncata																	
N. umbratica										X^{ns}	1/2						
N. velutina	X									$X^{1,2}$	X		1/2		15/45, 15/47		
N. wuttkei										10/28							
N. sylvestris								X, 1/2			X, [1/2]				X^{ij}		
N. kawakamii		2/13, 2/14									1/2						
N. otophora		X	X		1/7						1/2, 1/7				X^{ij}		
N. setchellii		2/9	X		1/7						11/30, 1/2			14/42			
N. tomentosa		X, 1/2	X		1/7						X, [1/2]				X^{i}, [15/49]		

N. tomentosif.	X^t	X, 2/14 2/15	X		1/7						11/31 1/2						
N. obtusifolia							X^0, 7/21			X^3	X, 1/2	12/39, 1/2					
N. palmeri											X	12/37					
N. arentsii			X		1/7		X^0, 7/21	1/2				12/39			X^i		17/50
N. glutinosa	X										X, 11/31	1/7, 12/35					X^{Gts}
N. thyrsiflora		X													X^i		
N. undulata	X, 1/1		X		1/7		$X^{0,1}$, 7/21	1/2	X^0			12/38, 1/2		14/42			
N. wigandioides	X^t, 1/2	X, 2/16			1/7		X^{0+}, 1/2	1/2				X, 1/2	1/2	X, 1/2			

Explanations:

Data arranged in rows (species) and diseases (columns). Abbreviated disease/pathogen names in line 2 are expanded in the footnotes. Throughout the table, uppercase "X" refers to the data published by Stavely (1979)

Uppercase X followed by a superscript refers to additional information not contained in the original Stavely's table: X^t – tolerance (symptomless host response), X^0, X^1, X^2, X^3 (one or more digits in superscript) denote race or strain number/s of the relevant pathogen; X^i,X^j, X^{ij} – (in column 15) superscripts mark resistance to *M. incognita*, *M. javanica* and both *M. incognita* and *N. javanica*, respectively; X^{Gts}, X^{Gtt} – (in column 17) mark resistance to *G.t. tabacum* and *G. t. solanacearum*, respectively

Explanations and references are provided in the footnotes separately for each column/ disease marked from 1 to 17 (line 1)

References to sources other than Stavely (1979) appear in the table as superscripts marked $^{x/y}$. (x) denotes a column in which the reference first appeared and (y) denotes its consecutive number as the references appear columnwise starting from column 1 and downwards within each column

Column 1 TMV (Tobacco Mosaic Virus)

[1/1]Yuan et al. (2015), hypersensitive localized necrosis response; [1/2]Tatemichi (1990); [1/3]implied by Palakarcheva (1984) and Dorossiev et al. (1990); [1/4]implied by Berbeć (1966); [1/5]temperature non-dependent hypersensitive localized lesion response (Depta et al., 2018); [1/6]Evans et al., 1981; [1/7]Wark's personal communication (Lucas, 1975)

Column 2 PVY (Potato Virus Y)

[2/8]Recessive 'va' factor and its alleles effective against some and ineffective against other necrotic strains, widely used in European cultivars (Julio et al., 2015; Korbecka-Glinka et al., 2017); symptomless host recessive factor deployed in some cultivars (Mazur, 1975; Michel et al., 2018); [2/9]resistant to some isolates, tolerant of others (Doroszewska & Depta, 2011); [2/10]susceptible (Doroszewska & Depta, 2011); [2/11]Lucas et al., 1980; [2/12]fully resistant to all isolates (Doroszewska & Depta, 2011); [2/13]tolerance conferred by a recessive gene non-allelic to *va* (Ohashi, 1985; Brandle et al., 1995); [2/14]resistant to a VAM non-breaking isolate only; [2/15]resistance of 'va' type reported (Gupton et al., 1982; Ano et al., 1995); [2/16]mildly susceptible (Doroszewska & Depta, 2011)

Column 3 TEV (Tobacco Etch Virus)

[1/7]Wark's personal communication (Lucas, 1975)

(continued)

Column 4 TSWV (Tomato Spotted Wilt Virus)
[4/17]Opoka, 1969, Laskowska et al., 2013; [4/18]Ivancheva-Gabrovska, 1978, Iancheva, 1989, Laskowska et al., 2013
Column 5 TRV (Tobacco Rattle Virus)
[1/7]Wark's personal communication (Lucas, 1975)
Column 6 STV (Tobacco Streak Virus)
[6/19]Burk and Heggestad (1966)
Column 7 WF (*Pseudomonas syringae* pv. *tabaci* Wildfire)
X^0: resistant to race 0; $X^{0,1}$: resistant to races 0 and 1; [7/20]race unspecified (Palakarcheva, 1976; Palakarcheva & Bailov, 1976); [7/21]: Ternovsky (after Stoyanova (1979), race unspecified); [1/2]Tatemichi, 1990, race unspecified
Column 8 ALS (*Pseudomonas syringae* pv. *angulata* Angular Leaf Spot)
[8/22]Mackenzie et al. (1986), resistance reported in var. *pumila* and var. *brasilea* (Woodend & Mudzengerere, 1992)
Column 9 BSh (*Phytophtora nicotianae* Black Shank)
[9/22]Furusato, 1960; [9/23]moderate resistance (Li et al., 2006); [9/24]resistance to race 0 reported in var. *pumila*, to race 1 in cv. Ky, from the reports by Woodend and Mudzengerere (1992), Bukuta (2002) and Drake et al. (2015) it can be inferred that var. *brasilea* is the source of resistance to races 0 and 1; resistance to race 1 from var. *brasilea* reported as subject to progressive erosion under pathogen pressure (McCorkle et al., 2018); [9/24a] resistant according to Li et al., 2006
Column 10 BM (*Peronospora hyoscyami* Blue mold)
[10/25]hypersensitive response reported, biotype unspecified (Zhang & Zaitlin, 2008); [10/26]lineage reported with superior resistance exceeding that of other Australian species including *N. debneyi* (Bolsunov, 1970); [10/27]Palakarcheva & Bailov, 1976; [10/28]: resistance reported, biotype unspecified (Laskowska & Berbeć, 2003)
Column 11 PM (*Erysiphe* (*Golovinomyces*) *cichoracearum* Powdery mildew)
[11/29]Furusato, 1960, Takenaka, 1960; [11/30]Shabanov et al., 1974; [11/31]Ternovsky (1934)
Column 12 BRR (*Thielaviopsis basicola* (*Chalara elegans*) Black root rot)
[12/32]Autotetraploid accession only (Doroszewska & Przybyś, 2007); [12/33]Stoyanova (1979); [12/34]Gajos (1984); [12/35]susceptible (Doroszewska & Przybyś, 2007); [12/36]moderately susceptible (Doroszewska & Przybyś, 2007); [12/37]moderately resistant (Ivancheva-Gabrovska & Kutova, 1981); [12/38]Trojak-Goluch and Berbeć (2005); [12/39]Ivancheva-Gabrovska & Kutova, 1981
Column 13 TBS (*Alternaria alternata* Tobacco Brown Spot)
[13/40] var. *acuminata* (Brandwagt et al., 2001)
Column 14 FE (*Cercospora nicotianae* Frogeye)
[14/41]var. *pavonii* and *brasilia* (Morgan, 1964); [14/42] symptoms after infection rated as very light (Morgan, 1964); [14/43] Stavely et al., 1973
Column 15 RK (*Meloidogyne incognita*, *M. javanica*, *M. arenaria* Root knot)
X^i,X^j, X^{ij} reported as resistant to *M. incognita*, *M. javanica* and both *M. incognita* and *N. javanica*, respectively; [15/44]high tolerance of *N. javanica* and *N. arenaria* (Davis et al., 1988); [15/45]: highly resistant and susceptible to *M. javanica*, respectively (Calitz & Milne, 1962); [15/46]tolerant of *N. javanica* and *N. arenaria* (Davis et al., 1988); [15/47]tolerant of *M. arenaria* (Davis et al., 1988); [15/48]tolerant of *M. javanica* (Davis et al. (1988); [15/49]highly resistant to *M. incognita* (Rufty et al., 1983)
Column 16 BRR (*Pratylenchus* ssp. Brown Root Rot)
Column 17 TCN (*Globodera tabacum tabacum*, *G. t. solanacearum* (Tobacco Cyst Nematodes)
[17/50]Resistant to *G. t. solanacearum* (Wark's personal communication (Lucas, 1975); [17/51] resistant to *G. t. tabacum* (Wark's personal communication (Lucas, 1975); [17/52] resistant to *G. t. solanacearum* (Hayes et al., 1997)

more recently reported species, some of them of uncertain status, and in *N. corymbosa*, *N. longibracteata* and *N. linearis*. It is very unlikely that any of these species have been tested for any resistance, however.

4.2 Goals and Types of Interspecific Introgression

Unlike the genetic exchange between different varieties of the same species, which is practically unrestricted and predictable as governed by the simple Mendelian principles of segregation, independent assortment and/or linkage, transmission and exchange of genetic material between different species is far more complicated, and there are no fixed rules that would extend over the whole range of potential introgression events. The next section will explain the basic mechanisms that govern the gene flow from species to species under conditions that essentially rely on natural processes in plant organisms without resorting to broad-sense genetic engineering.

By a curious coincidence, the hybrid *N. paniculata* × *N. rustica* was the first ever recorded to be artificially made in the genus *Nicotiana* (see Chap. 3) and the first that served to demonstrate that a potentially useful gene can be moved from one species and incorporated in another. A necrotic type of reaction to infection with tobacco mosaic was transferred from *N. rustica* to an apparently TMV-susceptible accession of *N. paniculata*[1] through repeated backcrossing of the hybrid to the paternal *N. paniculata* and selection for resistant individuals in each backcross generation (Holmes, 1936). See also section 4.4.1.

Save for some rare exceptions of mostly academic interest, e.g., the introgression of the mammoth gene from *N. tabacum* to *N. rustica* (Smith, 1950; Murthy & Swaminathan, 1957; Hillman & Smith, 1965), the further story of interspecific breeding in the genus was almost exclusively concerned with the cultivated tobacco *N. tabacum*.

The germplasm available in wild *Nicotiana* species can be potentially utilized for the improvement of cultivated *N. tabacum* in a twofold manner:

(a) polygenic introgression from wild species can broaden the germplasm basis of *N. tabacum*, which has tended to become increasingly narrow during the process of domestication and breeding manipulations
(b) specific usable traits can be transmitted to and incorporated in, the germplasm of *N. tabacum*

Polygenic Introgression The first set of possibilities has remained largely in the sphere of experiments. The closest relatives of *N. tabacum* were considered first because of relatively extensive opportunities for genetic interchange due to

[1] *N. paniculata* was later reported as TMV-resistant (see Table 4.1) but it appears to consist of both TMV-resistant and TMV-susceptible strains (Yuan et al., 2015)

intergenomic affinities in chromosome structure. Significant heterosis was found in hybrids of *N. tabacum* with *N. tomentosiformis* and *N. otophora* for yield, plant height, number of leaves and days to flower (Matzinger & Wernsman, 1967, see also Sect. 5.1). However, the hybrids had quality characteristics that would make them unusable for traditional commercial uses of the crop (Lewis, 2011). In the past, two papers that addressed this problem were concerned with the potential of *N. sylvestris* and *N. otophora* as sources of germplasm for the improvement of flue-cured tobacco. (Wernsman et al., 1976; Oupadissakon & Wernsman, 1977). Additive effects that might be interpreted as related to polygenic introgression from the alien species were found for quantitative traits such as plant height, days to flower, leaf number, contents of chemical constituents and general yield. In addition, yield also showed significant dominant effects. Possibilities for yield improvement were found but at the expense of other elements of agronomic performance, such as timely maturity and desirable quality of cured leaves. Similar experiments with the use of *N. sylvestris* and *N. tomentosiformis* were reported from Bulgaria (Manolov et al., 1978a, b). More recently, Hancock and Lewis (2017) re-examined the potential of the closest relatives of *Nicotiana tabacum* (*N. sylvestris*, *N. tomentosiformis* and *N. otophora*) for the improvement of *N. tabacum*. Synthetic tobacco 4x (*N. sylvestris* × *N. otophora*) and 4x (*N. sylvestris* × *N. tomentosiformis*) were found to be preferable to the direct hybrids with each of those species as vehicles for genetic exchange between tobacco and its wild progenitors. Heterotic effects were found for yield and growth rate in hybrids of synthetic tobacco with cultivated tobacco. Genetic recombination was found to be reduced, but germplasm exchange between chromosomes was relatively unrestricted due to the elimination of preferential pairing that hinders genetic flow between tobacco and its alien diploid relatives. The authors concluded that the synthetic tobacco offers a convenient system for introgressing genetic diversity into *N. tabacum*.

Nicotiana rustica is a rare example of successful general agronomic improvement of *N. tabacum* by the use of a species not directly involved in the ancestry of *N. tabacum*. That alien species was part of the pedigree of the flue-cured cultivars 'Delgold', 'Nordel' 'Newdel' and 'Delfield', each with a record of commercial cultivation. Delgold was originally developed from mating a tetraploid variant of the flue cured cv. V 115 to *N. rustica* (Pandeya & White, 1984). The new cultivar was found to be superior to V 115 for many traits, including leaf yield, money returns, alkaloid contents and improved resistance to some diseases. Nordel and Newdel were developed in a similar fashion with the 3x sesquidiploids *N. tabacum* × *N. rustica* as starting hybrids (Pandeya & White, 1981, 1984). Likewise, significant improvements over V 115 in earliness, alkaloid content, and tar-to-nicotine ratios were reported for Nordel and Newdel. Newdel, while exhibiting the desirable traits of Nordel, showed a significant improvement in yield. The development of cv. 'Delfield' involved protoplast fusion between *N. rustica* var. '*chlorotica*' and the *N. tabacum* mutant 'WS' (Pandeya et al., 1991). Delfield also outperformed the parental tobacco varieties for yield and for some other agronomic indices. However, in all these cases it is not possible to determine how much of this gain could be actually attributed to *N. rustica* germplasm and how much to gene reassortment within *N. tabacum* since in the breeding process of all four varieties two or more genotypes of *N. tabacum* were used. In yet another study on introgression from *N. rustica* to *N. tabacum* Pandeya et al. (1986) found blue mold-resistant segregants among BC_3 derivatives of protoplast fusion products between *N. tabacum* and *N. rustica*. Since neither *N. tabacum* nor *N. rustica* involved in the original somatic hybrid bore resistance to blue mold, the researchers attributed the appearance of resistant plants to genetic complementation between the two genomes, recombination of mitochondrial genes or interaction between mitochondrial and nuclear genes.

Oligogenic Introgression The majority of documented transfers of traits controlled by several genes seem to be confined to resistance to blue mold caused by *Peronospora hyoscyamii*. In a series of successive backcrosses of the amphidiploid 4x (*N. debneyi* × *N. tabacum*) to *N. tabacum,* Clayton et al. (1958) transferred resistance to blue mold from *N. debneyi*. In the backcross progenies, he would recover resistant plants that consistently contained the full chromosome complement of *N. tabacum* and varying numbers of univalents from *N. debneyi*. Further assiduous selection must have resulted in several alien translocations to the genome of *N. tabacum* since an agronomically acceptable stable resistant breeding line was ultimately developed (Smith, 1968). This apparently unique accomplishment, considering the distance between *N. debneyi* and *N. tabacum*, corresponds well with the much later findings by Kitamura et al. (2005), who found *N. debneyi* and *N. tabacum* to have more intergenomic affinities than their respective taxonomic positions would suggest. Oligogenic patterns of inheritance were also reported for blue mold resistance transferred from *Nicotiana goodspeedii* (Wuttke, 1969) and from *N. velutina* (Wark, 1963).

Chaplin (1977) and Chaplin and Sisson (1984) reported on transferring the genetic system controlling the levels of alkaloids from *N. rustica*. They used low-alkaloid flue-cured lines of *N. tabacum* as the recipient forms. The sesquidiploid hybrids RTT (2n *N. rustica* × 4n *N. tabacum*) were intermediate in alkaloid content between the two parents. Advanced introgression lines of *N. tabacum* with the alkaloid mechanism derived from *N. rustica* were developed by backcrossing and selfing. Those lines were not higher in alkaloids than the regular flue-cured varieties. An increase in alkaloid levels above that in the highest alkaloid cultivars could be identified in some progeny from crossing high-nicotine introgression lines with high-alkaloid *N. tabacum* flue cured varieties. The genetic alkaloid control system in *N. rustica* was not different from that in *N. tabacum,* but apparently, some additive effects of the two systems were possible. Interestingly, the Japanese scientist Furusato (1960) reported a more than threefold increase in nicotine content in sesquidiploids 3x (*N. rustica* × *N. tabacum*) involving two different *N. tabacum* varieties over the respective nicotine levels in *N. rustica* and in the recurrent *N. tabacum* parent.

Oligogenic inheritance was probably involved in attempts to transfer resistance to PVY from *N. africana* to *N. tabacum*. The work, carried out independently in two laboratories, resulted in the development of breeding lines that, although tolerant of the virus, did not express the full resistance of the wild species, the fact explained by more than one gene controlling the resistance in *N. africana* (Lewis, 2007; Korbecka-Glinka et al., 2017).

According to Tatemichi (1990), two dominant genes for resistance to powdery mildew were transferred from *N. tomentosiformis* to *N. tabacum*. However, Smeeton and Ternouth (1990) list the resistance to powdery mildew from *N. tomentosiformis* as controlled by a single dominant factor.

Breeding lines resistant to root knot disease caused by *Meloidogyne* nematodes are a unique case of oligogenic introgression that involves two genes, each from a different species (Shava et al., 2018; Jack, 2010; Raeber & Smeeton, 1980, 1984; Raeber & Schweppenhauser, 1964). Some Burley tobacco lines grown in Zimbabwe carry resistance to *Meloidogyne javanica* resulting from the R gene from *N. repanda* and the L gene from *N. longiflora*. The joint action of the two genes in the genetic milieu of cultivated tobacco proved to be superior to the resistance conferred by either species alone.

Monogenic Introgression Save for the above examples, controlled introgression in *Nicotiana* has been limited to simply inherited monogenic dominant traits, primarily to disease resistance factors. *Nicotiana* species have been screened for resistance to various diseases and pests. It appears from those studies that the genus is a vast pool of resistant germplasm that for various reasons remains largely unexploited. The major problem is that the reservoir is very hard to access for reasons discussed in many reviews, with those by Lewis (2011, 2020a) and Berbeć and Doroszewska (2020) being probably the latest ones. The following sections will take a brief look at how much of this potentially inexhaustible pool has been utilized or at least attempted to be used.

4.3 Basic Mechanisms of Interspecific Gene Transfer in *Nicotiana*

4.3.1 Alien Chromosome Substitution

Derivation of Samsun H The story of the transmission and incorporation of the gene conferring resistance to tobacco mosaic (TMV) from *N. glutinosa* to *N. tabacum* is the best example to illustrate and to help understand the mechanics and problems inherent to interspecific breeding in *Nicotiana*. This is at the same time the first successful transfer of a usable trait from an alien species to tobacco and one of the few that have continued to have a weighty and lasting impact, both academic and practical.

Holmes (1938) crossed *N. glutinosa*, already known for its resistance to TMV, with *N. tabacum*. The amphihaploid hybrid (TG) of *N. glutinosa* (GG) × *N. tabacum* (TT) that he produced was completely self and cross sterile, and he failed to produce seeds despite repeated efforts. Holmes gave up trying to convert his own sterile amphidiploid (TG) and made use of the ready-made fertile amphidiploid 4x (*N. tabacum* × *N. glutinosa*) (TTGG) also nicknamed '*N. digluta*' supplied to him by Dr Roy E. Clausen (Holmes, 1938). The transfer ran through the following stages:

(a) Amphidiploid generation**.** As a preliminary step, Holmes inoculated *N. digluta* plants (TTGG) with TMV and found that all of them produced localized necrotic lesions similar to those in TMV-inoculated *N. glutinosa.*
(b) Sesquidiploid generation**.** The amphidiploid TTGG plants were crossed to *N. tabacum* var. Connecticut Broadleaf. The resulting BC_1 progeny was composed of plants that contained two haploid genomes of *N. tabacum* and one haploid genome of *N. glutinosa* (TTG), so they had a chromosomal constitution of sesquidiploids. Additionally, in that generation, all plants were resistant to TMV.
(c) First breakdown generation**.** The sesquidiploids were crossed with three *N. tabacum* varieties, Burley 16', 'Samsun' and 'Connecticut Broadleaf', to produce BC_2 offspring. In interspecific breeding, the first two or three postsesquidiploid generations are often called 'breakdown generations'. This is because in those generations, the haploid set of chromosomes contributed by the donor species disintegrates into individual chromosomes that are included at random and in varying numbers in individual plants. In the BC_2 progenies derived from Connecticut Broadleaf, the ratio of resistant to susceptible plants was well below 50%, which reflected random loss or inclusion of *N. glutinosa* chromosomes. However, among some derivatives of Burley 16 and Samsun, approximately 50% of resistant plants were observed which was close to the expected Mendelian gamete segregation ratio for a dominant monogenic trait.
(d) Second and subsequent breakdown generations**.** A number of BC_2 plants involving all three cultivars were self-pollinated. The selfed BC_2F_2, BC_2F_3 and BC_2F_4 offspring derived from Connecticut Broadleaf kept segregating for resistance in a non-Mendelian fashion, and no stable resistant lines were obtained. Some of the selfed Burley 16 derivatives yielded 3:1 ratios, but no line that would breed true for resistance was obtained. It is only among the Samsun derivatives that a single BC_2 plant was found whose offspring yielded consistently 3:1 ratios when backcrossed to the recurrent parent and was entirely composed of resistant plants upon selfing. That homozygous TMV-resistant line developed by Holmes came to be known as 'Holmes Samsun'. In many papers and in germplasm collections, it is alternatively called 'Samsun H'. The spelling of the variety's name ('Samsun' or 'Samsoun') is dependent upon local language and usage and is not consistent.

Mechanism of the Incorporation of Resistance to TMV from *N. glutinosa* in the Genome of Holmes Samsun (Samsun H) The successful transfer of the genetic factor from *N. glutinosa* to *N. tabacum* raised the intriguing question of how it was actually accomplished. The chromosomes of *N. glutinosa* do not pair regularly with those of *N. tabacum* (Kostoff, 1943; Goodspeed, 1954), so there was little chance that the exchange of genetic material through recombination took place. A clue was provided by Mallach (1943), who found that while Samsun H showed 24 pairs in meiosis, its F_1 hybrids with other varieties of *N. tabacum* consistently showed 23 pairs and two univalents. These results indicated that considerable structural differences existed between one of the chromosomes contributed by Samsun H and its counterpart from other tobacco varieties, and these differences prevented the two chromosomes from normal pairing. Chromosome pairing and TMV resistance in reciprocal backcross progenies of *N. tabacum* var. Purpurea × Samsun H was analyzed by Gerstel (1943). In his study, the transmission of TMV resistance paralleled the distribution of the nonconjunctional pair of chromosomes and the factor for TMV resistance was found to be located on one of those two nonpairing chromosomes (see also Table 4.2). The ultimate conclusion was that Samsun H contained 23 pairs of *N. tabacum* chromosomes and a substituted pair of chromosomes from *N. glutinosa*. That pair of alien chromosomes in Samsun H was able to perform all the vital functions of their *N. tabacum* analogs in other varieties except that at Metaphase I, they were observed as univalents. Samsun H was in many ways similar to the original Samsun, was self and cross fertile and could be perpetuated by selfing like any other variety. In another study based on monosomic analysis, Gerstel (1945a) demonstrated that in Samsun H, the *N. glutinosa* chromosome had taken the place of the native H chromosome of *N. tabacum*.

A combination of *a priori* assumptions, experimental data and of the earlier account by Holmes himself (1938) were used by Gerstel (1943, 1946) to explain the chain of events that may have led to the incorporation of the pair of *N. glutinosa* chromosomes in the genome of Samsun H. According to Gerstel's reasoning, the substitution of an alien chromosome for a native chromosome may occur in the sesquidiploids and in further backcrosses to the recurrent parent as a result of the following events:

1. Formation of trivalents between two *N. tabacum* and one *N. glutinosa* chromosome in Metaphase I. This presupposes homeology between the *tabacum* chromosome and its *glutinosa* counterpart. The distribution during the subsequent meiotic stages might result in two types of gametes: one deficient for a tabacum chromosome replaced by its homoeologous glutinosa counterpart (23T+1G) and the other with a supernumerary tabacum chromosome (25T).
2. Nonconjunction, i.e., the failure of two tabacum homologs to pair at Metaphase I followed by random distribution of unpaired chromosomes at Anaphase I: both *N. tabacum* univalents go to the same daughter cell, the place of one of them in the opposite nucleus is taken by an *N. glutinosa* univalent.
3. Occasional pairing between an *N. tabacum* chromosome and its *N. glutinosa* homeologue with the other *N. tabacum* chromosome being left out of the association. At Anaphase I, two daughter nuclei may be formed: 23T+G and 24/25T.

To determine more about what had actually occurred, Gerstel (1946) crossed a tetraploid white-flowered mutant of *N. tabacum* cv. 'Cuba' with regular, pink-flowered *N. glutinosa*. Backcrossing the sesquidiploid thus obtained to white-

Table 4.2 Transmission of the non-conjunctional substituted chromosome of *N. glutinosa* from 'Samsun H' to BC_1 populations derived from backcrossing the F_1 hybrid (Samsun H × *N. tabacum* var.'Red Russian') to Red Russian (adapted after Gerstel, 1943)

Gametic output of the F_1 (Samsun H × Purpurea) used in BC_1 as male and female parent		Metaphase I configurations in the BC1 populations			Response to TMV	% Resistant plants
	Chromosome number		Chromosome type	Chromosome number		
female parent						
23T + G	24	23II + 2I	HG	48	HR	17
23T + (G+H)	25	24II + 1I	HHG	49	HR	2
23T + H	24	24II	HH	48	S	22
23T	23	23II + 1I	H	47	S	59
male parent						
23T + G	24	23II + 2I	HG	48	R	46
23T + H	25	24II	HH	48	S	54

Explanations: 23T – haploid complement of *N. tabacum* except chromosome H; H: chromosome H of *N.* tabacum; G: alien chromosome of *N. glutinosa* that substituted for the native chromosome H in Holmes Samsun resistant to TMV; Only one type of gametes contributed by Red Russian (23T + H); 23II: number of bivalent pairs; 2I, 1I: number of univalents, 2 or 1, respectively; HR –hypersensitive (resistant) response to TMV; S – susceptible response to TMV flowered tobacco and selfing the resultant offspring provided him with a convenient model that imitated the work of Holmes. The dominant pink factor from *N. glutinosa* and the recessive white factor from *N. tabacum*, both easy to track, were intended to mimic the behavior of resistance vs susceptibility observed by Holmes. Gerstel found evidence of irregular meiotic behavior in the sesquidiploids that resulted in rare genotypes with 23 pairs of *N. tabacum* chromosomes and additional *N. glutinosa* chromosomes in their offspring. Those plants could produce substitution gametes, but only if one of the *N. glutinosa* chromosomes could take over the function of the missing *N. tabacum* chromosome. Indeed, upon pollination of normal, white-flowered tobacco with one of the BC_1 offspring, a segregation of 1: 1 for pink vs white flowers was observed. This transmission rate indicated that the pink parent was heterozygous rather than trisomic for the pink factor and, consequently, contained a nonconjunctive pair of chromosomes from *N. glutinosa*. The results of further crosses corroborated the assumption that a substitution of the pair of alien chromosomes had actually occurred. The results fitted well with those reported by Holmes. Gerstel's ultimate conclusion was that the TMV-resistant plant that gave rise to Samsun H was a substitution heterozygote rather than a trisomic and that the substitution had taken place in the sesquidiploid as a result of its disturbed macrosporogenesis.

The investigations of Patel and Gerstel (1961) furnished evidence that it is not the sesquidiploid stage at which the substitution of an alien chromosome is most likely to occur. In their study of the amphidiploid 4x (*N. tabacum* × *N. glutinosa*), they found that in the sesqudiploid and breakdown generations, the meiotic divisions in the recurrent parent are usually very regular, and the chromosomes of the nonrecurrent parent are excluded from intergenomic associations since no nonconjunctions or trivalent formations were observed. However, the meiotic behavior of some other hybrids was somewhat different. In Metaphase I of the sesquidiploids obtained from mating tetraploid *N. tabacum* with *N. africana*, Doroszewska and Berbeć (2000) recorded a modal number of ca. 20 univalents rather than theoretically expected 24, indicating that at least four of them formed associations of higher valencies, possibly trivalents with the chromosomes of *N. tabacum*. Likewise, trivalents and pentavalents were found in the sesquidiploids from crossing autotetraploid *N. tabacum* with *N. benavidesii* (Berbeć & Głażewska, 1988). Reed and Collins (1980) reported pairing of nonhomologous chromosomes in the sesquidiploid progeny of the hybrids *N. stocktonii* × *N. tabacum* and *N. nesophila* × *N. tabacum*. It is also noteworthy that one of the best known cases of interspecific introgression in *Nicotiana tabacum*, the transfer of resistance to black shank from *N. plumbaginifolia*, involved the sesquidiploid from crossing 4n *N. tabacum* × 2n *N. plumbaginifolia* as the starting hybrid (Chaplin, 1962).

Transmission of Substituted Chromosomes The transmission of the substituted *N. glutinosa* chromosome that carried the N-gene for TMV resistance was investigated by Gerstel (1943) in the progenies obtained from backcrossing F_1 Samsun H × normal *N. tabacum* var. 'Purpurea' using the F_1 hybrid as both the male and the female parent (Table 4.2). Chromosome H of *N. tabacum* and chromosome G from

N. glutinosa that carries the N gene are further referred to as H and G, respectively. H and G are peculiar in that they are functionally equivalent, i.e., they can replace each other without affecting the development of an *N. tabacum* plant, yet they fail to pair with each other (Gerstel, 1943, 1945a). Nonconjunction determines the behavior of chromosomes G and H that is subject to poorly predictable rules governing the transmission of univalents during meiotic divisions.

Four types of viable egg cell gametes were produced by the F_1 plants owing to the random distribution of unpaired chromosomes G and H: 23T+G, 23T+H, 23T+G+H, and 23T (Table 4.2, column 1). However, only the first two classes of male gametes are competitive, i.e., 23T+G and 23T+H, because unbalanced 23- and 25-chromosome pollen grains are unable to compete with those having 24 chromosomes (Table 4.2, column 2). According to Gerstel (1946), pollen gametes deficient for 1 chromosome (23T) are rarely functional. On the other hand, single unpaired chromosomes tend to lag during meiosis and be left outside the daughter nuclei and are usually included in less than 25% of the gametes (Olmo, 1935). Thus, the preponderant class of viable microsporocytes would be of the 23T+1H and 23T +1G types (Table 4.2, column 1). Those types of gametes when fused with normal 23T+HH gametes from the recurrent Purpurea produced classes of offspring listed in columns 3, 4 and 5 of Table 4.2. As chromosome G carried the dominant factor for resistance to TMV, only those classes that contained that chromosome were resistant (Table 4.2, column 4). The percentages of particular plant classes within the BC_1 progeny reflected the gametic output of the F_1 hybrid Samsun H × Purpurea (Table 4.2, columns 4 and 7). Bearing in mind that unpaired chromosomes are largely eliminated from the gametes, the gametic output of the 23T condition (Table 4.2, column 1), as reflected by the percentage of the monosomic H plants, was surprisingly higher than theoretically expected. The very low frequency of trisomic HHG plants reflected the low chance of two univalents being included in one gamete. However, when F_1 was used as the male parent, nearly equal proportions of HH and HG (susceptible vs. resistant) plants were obtained because the two univalents had an equal chance of being included in the microsporocytes, and unbalanced genotypes were eliminated. The general conclusion from these results was that in backcrosses with normal varieties, substituted chromosomes tend to be quickly eliminated, especially when a substitution line is used as the maternal parent. Based on the latter finding, Gerstel postulated the use of heterozygous plants as pollen parents in the transfer of mosaic resistance from Samsun H to other varieties.

An extensive study of univalent transmission and alien substitution involving chromosomes from sources other than *N. glutinosa* was carried out by Chaplin and Mann (1961). They presented evidence for the substitution of chromosomes from *N. paniculata* (Pan), *N. plumbaginifolia* (Pl), and *N. rustica* (Rus) in the genome of *N. tabacum*. To facilitate the identification of a substitution event they used an autotetraploid *N. tabacum* stock homozygous for the recessive yellow–green gene (TTTT) to produce sesquidiploids with the wild species, each of them carrying the normal-green character (Pan, Pl, and Rus) dominant over the yellow-green mutation in the *N. tabacum* parent. The sesquidiploids (TTPan, TTPl and TTRus) were crossed as males with the female yellow green variety (TT) to produce the BC_1 breakdown generation. Evidence for substitution was sought in the BC_2 backcross and BC_2S_1 selfed generations by counting the segregation ratios of yellow–green to green plants. Green BC_1 plants were used as males in crosses with the recurrent yellow–green variety to facilitate the distinction between the 'trisomic' and the 'substituted' condition due to the expected much higher transmission rate of the latter. In the BC_2 and BC_2S_1 progenies, evidence for alien chromosome substitution from all three species was found. It was also established that the substituted chromosomes that carried the green factor did not pair with their *Nicotiana tabacum* counterpart.

It must be stressed that not all alien substitutions give rise to stable alien substitution lines. It is possible only when the substituted alien chromosome can take over all the vital functions of the displaced native chromosome. However, there are several reports on viable nullisomics in *N. tabacum*: nullisomic C (Gerstel & Parry, 1973; Mattingly & Collins, 1974); nullisomic D (Moore & Collins, 1982; Rufty et al., 1983); nullisomic E (Mattingly & Collins, 1974); nullisomic H (Kramer & Reed, 1988); nullisomic S (Lewis, 2020a); and an unidentified monosomic (Ramavarma et al., 1976). Of these, nullisomics E and S are even fertile enough to be maintained by self-pollination (Lewis, 2020a). These instances of viable nullisomics suggest that there may be exceptions to the requirement for the native and the introgressed chromosome to be mutually exchangeable.

4.3.2 Alien Chromosome Addition

Another means of introducing an alien gene from one species into the genome of another is through adding a pair of chromosomes from the genome of the former to that of the latter. Gerstel (1945b) first demonstrated the feasibility of the process by crossing autotetraploid *N. tabacum* (TTTT) with diploid *N. glutinosa* (GG) to obtain a sesquidiploid (TTG). Next, by backcrossing the sesquidiploid to the recurrent *N. tabacum,* he obtained a breakdown population composed of plants that, along with the whole TT genome, carried various numbers of additional chromosomes from the nonrecurrent *N. glutinosa* and segregated for the presence of the resistance factor. As a result of double selection for *the N. tabacum* phenotype and resistance to TMV, two lines were developed that contained 50 chromosomes in their somatic cells, 24 pairs from *N. tabacum* and one supernumerary pair from *N. glutinosa*. The simplest explanation of the origin of those alien substitution plants was the fusion of two gametes, each containing a haploid complement of 24 chromosomes from *N. tabacum* and an additional monosome from *N. glutinosa*.

A 50-chromosome alien addition line was a transition stage in integrating the factor of resistance to PVY from *N. africana* in the genome of *N. tabacum* (Lewis, 2005, see also Sect. 4.6.5).

Black root rot-resistant addition lines of *N. tabacum* were obtained by Bai et al. (1996) among backcross derivatives of the somatic hybrid *N. tabacum* + *N. debneyi*.

Addition lines of *N. tabacum* containing two pairs of alien chromosomes from *N. quadrivalvis* were also developed (Burk, 1960).

An alien addition line can be produced if the amphidiploid hybrid and its sesquidiploid derivative show some degree of fertility. Given that this prerequisite is met, the mode of origin of alien addition lines can be explained by one of two phenomena:

(a) Fusion of two gametes, both having a full chromosome complement of one species and an added chromosome from another. This should be considered a relatively rare event, as extra chromosomes tend to be eliminated from pollen mother cells (Olmo, 1935; Gerstel, 1945b), and microsporocytes with supernumerary chromosomes are not competitive with normal 24-chromosome gametes (Gerstel, 1943).
(b) Division of the univalent from an alien species in the female meiocyte followed by the inclusion of both chromatids in one daughter nucleus. Splitting of unpaired chromosomes during the first reduction division occurs frequently in *Nicotiana* (Gerstel, 1946). This is reflected by the fact that the sum of the chromosomes contained in both daughter nuclei after the first meiotic division often exceeds the somatic chromosome number of a given plant (Kostoff, 1943).

4.3.3 Segmental Alien Substitution

Alien substitution or addition lines that involve whole foreign chromosomes are a poor choice in tobacco breeding. Traits introgressed in this manner are difficult to transfer to other varieties because of irregular inheritance, and the varieties thus modified are generally below the agronomic standards because they carry a heavy load of alien chromatin derived from the wild donor species, a phenomenon termed 'genetic drag' or 'linkage drag' (see Sect. 4.6.3). It practically makes them useless as cultivars. Addition lines are even more problematic if introgression of only a small chromosome fragment carrying the gene of interest is intended. Supernumerary alien chromosomes are unlikely to recombine with the chromosomes of the recipient genome (Lewis, 2005).

Nicotiana tabacum can hybridize with nearly sixty of its sister *Nicotianae* (Table 3.10), which are now known to be nearly 100 in number (go back to Table 2.1). Of these, only seven directly or indirectly involved in the parentage of *N. tabacum,* i.e., *N. sylvestris* and *N. tomentosiformis*, *N. tomentosa*, *N. setchellii*, *N. otophora*, *N. kawakamii* of the section Tomentosae, show no particularly difficult restrictions in the gene flow between them and *N. tabacum* (Clausen & Cameron, 1957; Gerstel, 1960; Wernsman et al., 1976; Oupadissakon & Wernsman, 1977; Ohashi, 1985; Hancock & Lewis, 2017). This is due to intergenomic affinities in chromosome structure that allow relatively extensive opportunities for genetic exchange via recombination.

However, with the remaining *Nicotiana* species, those opportunities are severely restricted, including the paradigmatic introgression of TMV resistance from

N. glutinosa. The original resistant variety, Samsun H, was flawed with undesirable agronomic characteristics associated with the presence of a pair of whole alien chromosomes. Nonetheless, the acceptable TMV-resistant variety Burley 21 was ultimately developed by W. D. Valleau through persistent backcross and selection cycles that involved recipient commercial varieties and Samsun H as the donor (Gerstel & Burk, 1960). The cytological analysis of how only a fragment of an alien chromosome can be integrated with the genome of the recipient species was presented and discussed in detail by Clausen and Cameron (1957). At least some residual homology between the chromosomes involved in the transfer is recommended to facilitate the substitution of a whole chromosome, e.g., by trivalent formation (Gerstel, 1946), but at least occasional pairing between an *N. tabacum chromosome* and its wild species counterpart was assumed to be the prerequisite for the exchange of chromosome fragments, a process analogous to recombination but which became known as segmental substitution in introgressive breeding (Clausen & Cameron, 1957). Gerstel and Burk (1960) demonstrated that the F_1 hybrids obtained by crossing Valleau's new resistant variety with Samsun H exhibited a higher degree of complete pairing during meiosis than did the hybrids between Samsun H and normal *N. tabacum*. The results indicated that a fragment of the *N. glutinosa* chromosome carrying the resistance factor was translocated to a native Burley 21 chromosome. The fragment was long enough to permit frequent formation of chiasmata with the substituted whole *N. glutinosa* chromosome from Samsun H. Although chromosome H from *N. tabacum* and its analog from *N. glutinosa* appeared to be nonconjunctional in Metaphase I (Mallach, 1943), the meiotic configurations studied by Gerstel and Burk (1960) suggested that the two chromosomes may have been paired in Prophase I. The affinity between the two chromosomes helped demonstrate that an alien segmental substitution had actually taken place.

Subsequent experiments yielded much evidence for recombination within a pair of homeologous or partly homeologous chromosomes as probably the most frequent mechanism of chromosomal interchange between *N. tabacum* and an alien species. Experimental results of the transfer of a marker locus from *N. plumbaginifolia* to *N. tabacum* performed by Moav (1958) showed that the incorporation of the *N. plumbaginifolia* locus in *N. tabacum* was clearly nonrandom. Out of fourteen transfers, eight involved the same chromosome, and six were equally distributed among three others. This proved the existence of residual homology between the *N. tabacum* chromosomes and their *N. plumbaginifolia* counterparts and that the homology was responsible for the interspecific interchanges. Likewise, Lewis (2002) found that the introgression of resistance to PVY from *N. africana* to *N. tabacum* was nonrandom. In three out of seven introgression events, the resistance locus was integrated with the same *N. tabacum* chromosome. The factor for resistance to TMV from *N. glutinosa*, the N gene, was found to be integrated with two chromosomes of *N. tabacum*: chromosome H in the TMV-resistant varieties grown in the US and with another unidentified chromosome in other resistant accessions (Lewis, 2005; Lewis & Rose, 2010).

However, in each of the reports cited in the preceding paragraph (Moav, 1958; Lewis, 2002; Lewis & Rose, 2010) more than one chromosome were reported as the sites of the introgressed loci. Although not proved conclusively, the distribution among different chromosomes may indicate that, in some of those cases, the segmental substitution took place through mechanisms other than homeologous recombination e.g. by translocation. Random breakage and reunion of chromosomes, either spontaneous as postulated by Moav (1958) or induced, e.g., by irradiation as performed by Niwa (1969), may play a significant part in translocating chromosome fragments from chromosome to chromosome. However, because of the random nature of such translocations, they usually result in biologically ill-compensated and agronomically inferior products (Lewis, 2011).

The opportunities for recombination or translocation between alien chromosomes are the greatest in the amphidiploid generation and diminish with subsequent transfer generations (Patel & Gerstel, 1961). A translocated segment can be large enough to make the rearranged chromosome visually recognizable as a heteromorphic bivalent when it pairs in meiosis with its normal unchanged homolog (Ramavarma et al., 1991). The translocated fragment itself could be visualized using the GISH method, an event demonstrated by Laskowska et al. (2015) in their study of the amphidiploid 4x (*N. wuttkei* × *N. tabacum*).

Resistance to *parasitica* var. *nicotiana* (now *Phytophtora nicotianae*) was transferred to *N. tabacum* from two species, *N. longiflora* (Valleau et al., 1960) and *N. plumbaginifolia* (Chaplin, 1962). Recently, the root knot resistance gene was found to be originally located on chromosome 9 of *N. plumbaginifolia* (Dang et al., 2019). The single dominant genes for resistance to black shank disease from *N. plumbaginifolia* (*Php*) and *N. longiflora* (*Phl*) were found to be inserted on the same chromosome of *N. tabacum* but were not allelic. However, the two loci were at such a close distance from each other that the double recessive recombinants were produced in F_2 *Php* × *Phl* at a frequency just slightly higher than 0.05%, and test crosses gave a frequency of ca. 1.5% (Johnson et al., 2002b).

Interestingly, a similar situation was found with two single dominant genes for resistance to *Thielaviopsis basicola* (black root rot BRR), one originally from *N. debneyi* and transferred to local cultivars via AC Gayed, a Canadian flue-cured cultivar (Brandle et al. (1997) and the other introgressed directly from *N. glauca* (Trojak-Goluch & Berbeć, 2009). The F_2 population of ca. 600 plants from selfing the F_1 hybrid between the two resistant genotypes contained no susceptible double recessive recombinants (author of this review, unpublished observation). Unlike *N. longiflora* and *N. plumbaginifolia,* which are the closest relatives within the same section, *N. debneyi* and *N. glauca* have been classified into two distantly related sections (Chase et al., 2003). However, the analysis of 5S rDNA spacer sequences (Kitamura et al., 2005) revealed that half of the 48 chromosomes of *N. debneyi* showed a high degree of homology with the genomic DNA of *N. glauca*. A SCAR marker linked to the black root rot resistance region from *N. debneyi* (Julio et al., 2006) was amplified in accessions of both *N. debneyi* and *N. glauca* as well as in two BRR-resistant lines of *N. tabacum,* one with BRR resistance from *N. debneyi* and the other from *N. glauca* (communicated by Dr G. Korbecka of this laboratory). All these results suggest a close affinity between the BRR resistance loci from *N. debneyi* and *N. glauca*. Paradoxically, the detection of the same amplified region in the two studied resistant lines and in both *N. debneyi* and *N. glauca* reduces the likelihood that the resistance to black root in the two resistant lines may have actually come from the same species (compare Milla et al., 2005).

Table 4.3 lists some cases of controlled introgression from several *Nicotiana* species to *N. tabacum* that resulted in different types of introgression: additions and substitutions of whole chromosomes and segmental substitutions.

4.4 Optional Routes of Introgression from *Nicotiana* Species to *N. tabacum*

4.4.1 Bridge-Cross Method

As discussed in the previous chapter, chromosomal and genic incongruities between the donor and the recipient, both pre- and postfertilization, were the primary hindrances that the breeders had to surmount to accomplish the interspecific transfer of a desired trait, and different tools were used to this end. Failing this, there was yet another agent left to be deployed: an intermediary between two incongruent species. The main property expected of such an intermediary species was that it ought to cross readily with both the recipient and the donor. Interestingly, the first documented interspecific gene transfer from a wild *Nicotiana* to *N. tabacum* also involved a bridging species. F.O. Holmes, an American geneticist and breeder used *N. paniculata* as an intermediary to introgress a hypersensitive, age-dependent response to TMV from *N. rustica* var. *jamaicensis* into *N. tabacum* var. Samsoun (Holmes, 1936, 1937a, b; Scholthof, 2017). First, Holmes crossed *N. paniculata* with *N. rustica,* successfully overcame the sterilty of the F_1 hybrid by pollinating it with pollen of *N. paniculata* and after several further backcrosses to the female parent recreated a TMV resistant strain of *N. paniculata.* As the next step, he crossed the TMV-resistant *N. paniculata* with *N. tabacum* and after several backcrosses transferred the resistance factor to *N. tabacum*. This approach was repeated by Burk (1967). *N. repanda*, a valuable source of germplasm resistant to many diseases of tobacco, is very difficult to hybridize directly with *N. tabacum*. Burk's goal was to transfer resistance to TMV from *N. repanda* to *N. tabacum*. The following procedure was devised to transfer chromosomes from *N. repanda* to *N. tabacum*. First, the amphihaploid *N. repanda* × *N. sylvestris* was created, a hybrid that is easy to make but sterile (Kostoff, 1943). The amphihaploid was converted to a fertile amphidiploid and backcrossed twice to *N. sylvestris* to create first a sesquidiploid (SSR) and next, by backcrossing to *N. sylvestris* to produce a highly variable breakdown population [(SS+(r)]. It was screened for *N. sylvestris*-like phenotypes that combined resistance to TMV with an ease of hybridization with *N. tabacum*. The [(SS+(r)] × *N. tabacum* allohaploids were successfully backcrossed to the recurrent *N. tabacum*, possibly because the former produced a certain amount of unreduced

Table 4.3 Instances of alien chromosome additions and substitutions in gene transfers from different *Nicotiana* species to *N. tabacum*

Source species of introgressed trait	Gene transfer mechanism	Transferred trait	Reported by
N. alata (*n* = 9)	SS[1]	Resistance to TMV	Gajos (1981)
N. plumbaginifolia (*n* = 10)	A SS	Pollen killer locus	Cameron and Moav (1957)
		Marker gene substituting for white seedling locus (Ws(pbg)	Clausen (1952), Moav (1958), and Niwa (1969)
	CS, SS A	Resistance to *Phytophtora nicotianae* race 0	Apple (1967), Chaplin (1962), and Dang et al. (2019)[8]
	SS	Purple flower color	Ar-Rushdi (1957)
N. longiflora (*n* = 10)	SS CS	Resistance to *Meloidogyne javanica*	Schweppenhauser (1975)
	SS	Resistance to *Phytophtora nicotianae* race 0	Valleau et al. (1960) and Johnson et al. (2009)
N. glauca (*n* = 12)	SS	Resistance to *Thielaviopsis basicola*	Trojak-Goluch and Berbeć (2002)
N. benavidesii	S/SS	Gene substituting for pink flower locus	Berbeć (1980)
N. paniculata (*n* = 12)	S, SS	Dominant green chlorophyll markers	Chaplin and Mann (1961) and Lucov et al. (1970)
N. quadrivalvis (*n* = 24)	A[2]	Male fertility restorer locus/ loci	Burk (1960), Gerstel and Burns (1983)[2]
N. repanda	A	Male fertility restorer locus	Gerstel et al. (1978)
N. repanda	CS[3]	Resistance to TMV	Gwynn et al. (1986)
N. repanda	A, SS	Resistance to *Meloidogyne javanica*	Schweppenhauser (1968, 1974)
N. rustica (*n* = 24)	SS	Resistance to *Phytophtora nicotianae* Race 1 (black shank)	Drake and Lewis (2013)
N. africana (*n* = 23)	A	Resistance to PVY and TEV	Wernsman (1992) and Lewis and Wernsman (2001)
N. benthamiana (*n* = 19)	SS	Resistance to *Spodoptera litura* (tobacco caterpillar)	Ramavarma et al. (1991)
N. debneyi (24)	AA	Resistance to *Thielaviopsis basicola*	Bai et al. (1996)
N. gossei (36)	A, CS	Resistance to *Spodoptera litura*	Rao et al. (1980)
N. suaveolens (*n* = 16)	A, SS	Male fertility restorer locus	Schweppenhauser and Mann (1968) and Hosfield & Wernsman (1974)
N. sylvestris (*n* = 12)	CS, SS	Marker gene substituting for white seedling locus Ws(syl)	Chaplin and Mann (1961)

(continued)

Table 4.3 (continued)

Source species of introgressed trait	Gene transfer mechanism	Transferred trait	Reported by
N. otophora (*n* = 12)	CS, SS	Marker gene substituting for white seedling locus (Ws(oto)	Chaplin and Mann (1961)
N. tomentosiformis (*n* = 12	CS, SS	Marker gene substituting for white seedling locus (Ws(tmf)	Chaplin and Mann (1961)
N. glutinosa (*n* = 12)	S[4], SS[5], AA[6], AAAA[7]	Resistance to TMV, reduced flower size, gene substituting for white flower locus	Holmes (1938), Mallach (1943), and Gerstel (1943, 1945b, 1946)

Abbreviations: *CS* substitution of a pair of chromosomes, *SS* substitution of a chromosome fragment (segmental substitution), *AA* addition of a pair of chromosomes, *A* addition of a single chromosome

[1]Segmental substitution manifested as a unique linkage in 48-chromosome *N. tabacum* lines (leaf malformations and tumors);

[2]Addition of two pairs of chromosomes (52-chromosome line);

[3] heterozygous substitution (23 chromosome pairs from *N. tabacum* + 2 univalents (from *N. tabacum* and from *N. repanda*, respectively);

[4]Resistance to TMV (Holmes, 1938; Mallach, 1943; Gerstel, 1943); pink flower locus (Gerstel, 1946);

[5]Resistance to TMV (Clausen & Cameron, 1957; Gerstel & Burk, 1960; Patel & Gerstel, 1961);

[6]Resistance to TMV (Gerstel, 1945b);

[7]Addition of two pairs of chromosomes conferring resistance to TMV and reduced flower size, respectively (Gerstel, 1945b);

[8]Monosomic alien addition of chromosome 9 carrying the *Php* gene

restitution gametes. Through repeated backcrossing and screening for resistant phenotypes, the hypersensitive reaction to TMV was transferred to *N. tabacum*.

The procedure that came to be called the bridge cross method was used by several other investigators (Table 4.4). They used different combinations of donors, recipients and bridging species to achieve their goals. *N. otophora* was used as a bridging species in a successful attempt to transfer a hypersensitive response to TSWV from *N. alata* to *N. tabacum* (Gajos, 1976, 1981). The amphihaploid *N. tabacum* × *N. alata* (TA) was treated with colchicine, and one male fertile plant, most likely an amphidiploid (TTAA), was selected among the treated plants (Gajos, 1976, 1981). Sesquidiploids obtained by backcrossing TTAA × TT (sesquidiploids (TTA)) were self- and cross-sterile. However, seeds were obtained by crossing the colchicine-treated fertile plant (TTAA) with another amphidiploid hybrid, 4x (*N. tabacum* × *N. otophora*), obtained earlier (Gajos, 1979). In the offspring thus obtained, one plant showed a hypersensitive reaction to TSWV inoculation. The plant was self-sterile due to pollen sterility but produced offspring when backcrossed as female to *N. tabacum*. Self-fertility was restored, and several lines that showed hypersensitivity to TSWV, characteristic of the donor parent (*N. alata*), were developed. The hypersensitivity was linked to morphological disorders, however (see relevant part of Sect. 4.6.3). In another case, two rather than one bridging species (*N. sylvestris* and *N. longiflora*) were used in transferring resistance to root

Table 4.4 Instances of bridge crosses in attempted interspecific gene transfers to *N. tabacum*

Source species	Bridging species	Bridge cross	Transferred trait	Authors
N. alata	*N. otophora*	4x (*N. tabacum* × *N. alata*) × 4x (*N. tabacum* × *N. otophora*)	Resistance to TSWV	Gajos (1979, 1981)
N. repanda	*N. sylvestris*	4x (*N. repanda* × *N. sylvestris*) × *N. tabacum*	Resistance to TMV	Burk (1967), Chaplin (1978)
N. repanda	*N. sylvestris*	triple allopolyploid *N. tabacum* × *N. repanda* × *N. sylvestris* backcrosssed to *N. tabacum*	Resistances to *Meloidogyne javanica* and *Cercospora sp.*	Stavely et al. (1973)
		[4x (*N. repanda* × *N. sylvestris*) × *N. sylvestris*] × *N. tabacum*	Resistances to *Meloidogyne incognita* race 3 and 4; *Globodera tabacum*, *Pseudomonas syringae* pv. *tabaci*, TMV	Gwynn et al. (1986)
N. repanda	*N. sylvestris*, *N. longiflora*	4x (*N. repanda* × *N. sylvestris*) × (*N. longiflora* × *N. sylvestris*) × *N. tabacum*	Resistance to *Meloidogyne* sp.	Schweppenhauser (1974)
N. repanda	*N. sylvestris*	4x (*N. repanda* × *N. sylvestris*) × *N.* sylvestris[1] × *N. tabacum*	Resistance to *Cercospora nicotianae*	Wan et al. (1971)
N. rustica	*N. alata*	4x (*N. rustica* × *N. alata*) × *N. tabacum*	CMS factors	Shabanov and Nikova (1984), Nikova et al. (1997), and Nikova and Vladova (2002)
	N. paniculata	N. paniculata × N. rustica backcrossed to N. paniculata; N. paniculata × N. tabacum backcrossed to N. tabacum	transfer of resistance to TMV from N. rustica to N. tabacum via N. paniculata	Holmes (1936, 1937a, b)
N. benthamiana	*N. glutinosa*	6x (*N. benthamiana* × *N. glutinosa* × *N. tabacum*) × *N. tabacum*	CMS factors	Ramavarma et al. (1977)
	N. repanda	4x (*N. benthamiana* × *N. repanda*) × *N. tabacum*	Resistance to aphids	Murthy et al. (2014)
N. kawakamii	*N. sylvestris*	4x (*N. sylvestris* × *N. kawakamii*) × *N. tabacum*	Resistance to PVY	Ohashi (1985)
N. tomentosiformis	*N. sylvestris*	No details available	Resistance to *Erysiphe cichoracearum*	Tatemichi, 1990
N. exigua	*N. rustica*	4x (*N. rustica* × *N. exigua*) *N.* 'diruex' × *N. tabacum*	Resistance to *Peronospora* (project not completed)	Bolsunov (1970)
N. glutinosa	*N. sylvestris*	*N. glutinosa* × 4x (*N. sylvestris* × *N. tabacum*)	Resistance to TMV	Hitier and Izard (1958)

[1]Several backcrosses to *N. sylvestris*, *N. sylvestris*-like individuals resistant to *C. nicotianae* crossed with *N. tabacum*

knot disease from *N. repanda* to *N. tabacum* (Schweppenhauser, 1968, 1974). It may be added that in this particular instance, Schweppenhauser used a susceptible strain of *N. longiflora* as a bridging species. In several other projects, Schweppenhauser and other investigators used resistant accessions of *N. longiflora* as the direct source of resistance to that pathogen (Schweppenhauser, 1975; Raeber & Smeeton, 1980; Mudzengerere, 1994). In a few of the cases listed in Table 4.4, the construction of a hybrid that united the genomes of the donor, the ultimate host and the bridging species was the first step of the bridge transfer process: 4x (*N. rustica* × *N. alata*) × *N. tabacum*, 4x (*N. benthamiana* × *N. repanda*) × *N. tabacum*, 4x (*N. rustica* × *N. exigua*) × *N. tabacum*. Such multispecific hybrids will be briefly discussed in the next Sect. (4.4.2).

4.4.2 Multiple or Polygenomic Hybrids as Starting Hybrids in Gene Transfer

Simple digenomic unions or their derivatives of a higher ploidy level are the preponderant class of *Nicotiana* hybrids that have been investigated and utilized in academic and applied studies.

However, some researchers moved a step further. For various reasons, including simple curiosity, they made attempts to cross such a digenomic hybrid with a third species or with another hybrid. Such matings gave rise to complex hybrids that united genomes supplied by three or even four species.

Trispecific and tetraspecific hybrids and even some rare multiple combinations of higher order are the general terms used to describe hybrid combinations that involve more than two species. However, the actual cytogenetic character of such hybrids is highly dependent on to what extent the genetic exchange is restricted between the genomes of the constituent species. The best examples of multiple hybrids within which the interspecific genetic flow is almost unobstructed are those that involve the 9-chromosome species of the section Alatae: *N. alata*, *N. forgetiana*, *N. mutabilis*, *N. langsdorffii*, *N. sanderae*, the putative species *N.* 'Rastroensis' and, probably to some lesser extent, *N. bonariensis* (Williams & Pandey, 1975; Lee et al., 2008). Due to extensive chromosome segregation and recombination within the constituent homoeologous genomes of such multiple hybrids, their integrity can hardly be preserved in the resulting multiploids, so they should not be regarded as truly polygenomic. The same reservation applies to trispecific combinations within the sections Tomentosae, some hybrids of the section Suaveolentes and the two species of the section Trigonophyllae.

Trigenomic hybrids normally result from mating an amphidiploid or amphihaploid hybrid to a third species. In the latter case, the amphihaploid parent must retain at least some ability to produce unreduced gametes. (*N. rustica* × *N. paniculata*) × *N. glutinosa*, probably the first hybrid of this cathegory was reported in the mid-eighteenth century by Koelreuter, the father of interspecific hybridization in *Nicotiana* (Kostoff, 1943). Trigenomic combinations not involving *N. tabacum* that have been reported to date are not very numerous: *N. rustica-paniculata-pauciflora* (Kostoff, 1935, 1943), *N. rustica-paniculata-undulata* (Kostoff, 1936), *N. glauca-langsdorffii-sanderae*, *N. glauca-langsdorffii-alata* (Kostoff, 1943), *N. longiflora-glauca-forgetiana* (Apparao & Gopinath, 1968), *N. alata-rustica-glauca* (Butenko & Luneva, 1966), *N. glutinosa-obtusifolia-megalosiphon* (Gopinath et al., 1965), *N. debneyi-glauca-alata*, *N. debneyi-glauca-langsdorffii*, *N. debneyi-glauca-plumbaginifolia*, *N. debneyi-glauca-sanderae* (Kehr, 1951), *N. quadrivalvis-debneyi-glauca*, *N. debneyi-glutinosa-glauca*, *N. rustica-suaveolens-langsdorffii* (Kehr & Smith, 1952), *N. rustica-suaveolens-langsdorffii, glauca-langsdorffii-sanderae* (Brieger & Forster, 1942; Izard, 1952), *N. rustica-suaveolens-sanderae* (Izard, 1950), *N. rustica-paniculata-langsdorffii* (McCray, 1932), *N. rustica-quadrivalvis-suaveolens*, *N. quadrivalvis-suaveolens-glauca*, *N. quadrivalvis-suaveolens-langsdorffii* (Modilevsky, 1939), *N. suaveolens-langsdorffii-sanderae* (Smith, 1958), *N. quadrivalvis-debneyi-rustica*, *N. quadrivalvis-glutinosa-paniculata*, *N. quadrivalvis-glutinosa-rustica*, *N. quadrivalvis-suaveolens-glutinosa*, *N. debneyi-plumbaginifolia-sanderae*, *N. suaveolens-langsdorffii-sanderae* (Smith & Abashian, 1963), *N. paniculata-langsdorffii-africana* (Kitamura et al., 2005). A few quadruple hybrids involving the whole genomes of four non-tabacum species were also reported: *N. quadrivalvis-suaveolens-glauca-langsdorfii*, *N. rustica-glauca-quadrivalvis-suaveolens*, *N. rustica-paniculata-quadrivalvis-suaveolens* (Kostoff, 1943).

The majority of multiple hybrids were made and studied to gather additional information on phylogenetic relations between individual species (McCray, 1932; Greenleaf, 1941; Krishnamurthy et al., 1960) to explain the physiological, biochemical and genetic mechanisms behind the formation of genetic tumors in *Nicotiana* (Kehr, 1951; Kehr & Smith, 1954; Ahuja, 1962; Ahuja, 1998). Lastly, they were synthetized to examine the limits to feasibility of experimentally synthetized multiple allopolyploids and to their survival (Kostoff, 1943; Kehr & Smith, 1952; Smith, 1958).

The potential of trigenomic hybrids as a starting material in bridge-cross gene transfers from a wild *Nicotiana* to *N. tabacum* was indicated by Ramavarma et al. (1977, 1978). A trigenomic that made history was the hybrid combination (*tabacum-otophora-alata*), the most likely product of crossing two amphidiploids 4x (*N. tabacum* × *N. otophora*) and 4x (*N. tabacum* × *N. alata*). It was the staring allopolyploid in the transfer of resistance to TSWV from *N. alata* to *N. tabacum* (Gajos, 1976, 1981, 1984). Of the hybrids listed in Table 4.5, those that unite the genome of *N. tabacum* with that of *N. alata* (Kehr & Smith, 1952; Larkina, 2015; Berbeć, unpublished), *N. bonariensis* (Ahuja, 1962) and *N. langsdorffii* (Kehr & Smith, 1952) may be of particular interest. The direct gene transfer from *N. alata* to *N. tabacum* is known to become stalled at the sesquidiploid stage (Chaplin & Mann, 1961; Gajos, 1981; Berbeć, 1987). Gajos (1981) was able to bypass that stage by using *N. otophora* as the bridging species, but the transferred TSWV resistance factor was found to be linked to a serious defect (Moon & Nicholson, 2007; Laskowska & Berbeć, 2010). Optional transfer routes from *N. alata* to *N. tabacum* seem to be available via *N. debneyi* (Kehr & Smith, 1952; Larkina, 2015) and *N. raimondii*. The amphidiploid 4x (*N. raimondii* × *N tabacum*) is easily crossable

Table 4.5 Trispecific (trigenomic) hybrids involving *N. tabacum*

No.	Female parent	Male parent	Reported by
1	4x (*N. benthamiana* × *N. africana*)	*N. tabacum*	Iizuka et al. (2012)
2	4x (*N. benthamiana* × *N. glutinosa*)	*N. tabacum*	Ramavarma et al. (1977)
3	4x (*N. benthamiana* × *N. repanda*)	*N. tabacum*	Murthy et al. (2014)
4	4x (*N. debneyi* × *N. fragrans*)	*N. tabacum*	Iizuka et al. (2012)
5	4x (*N. debneyi* × *N. tabacum*)	*N. alata*	Kehr (1950, 1951) and Kehr and Smith (1952)
6	4x (*N. debneyi* × *N. tabacum*)	*N. bonariensis*	Ahuja (1962)
7	4x (*N. debneyi* × *N. tabacum*)	*N. forgetiana*	Ahuja (1962)
8	4x (*N. debneyi* × *N. tabacum*)	*N. glauca*	Kehr and Smith (1952) and Luneva and Poddubnaya Arnoldi (1979)
9	4x (*N. debneyi* × *N. tabacum*)	*N. glutinosa*	Kehr and Smith (1952)
10	4x (*N. debneyi* × *N. tabacum*)	*N. langsdorffii*	Kehr (1950, 1951) and Kehr and Smith (1952)
11	4x (*N. debneyi* × *N. tabacum*)	*N. longiflora*	Ahuja (1962, 1974, 1998)
12	4x (*N. debneyi* × *N. tabacum*)	N. *plumbaginifolia*	Kehr (1950, 1951) and Kehr and Smith (1952)
13	4x (*N. debneyi* × *N. tabacum*)	*N. quadrivalvis*	Smith (1958)
14	4x (*N. debneyi* × *N. tabacum*)	*N. sanderae*	Kehr (1950, 1951) and Kehr and Smith (1952)
15	4x (*N. debneyi* × *N. tabacum*)	*N. rustica*	Kehr and Smith (1952)
16	4x (*N. debneyi* × *N. tabacum*)	*N. sylvestris*	Kehr and Smith (1952)
17	4x (*N. goodspeedii* × *N. tabacum*)	*N. benthamiana*	Dorossiev et al. (1990)
18	4x (*N. goodspeedii* × *N. tabacum*)	*N. excelsior*	Dorossiev et al. (1990)
19	? (*N. glutinosa* × *N. sylvestris*)[1]	*N. tabacum*	Fardy and Hitier (1945)
20	4x (*N. quadrivalvis* × *debneyi*)	*N. tabacum*	Kehr and Smith (1952), Smith et al. (1958), and Morgan (1964)
21	4x (*N. raimondii* × *N. tabacum*)	*N. alata*	author of this volume, unpublished
22	4x (*N.repanda* × *N. longiflora*)	*N. tabacum*	Davis et al. (1988)
23	4x (*N. repanda* × *N. sylvestris*)	*N. tabacum*	Burk (1967)
24	4x (*N. rustica* × *N. alata*)	*N. tabacum*	Shabanov and Nikova (1984), Nikova et al. (1997) and Nikova and Vladova (2002)
25	4x (*N. rustica* × *N. debneyi*)	*N. tabacum*	Sabour et al. (1986)

26	4x (*N. rustica* × *N. exigua*)	*N. tabacum*	Bolsunov (1970)
27	4x (*N. rustica* × *N. suaveolens*)	*N.tabacum*	Hitier and Izard (1950)
28	4x (*N. rustica* × *N. tabacum*)	*N. glauca*	Kostoff and Radjably (1935) and Kehr and Smith (1952)
29	4x (*N. rustica* × *N. tabacum*)	4x (*N.suaveolens* × *N. tabacum*)	Moav and Cameron (1961)
30	4x (*N. rustica* × *N. tabacum*)	*N.glutinosa*	Moav and Cameron (1961)
31	4x (*N. rustica* × *N. tabacum*)	*N. otophora*	Moav and Cameron (1961)
32	4x (*N. rustica* × *N. tabacum*)	*N. sylvestris*	Kostoff and Radjably (1935) and Moav and Cameron (1961)
33	4x (*N. rustica* × *N. tabacum*)	*N. tomentosiformis*	Moav and Cameron (1961)
34	4x (*N. setchellii* × *N. otophora*)	*N. tabacum*	author of this volume, unpublished
35	? (*N. sylvestris* × *N. kawakamii*)	*N. tabacum*	Ohashi (1985)
36	4x (*N. sylvestris* × *N. tomentosiformis*)	*N. tabacum*	Kostoff (1931)[3]
37	*N. tabacum*	4x (*N. glutinosa* × *N. obtusifolia*)	Krishnamurthy et al. (1960)[4] and Appa Rao and Krishnamurthy (1963)[4]
38	*N. tabacum*	*N. glutinosa* × *N. gossei* (?)[2]	Anon (1999)
39	*N. tabacum*	2x (*N. sylvestris* × *N. setchellii*)	Greenleaf (1941)
40	*N. tabacum*	4x (*N. sylvestris* × *N. otophora*)	Ternovsky (1963)
41	*N. tabacum*	4x (*N. sylvestris* × *N. tomentosiformis*)	Kostoff (1931)[3] and Greenleaf (1941)
42	*N. tabacum*	4x (*N. tabacum* × *N. otophora*)	Greenleaf (1941)
43	4x *N. tabacum*	4x (*N. repanda* × *N. sylvestris*)	Stavely et al. (1973)
44	4x (*N. tabacum* × *N. alata*)	4x (*N. tabacum* × *N. otophora*)	Gajos (1981, 1984)[5]
45	4x (*N. tabacum* × *debneyi*)	*N. alata*	Larkina (2015)
46	4x (*N. tabacum* × *N. glutinosa*)	*N. glauca*	Fardy and Hitier (1945)
47	4x (*N. tabacum* × *N. glauca*)	*N. alata*	Kehr (1950, 1951) and Kehr and Smith (1952)
48	4x (*N. tabacum* × *N. glauca*)	*N. langsdorffii*	author of this volume, unpublished
49	4x (*N. tabacum* × *N. glauca*)	*N. sanderae*	Kehr (1950, 1951) and Kehr and Smith (1952)
50	4x (*N. tabacum* × *N. sylvestris*)	*N. tomentosiformis*	Greenleaf (1941)

(continued)

Table 4.5 (continued)

No.	Female parent	Male parent	Reported by
51	4x (*N. tabacum* × *N. tomentosiformis*)	*N. glauca*	Lehmann (1936) and Fardy and Hitier (1945)
52	4x (*N. undulata* × *N. tabacum*)	*N. alata*	Ahuja (1962)
53	4x (*N. undulata* × *N. tabacum*)	*N. bonariensis*	Ahuja (1962)[6]
54	4x (*N. undulata* × *N. tabacum*)	*N. forgetiana*	Clausen after Kehr and Smith (1954)
55	4x (*N. glutinosa* × *N. obtusifolia*)	*N. tabacum*	Krishnamurthy et al. (1960)
56	4x (*N. quadrivalvis* × *N. debneyi*)	*N. tabacum*	Smith and Abashian (1963)

[1]Ploidy level of the maternal parent not known;
[2]Ploidy level of the paternal parent not known;
[3]Very similar to Kostoff's 4x (*N. sylvestris* × *N. tomentosiformis*) known in literature as 'Kostoff's hybrid'. Kostoff's hybrid is now regarded to be the trigenomic hybrid *sylvestris-tomentosiformis-tabacum* (Lim et al. 2006). This trigenomic combination but with *N. tabacum* as the maternal third species is also known as '*Nicotiana triplex*' (Kostoff, 1933; McRae, 1941); the triple hybrid by Kostoff (1931) named '*N. triplex*' originated from pollinating *N. tabacum* with partly fertile pollen of 2x (*N. sylvestris* × *N. tomentosiformis*), due to a relatively large number of unreduced gametes produced by the latter;
[4]Reciprocal triple hybrid also obtained
[5]No cytological data are available on the resulting hybrid. Barring chromosomal interchanges between the tomentosoid subgenomes of *N. tabacum* and *N. otophora* in the paternal amphidiploid, an allotetraploid uniting the diploid genome of *N. tabacum* and two haploid genomes of the remaining constituent species (tab-tab-al-oto) was the most likely product of that cross.
[6]Six-genomic allohexaploid also obtained

with *N. alata* and produces viable trispecific offspring (Berbeć, unpublished). *N. bonariensis* is a potential source of resistance to brown leaf spot (Stavely, 1979; Brandwagt et al., 2001), but the species is highly incompatible with *N. tabacum,* and there is only one report on the successful crossing of the two species (Busconi et al., 2010). The deployment of the trispecific hybrid 4x (*N. undulata* × *N. tabacum*) × *N. bonariensis* (Ahuja, 1962) may be an option to bridge *N. tabacum* with *N. bonariensis*. To bridge *N. langsdorffii* with *N. tabacum,* the trispecific hybrid 4x (*N. debneyi* × *N. tabacum*) × *N. langsdorffii* (Kehr, 1951; Kehr & Smith, 1952) can be used. An accession of *N. langsdorffii* was found that carries a single dominant gene conferring a localized necrosis response, thereby containing the spread of the blue mold pathogen throughout the life cycle of the plant (Zhang & Zaitlin, 2008).

Trigenomic hybrids produced with the involvement of *N. tabacum* as part of their genomic makeup are listed in Table 4.5.

Quadruple tetraspecific hybrids that combined the genomes of four different species were also obtained to study the effects of multiple alloploidy. They usually originated from crossing two amphidiploids, each involving two different species, e.g., 4x (*N. quadrivalvis* × *N. suaveolens*) × 4x (*N. glauca* × *N. langsdorffii*), 4x (*N. rustica* × *N. paniculata*) × 4x (*N. quadrivalvis* × *N.* suaveolens), and 4x (*N. rustica* × *N. glauca*) × 4x (*N. quadrivalvis* × *N. suaveolens*) – Kostoff, 1943. Tetraspecific hybrids that include *N. tabacum* as one of their constituent genomes are listed in Table 4.6.

Table 4.6 Multiple tetraspecific hybrids involving *N. tabacum*

No.	Female allopolyploid	Male allopolyploid/diploid	Reported by
1	4x (*N. rustica* × *N. tabacum*)	(*N. quadrivalvis* × *N. suaveolens*)	Kostoff (1943)
2	4x (*N. rustica* × *N. tabacum*)	4x (*N. glauca* × *N. langsdorffii*)	Kostoff (1943)
3	4x (*N. undulata* × *N. tabacum*)	4x (*N. debneyi* × *N. plumbaginifolia*)	Kehr and Smith (1954)[1]
4	4x (*N. glutinosa* × *N. glauca*)	4x (*N. debneyi* × *N. tabacum*)	Kehr and Smith (1952)
5	6x (*N. glutinosa-obtusifolia-tabacum*)	*N. glauca*	Appa Rao and Krishnamurthy (1963)
6	6x (*N. glutinosa-obtusifolia-tabacum*)	*N. longiflora*	Appa Rao and Krishnamurthy (1963)
7	6x (*N. glutinosa-obtusifolia-tabacum*)	*N. forgetiana*	Appa Rao and Krishnamurthy (1963)
8	6x (*N. quadrivalvis-debneyi-tabacum*)	*N. glutinosa*	Kehr and Smith (1952)
9	4x (N. *quadrivalvis* × *N. suaveolens*)	4x (*N. tabacum* × *N. glauca*)	Smith and Abashian (1963)
10	4x (*N. rustica* × *N. quadrivalvis*)	4x (*N. debneyi* × *N. tabacum*)	Smith and Abashian (1963)

[1]Semi lethal dwarf

4.4.3 The Use of Autotetraploid Variants of Parental Species to Start Gene Transfer

Interspecific autotetraploid × diploid crosses to produce partly fertile sesquidiploids as a means of bypassing the sterility of the amphihaploid hybrid are briefly discussed in the section on overcoming hybrid sterility (Sect. 3.6.3). Occasionally, that approach may also circumvent the crossability barrier between two species. It is also effective from the standpoint of gene transfer. The potential gene donor in that scheme, a wild *Nicotiana* species, is deployed as a natural diploid and is thus very unlikely to produce imbalanced gametes deficient for genes of interest.

The use of autotetraploid *N. tabacum* in crosses with diploid *Nicotiana* species was explored extensively by Chaplin and Mann (1961), who investigated the crossability of 24 wild *Nicotianae* with the tetraploid form of *N. tabacum*. The researchers found that using autotetraploids of *N. tabacum* helped overcome the incompatibility that existed between the cultivated species on the one hand and *N. rustica* and *N. alata* on the other. The method was also found to be effective in producing viable hybrids of *N. tabacum* with *N. alata* (Berbeć, 1987; Laskowska & Berbeć, 2005). Takenaka (1962) and Burk (1972) reported success from crossing autotetraploid *N. tabacum* with incompatible *N. langsdorffii*.

Another benefit accruing from using an autotetraploid *N. tabacum* parent to produce the starting hybrid is that it bypasses the amphidiploid stage by directly yielding sesquidiploid hybrids. A possible trade-off is the lessened likelihood for chromosomal interchange in sesquidiploids compared to that in amphidiploids (Patel & Gerstel, 1961).

As already indicated in Sect. 4.3.2, sesquidiploids may provide a very convenient starting material for producing alien addition lines. As the chromosomes of the nonrecurrent parent are quickly and selectively eliminated when these sesquidiploids are backcrossed to *N. tabacum* individuals with a single added chromosome from the alien species have a relatively high chance of appearance in early backcross generations. A plant monosomic for chromosome 9 of *N. lumbaginifolia* was picked up by Dang et al. (2019) already in first breakdown generation of the cross 4n *N. tabacum* × 2n *N. plumbaginifolia*) to 2n *N. tabacum* (see Sect. 4.3.2)

Clayton (after Valleau, 1952) produced a fertile hybrid from crossing *N. tabacum* with autotetraploid *N. debneyi*. The hybrid could be repeatedly backcrossed to *N. tabacum* until *N. tabacum*-like phenotypes that resisted blue mold were recovered. The account is surprising enough, given the cytogenetics of such a transfer that most likely must have involved a reversion to allohaploid, or more likely, to near-allohaploid situation in its first stage. In a similar attempt to transfer the hypersensitive response to TMV from *N. repanda* to *N. tabacum,* Pittarelli and Stavely (1975) successfully overcame the crossability barrier between the two species by using the autotetraploid variant of *N. repanda* as the female parent in the cross with *N. tabacum*. They backcrossed the resulting sesquidiploid TRR to *N. tabacum*. The BC_1 (breakdown generation) was composed of viable plants with a haploid complement of *N. tabacum* and varying numbers of extra single chromosomes from

N. repanda. These aneuploid plants had enough fertility to be backcrossed to *N. tabacum*. From these results, it can be further inferred that the aneuploids of *N. tabacum* × *N. repanda* must have produced a sufficient number of unreduced gametes to retain vestigial fertility.

The sesquidiploids of the SST type were reported by Wong (1975) in the F_1 progeny from mating 2n *N. stocktonii* to 2n *N. tabacum*, probably due to unreduced gametes produced by the wild parent. In the experience of the author of this review, the mating of *N. tabacum* with *N. palmeri* failed to produce hybrid plants when both parents were diploids, but it was easily done when autotetraploid *N. palmeri* was used as the maternal parent. However, the sesquidiploid TPP thus obtained was highly sterile, and backcrosses to *N. tabacum* yielded no offspring.

The use of chromosome-doubled parental forms to bypass the incompatibility barrier between *N. tabacum* and its wild *Nicotiana* relatives is also discussed in Chap. 3 (Sects. 3.2.3 and 3.4.1).

4.4.4 "Egg Transformation" Without Gametic Fusion

One of the major controversies regarding routes of interspecific gene transfer came about as a spinoff from studies on the use of irradiated pollen to overcome self-compatibility barriers in different plants, including *Nicotiana* species (Pandey, 1974). In a series of articles, Kamla Pandey, an Indian geneticist settled in New Zealand, reported transferring minute chromosome fragments from the male gametophyte of one species to the embryo sac of another by using the pollen of the donor species that had been exposed to a high dose (100 Kr) of ionizing irradiation prior to fertilization (Pandey (1975, 1978, 1980a, b). The author claimed that a transfer of limited amounts of "pulverized" DNA from the pollen parent to the egg nucleus can occasionally be achieved, thus producing genetically transformed maternal plants. According to that hypothesis, pseudofertilization and induction of parthenogenetic diploidy, essential for the process to occur, were greatly facilitated by the close linkage of the self-incompatibility locus with genes that stimulate egg cells-to-diploid embryo development (Pandey, 1980a, 1980b). The latter finding also provided an explanation for the selective transfer of self-incompatibility genes through parthenogenetic induction. This phenomenon was called "egg transformation" by Pandey (1975). Pandey's experiments were mainly concerned with the offspring of crosses within the section Alatae: *N. alata*, *N. forgetiana*, *N. langsdorffii*, *N. bonariensis* but also involved the hybrid *N. bonariensis* × *N. glauca* (Pandey, 1980b). Some phenotypic characteristics peculiar to the male parent (flower color, pollen color, self-incompatibility) were recovered in the nonhybrid maternal progeny after using heavily irradiated donor pollen either alone or as a mixture with self-incompatible maternal pollen. Thus, white flower color and a self-incompatibility allele were allegedly transferred from *N. alata* to *N. forgetiana* (Pandey, 1975, 1978, 1980b), white pollen color transferred from *N. alata* to *N. langsdorffii*, blue pollen color originally from *N. langsdorffii* transferred to *N. alata* via *N. forgetiana* as the donor of irradiated pollen (Pandey, 1980b), and self-incompatibility alleles were transferred via irradiated pollen involving different crosses, including the intersectional combination *N. bonariensis* × *N. glauca* (Pandey, 1980b).

Pandey envisaged a promising future for the phenomenon of egg transformation as a novel tool for crop improvement that bypasses the limits of conventional introgressive breeding without resorting to sophisticated manipulations at the cell or molecular levels. He even called it "poor man's genetic engineering" (Pandey, 1981), an obvious allusion to the first big steps in *Agrobacterium*-mediated gene transfer that were being made just then.

However, within a few years following the above reports, Engvild (1985) and Chyi and Sanford (1985) put Pandey's data to re-examination. Engvild (1985) attempted to repeat Pandey's experiments with the latter scholar's own material but failed to obtain any seed set. Maternal plants that may have originated through diploid parthenogenesis were obtained from pollinating *N. paniculata* and *N. plumbaginifolia* with pollen of *N. alata*. In neither case did the progeny of these maternal diploids show any segregation for the traits of the pollen donor. The author concluded that some gene transfer by pollen irradiation is possible between species that produce fertile hybrids, but such transfer between species separated by crossability or fertility barriers is probably a rare phenomenon.

Likewise, Chyi and Sanford (1985) closely mimicked Pandey's experiments and found no evidence for gene transfer by irradiated pollen. They concluded that Pandey's results are not reproducible but, at the same time, made a reservation that such transformation events are not entirely impossible, but they may occur as extremely rare events.

Finally, it can be observed that the above dispute is of limited relevance to tobacco breeding since neither Pandey nor his opponents included *N. tabacum* in their experiments.

4.5 Barriers to Introgressive Breeding in *Nicotiana*

4.5.1 First Decisions Aimed at Making Interspecific Gene Transfer More Likely to Be Accomplished

As early as the 1950s, Clayton (1954) laid down some basic rules that ought to be applied in identifying the species most suited to be deployed in interspecific transfer. The species of choice ought to express the desired trait uniformly and to a high degree, the degree of expression should be comparably well expressed in the donor species and in its hybrid with the recipient, and the work should be started with an amphidiploid stock that fully expresses the trait of interest. The expediency of the two last suggestions is readily seen when one considers the mechanics of conventional sexual gene transfer. Unlike intervarietal gene exchange, in which recessive traits are obviously easier to work with than dominant characters, the interspecific introgression of a recessive trait, if not assisted, e.g., by sophisticated molecular technologies, is in most cases technically unfeasible. Because of unpredictable and sometimes very low transmission rates, identification of the carrier of the trait of interest would require creating and screening a very large number of selfed populations in each transfer generation, provided the tested individuals would have enough fertility to be selfed, in the first place.

An interesting case when the rule concerning the dominant expression of the trait of interest did not seem to apply was the resistance to PVY from *N. raimondii*. The resistance of that species to PVY, although fully expressed against all PVY strains by which it was challenged, was declared unusable because of its recessive character in crosses with *N. tabacum* (Burk et al., 1982). Paradoxically, due to a specific epistatic interaction, the fate of the factor from *N. raimondii* responsible for the resistance could be tracked as severe susceptibility that manifested itself in segregating populations derived from the amphidiploid 4x (*N. raimondii* × *N. tabacum*) (Berbeć, 1988, see also Sect. 4.6.1).

In many cases, introgressive breeding in tobacco is probably facilitated by the fact that *N. tabacum* originated as an amphidiploid and, genome downsizing processes notwithstanding, has retained many, if not most, of its loci in duplicate (Clausen & Cameron, 1944). This makes it more forgiving to chromosomal and genetic imbalances that the introgression process is likely to produce but not quite immune to them. 'Synthetic tobaccos', such 4x *sylvestris-tomentosiformis* or 4x *sylvestris-otophora,* in which the most important loci are duplicated, were advocated as a substitute for *N. tabacum* to facilitate and increase the interspecific flow of genes (Hancock & Lewis, 2017, see also Sect. 4.6.4).

4.5.2 Extent of Chromosome Homology/Homeology

Recombination involving exchange of a chromosome segment (alien substitution) in the first hybrid generations was reported for crosses with the progenitor species of *N. tabacum* or its close relatives, i.e., *N. otophora*, *N. tomentosiformis* and *N. sylvestris*, which have retained a considerable degree of chromosome homeology with their cultivated descendant (Chaplin & Mann, 1961). As was recently demonstrated by Laskowska et al. (2015), translocations of chromosome fragments can also occur in the amphidiploid stage of the considerably more remotely related combination, *N. wuttkei* × *N. tabacum*, in which only some vestigial homeology of parental chromosomes was conserved. In some other closely tracked introgression events, the transfer of a whole chromosome either as an addition or as a substitution was the first stage of introgression (Gerstel, 1945a; Burk, 1960; Lewis, 2005).

4.5.3 Preferential Pairing

Preferential pairing in interspecific hybrids is the propensity of a chromosome to associate with its counterpart from the same species with the exclusion of an alien chromosome even if the latter shows some degree of homology with the former (Gerstel, 1961). Preferential pairing further restricts already scant opportunities for chromosomal interchanges between alien species, and the condition becomes increasingly worse in sesquidiploid and breakdown generations. In contrast to many other genera where preferential pairing is controlled by specific genes, in *Nicotiana,* it is controlled chromosomally, mainly due to substantial structural differences among the chromosomes contributed to *Nicotiana* allopolyploids by their respective diploid parents (Gerstel, 1961).

4.5.4 Disturbed Gametogenesis and Sterility in Allopolyploid Hybrids

Although chromosome doubling is the usual treatment to make sterile amphihaploids amenable to further breeding manipulations, amphidiploid or near-amphidiploid condition does not always mean restored fertility. Notorious in this respect is the amphidiploid *N. sylvestris* × *N. tomentosiformis*, the putative progenitor of the present-day *Nicotiana tabacum*, which is either completely female sterile or shows only vestigial self-fertility (Greenleaf, 1941; Burk, 1973; Lilienfeld, 1952, 1953; Lim et al., 2006). That kind of female sterility due to arrested development of embryo sacs during meiosis is common to several hybrid combinations that involve some species from the section Tomentosae (Greenleaf, 1941, Ternovsky, 1963, Table 4.7). A different type was reported in the sesquidiploid hybrids 3x (*N. tabacum* × *N. setchellii* and 3x *N. tabacum* × *N. tomentosa* var. 'Acomayo' (Ar-Rushdi, 1955). The female sterility in those hybrids was due to arrested pollen tube growth in the styles of the hybrid plants. It was found to be controlled by a recessive gene system complementary to the sterility gene located on chromosome H of the tomentosoid subgenome of *N. tabacum* (see, however, Sect. 3.3.4 on the identity of chromosome H in that species). It is unclear, however, which of the two systems was responsible for the expression of female sterility in the trispecific hybrid *N. tabacum* × 4x (*N. otophora* × *N. setchellii*) (Berbeć, unpublished).

The term desynapsis is usually applied to situations in which chromosomes pair normally at pachytene but begin to fall apart thereafter. By Metaphase I, the number of bivalents is greatly reduced, and they are randomly distributed to daughter nuclei, which results in imbalanced, unviable gametes. Limited desynapsis is probably quite common in amphihaploids and may account for the variable number of bivalents observed in the same hybrid plant. Swaminathan and Murthy (1959) explained the phenomenon as caused by the precocious terminalization of chiasmata. Desynapsis of homologous chromosomes may also account for at least part of the instability observed in some amphidiploids, although it has never been indicated as the principal cause of amphidiploid sterility, save for one instance described in the next paragraph.

Table 4.7 Instances of sterility and instability in amphidiploids and some other allopolyploids of *N. tabacum* with other *Nicotiana* species

Species involved with the allopolyploid hybrid with *N. tabacum*	Type of abnormal behavior in allopolyploid	Reported by:
N. sanderae	Highly unstable mixoploids (44-82 chromosomes)	Nikova et al. (2003)
N. glauca	Instability caused by preferential loss of *N. tabacum* chromosomes	Szilagyi (1975)
	Male sterility caused by collapse of microsporogenesis at tetrad stage	Trojak-Goluch and Berbeć (2003)
N. paniculata	Probably preferential loss of *N. paniculata* chromosomes in successive passages of the amphidiploid cultured in vitro that was accompanied by the onset of cms in regenerants that had from 42 to 64 somatic chromosomes	Nikova and Vladova (2002)
N. raimondii	Instability caused by preferential loss of *N. raimondii* chromosomes	Berbeć (1988)
N. rustica	High chromosomal instability self-sterility	Kostoff (1937, 1943), Takenaka (1963), Furusato (1960)[4], and Moav and Cameron (1961)[4]
	Reduced vigor, dwarfism	Moav and Cameron (1961)
N. nudicaulis	Reduced self-fertility, morphological variation	Burk and Neas (1964)
N. africana	Stability varied with *N. tabacum* genotype from fairly stable to unstable due to meiotic disturbances	Doroszewska and Berbeć (2000)
	Segregation for PVY resistance in F_2 generation of the amphidiploid	Keum et al. (1991)
N. amplexicaulis	Self- and cross-sterility caused by precocious chromosome disjunction in meiosis (desynapsis)	Berbeć and Doroszewska (1981)
N. exigua	Instability due to chromosome loss	Kobus (1973)
N. maritima	Instability due to chromosome loss or chromosome interchange	Nikova and Zagorska (1987)
N. rosulata	Highly instable due to asynchronous meiosis	Ternovsky and Larkina (1978)

(continued)

Table 4.7 (continued)

Species involved with the allopolyploid hybrid with *N. tabacum*	Type of abnormal behavior in allopolyploid	Reported by:
N. suaveolens	High cytological instability and reduced seed set	Shinkareva (1979)
4x (*N. sylvestris* × *N. setchellii*)	Female sterility	Ar-Rushdi (1955)
4x (*N. sylvestris* × *N. tomentosa* var. 'Acomayo')	Female sterility	Ar-Rushdi (1955)
N. otophora	Chromosome elimination in gametes	Gerstel (1960) and Yang (1960)
N. setchellii × *N. otophora*[1]	Female sterility	Berbeć et al. (1982)
N. setchellii[2]	Female sterility	Berbeć (unpublished)
N. setchellii[3]	Female sterility	Ar-Rushdi (1955)
N. tomentosa[4]	Female sterility	Ar-Rushdi (1955)
	Chromosome elimination in gametes	Yang (1960)
N. obtusifolia	Female sterility with some occasional selfed seeds produced	Chung et al. (1996)
N. glutinosa	Instability due to non-preferential chromosome loss	Patel and Gerstel (1961)
	Moderately unstable	Ternovsky (1962)

'Allopolyploids' in this table generally means amphidiploid hybrids. If otherwise, the relevant information is supplied below as a footnote and indicated by numerical superscript after the alien species name in column 1

[1]Trispecific allopolyploid (hexaploid): 6n *N. tabacum* × (*N. setchellii* × *N. otophora*);

[2]Amphidiploid 4x (*N. tabacum* × *N. setchellii*);

[3]Sesquidiploid from crossing 4n *N. tabacum* with 2n *N. tomentosa* var 'Acomayo'

[4]Self-sterility in spite of high pollen fertility (Furusato, 1960, variegation, tumors (Moav & Cameron, 1960)

The single case in which desynapsis was the sole cause of the complete self and cross sterility of the allopolyploid *Nicotiana* hybrid was reported by the author of this review. The crosses of *N. amplexicaulis* with true-breeding *N. tabacum* varieties consistently failed to produce surviving hybrids, but the situation was improved substantially when 48-chromosome *N. tabacum*-like derivatives (Tg) from another interspecific hybrid, *N. tabacum* × *N. glauca*, were used as pollen parents (Berbeć & Doroszewska, 1981). Among colchicine-treated *N. amplexicaulis* × Tg plants, a single individual was found with doubled (84) mitotic chromosomes. It showed nearly complete pairing up to the early diakinesis stage, indicating that cytologically, it was an amphidiploid 4x (*N. amplexicaulis* × *N. tabacum*). Starting with late diakinesis to early Metaphase I, most of the bivalents would fall apart, resulting in a random distribution of univalents to the daughter nuclei and rendering the amphidiploid completely female and male sterile. It is not known what exactly caused the desynaptic behavior of this particular amphidiploid plant since desynapsis was not reported for amphidiploids 4x (*N. amplexicaulis* × *N. tabacum*) obtained elsewhere (Wark, 1970; DeVerna et al., 1987; Berbeć & Doroszewska, 1992, Venkateswarlu et al., 1998). Hypothetically, alien genetic material from the third species, *N. glauca*, may have caused genic or chromosomal imbalances that resulted in a disturbed conjugation process. The desynaptic amphidiploid 4x (*N. amplexicaulis* × *N. tabacum*) showed signs of introgression from *N. glauca* expressed as anthocyanin coloration of the stem and upper leaves. In this context, it is also interesting that partial fertility was restored to another survivor of the cross *N. amplexicaulis* × Tg by cochicine treatment of its lateral shoot. The fertile shoot of undetermined cytological status was pollinated by *N. tabacum* cv. 'Zamojska 4'. The resulting progeny was of the sesquidiploid type (Berbeć & Doroszewska, 1992) indicating that the colchicine-treated branch of the original hybrid produced amphidiploid flowers. Further backcrossing to Zamojska 4 resulted in the development of a cytoplasmically male sterile alloplasmic analog of that cultivar (Zamojska 4 cms *amplexicaulis*).

Spontaneous chromosome elimination is frequently observed in amphidiploid hybrids and other allopolyploids of *Nicotiana*. It was reported, e.g., in the amphidiploids of *N. tabacum* × *N. africana* (Doroszewska & Berbeć, 2000), *N. tabacum* × *N. glauca* (Szilagyi, 1975), *N. raimondii* × *N. tabacum* (Berbeć, 1988), *N. tabacum* × *N. glutinosa* (Patel and Gerstel (1961), *N. tabacum* × *N. sylvestris*, *N. tabacum* × *N. otophora*, *N. tabacum* × *N. tomentosiformis* (Yang, 1960), *N. knightiana* × *N. tabacum* (author's unpublished observations) and several other amphidiploids involving *N. tabacum*. The elimination of chromosomes contributed by two parental species in the selfed offspring of an amphidiploid may be random and not distinctly biased toward either parental chromosome set. However, in at least two amphidiploids, 4x (*N. raimondii* × *N. tabacum*) (Berbeć, 1988) and 4x (*N. tabacum* × *N. glauca*) (Szilagyi, 1975), a clear nonrandom, unidirectional chromosome loss was observed. In the successive selfed generations of the 4x (*N. raimondii* × *N. tabacum*), a progressive shift and, finally, reversion to *N. tabacum*-like phenotypes was observed accompanied by other phenomena, e.g., the appearance of male sterile segregants phenotypically close to *N. tabacum* (Berbeć, 1988, see Sect. 4.6.1). This is also true of amphidiploids produced via somatic protoplast fusion. The self-pollinated progeny of a true somatic amphidiploid 4x (*N. tabacum* + *N. debneyi*) ($2n = 96$) ranged from an amphidiploid to an aneuploid with 60 somatic chromosomes (Sproule et al., 1991).

Instances of sterile and unstable amphidiploids involving *N. tabacum* are given in Table 4.7.

4.5.5 Elimination of Somatic Chromosomes

Elimination of somatic chromosomes can also be the source of unpredictable inheritance patterns. It commonly occurs in somatic hybrids, as discussed previously, and is the source of somaclonal variability in tissue culture. Variable chromosome numbers were recorded in regenerants from embryo rescue cultures of *N. glutinosa* × *N. megalosiphon* (Subhashini et al., 1986), *N. tabacum* × *N. knightiana* (Slusarkiewicz-Jarzina & Zenkteler, 1983) and *N. repanda* × *N. tabacum* (Iwai et al., 1985). Explant cultures used to induce chromosome doubling also resulted in chromosome loss, aneuploidy and mixoploidy of the regenerants of *N. velutina* × *N. tabacum*, *N. maritima* × *N. tabacum*, and *N. benthamiana* × *N. tabacum* (Nikova et al., 1991). A particularly highly variable population of regenerants whose chromosome numbers ranged from 42 to 64 was produced in the stem pith culture of *N. paniculata* × *N. tabacum* (Nikova et al., 1991).

Somatic instability also occurs in sexually produced hybrids grown *in vivo*. Moav and Cameron (1960) found that the chromosomes of *N. plumbaginifolia* in the hybrid *N. tabacum* × *N. plumbaginifolia* and in its derivatives exhibited instability, which led to chromosome loss during reductional divisions. The phenotypic manifestation of that instability was readily discernible as a multitude of white spots when the *N. plumbaginifolia* chromosome carrying a dominant factor for chlorophyll production was introduced in the background of albinotic chlorophyll-deficient *N. tabacum*. The chromosomes of *N. plumbaginifolia* tended to be more stable when they were associated with those of *N. tabacum* as intact chromosome sets (amphihaploids, amphidiploids, and sesquidiploids), whereas in breakdown generations, the rate of chromosome loss greatly increased.

4.5.6 Sterility of the Sesquidiploid Generation

At least partial fertility of sesquidiploids or near-sesquidiploids is essential for conventional introgression to be accomplished. The sesquidiploid condition is the bottleneck of introgression, as pointed out by Clausen and Cameron (1957). In theory, at least all interspecific gene transfers have to pass through that stage regardless of how the transfer is actually started: by backcrossing of the amphihaploid hybrid to the recipient parent taking a chance of restitution gametes being formed by the former, by converting the sterile amphihaploid to fertile amphidiploid and backcrossing it to the recurrent parent, or by crossing the autotetraploid recipient with the diploid donor. Sesquidiploids are also the usual direct product of gametosomatic fusion (Giddings and Rees, 1992). Sesquidiploid generation may be avoided, however, by fusing somatic protoplasts as the initial stage of gene transfer. Asexually produced interspecific hybrids and their sexual progeny usually cover a range of products that include both true allopolyploids and aneuploids of different chromosomal and genomic makeup (Pandeya et al., 1991; Brandle et al., 1992; Bai et al., 1996).

Fortunately, fully sterile sesquidiploids occur relatively rarely. The most known case of this category is the sesquidiploid involving *N. tabacum* and *N. alata*. It has been consistently reported as sterile by several investigators regardless of how it was arrived at (Chaplin & Mann, 1961; Gajos, 1975, 1981; Berbeć, 1987). It is only Stoyanova (1979) who did not report any difficulty in backcrossing

N. tabacum × *N. alata* sesquidiploids (TTA) to *N. tabacum*. Additionally, Ivancheva and Manolov (1982) reported obtaining a few plants from pollinating TTA and TTS (*tabacum-sanderae*) with *N. tabacum*. The two authors also succeeded in obtaining postsesquidiploid progeny from TTS plants by selfing. Interestingly, Gajos (1981) was able to surmount that obstacle by deploying a third species, *Nicotiana otophora*, in the transfer of TSWV resistance from *N. alata* to N. *tabacum* (see Sect. 4.6.3).

The sesquidiploid obtained from direct crossing 4n *N. tabacum* × 2n *N. rustica* was also highly self-and cross sterile, but the sterility was finally overcome by persistent backcrossing to *N. tabacum* (Pandeya & White, 1981, 1984). Nifong (2008) found the sesquidiploids from crossing autotetraploid *N. tabacum* cv. 'K326' with *N. rustica* to be self-sterile and to vary substantially for the degree of female fertility depending on the *N. rustica* accession used as their pollen parent. Likewise, the sesquidiploid 2n *N. rustica* × 4n *N. tabacum,* in which the tetraploid form was the pollen parent, was practically sterile (5% viable pollen). In that case, fertility was restored to the sesquidiploid by doubling its chromosome number with colchicine (Pittarelli & Sisson, 1989).

Female sterility similar to that observed in amphidiploids and some multiple allopolyploids (Sect. 4.4.2) was also observed in sesquidiploids (Ar-Rushdi, 1955, Table 4.7).

4.5.7 Erratic Inheritance in Early Transfer Generations

Transmission Rate of Extra Alien Univalents Lack of homology between the species involved in interspecific transfer not only blocks the necessary chromosome interchanges and recombination but also results in a substantial number of unassociated chromosomes (univalents) and their random elimination or/and uneven distribution to the gametes. Male gametophytes with additional chromosomes or deficient for some chromosomes show reduced competitive ability versus their chromosomally balanced counterparts and rarely reach the ovules, which results in a loss of unassociated chromosomes and discordant segregation ratios for the introgressed trait, a common observation by many investigators. The unpaired chromosome H of *N. tabacum* and its homeologue from *N. glutinosa* were included in only 25% to 20% of the female gametes (Gerstel, 1943; Chu, 1954) and, on the male side, the transmission rate was much lower in a study of the transfer of TMV resistance factor from the wild to the cultivated species (Gerstel, 1943). In an attempt to transfer the purple flower factor from *N. plumbaginifolia* to *N. tabacum,* Ar-Rushdi (1957) observed that the additional univalent from *N. plumbaginifolia* that carried the trait was transmitted at 20% through the egg and at 0% through the pollen. In a similar study involving *N. plumbaginifolia* and *N. tabacum*, an additional chromosome that carried the white seedling factor was included in 15% of female and 3% of male gametes (Moav, 1958). In his study on the genetic control of tumor formation in *Nicotiana* hybrids, Ahuja (1968) reported that an *N. longiflora* chromosome fragment was transmitted to the amphidiploid 4x (*N. debneyi-tabacum*) at a rate of 40% through the egg and 25% through pollen. Ovular and pollen transmission rates of 25–52% and 0–39%, respectively, of an added chromosome from *N. africana* were recorded in the backcross progenies of an alien addition line of *N. tabacum* segregating for resistance to PVY (Lewis, 2005). For the same alien addition, Campbell et al. (1994) reported 10% female transmission rate of the added chromosome from *N. africana*. On the other hand, ovular and pollen transmission rates of the extra *N. debneyi* chromosome in an alien addition line of *N. tabacum* resistant to black root rot were slightly higher through pollen than through the egg (10.1% and 8.1%, respectively) (Bai et al., 1996).

Another departure from the rule of higher ovular vs. pollen transmission of alien univalents was also reported by Smith (1988) for a case outside the involvement of *N. tabacum*. An extra univalent from *N. glauca* added to the genome of *N. langsdorffii* was transmitted by selfing at a rate of 20% through pollen and as little as 1% through the egg.

Other Causes of Discordant Inheritance An unusual case of the genetic locus called pollen-killer was found by Cameron and Moav (1957). In the hybrid derivatives of *N. plumbaginifolia* × *N. tabacum,* the tetrad stage nuclei that contained 24 chromosomes of *N. tabacum* plus the pollen killer gene-carrying chromosome from *N. plumbaginifolia* developed into normal functional pollen grains, whereas those that did not were aborted. The interference from the pollen killer gene was recorded by Chaplin (1962) in his ultimately successful transfer of resistance to black shank (*Phytophtora nicotianae*) from *N. plumbaginifolia* to *N. tabacum.*

Alterations in chromosome structure and number may occur as the result of the action of mitotic drugs used for the doubling of chromosomes. They are frequently observed in *in vitro* cultures, especially if the culture is prolonged and goes through several passages (go back to Sect. 3.6.5).

4.5.8 Erratic Inheritance in Advanced Introgression Stages

The hypersensitive response to TSWV transferred from *N. alata* and incorporated into the tobacco variety 'Polalta' (Gajos, 1981), the sole known and confirmed source of resistance to that virus, has become notorious in introgressive breeding. A peculiar kind of genetic drag is the main issue, and this will be addressed in a subsequent section. Apart from that, the resistance has also other flaws related to inheritance. In some cases, it showed perfect ratios for the monogenic dominant pattern (Moon & Nicholson, 2007). In some others, the proportion of resistant to susceptible plants was close to that expected for the digenic type of inheritance (Kennedy & Nielsen, 1993). Currently, the resistance to TSWV from *N. alata* is considered to be controlled by the single dominant *RSTV-al* gene (Moon & Nicholson, 2007; Laskowska et al., 2013). However, the resistant vs. susceptible ratios tend to be skewed toward the susceptible genotypes if male gametes are involved in the transmission of the resistance factor (Laskowska & Berbeć, 2010). The *RSTV-al*-mediated hypersensitive response to TSWV is known to have become highly unstable once introgressed in a new genetic background. The case in point is cv. 'Wiktoria', from which the resistance is quickly lost, presumably through the elimination of the resistance gene (Laskowska et al., 2013). Moon and Nicholson (2007) found Wiktoria and another resistant genotype ('ZG-8') to uniformly amplify all SCAR markers linked to *RSTV-al.* This finding may suggest the existence of different lineages of Wiktoria, some of which are stably resistant, although it is not clear whether Wiktoria' and ZG-8 were prechecked for resistance before having been tested for the presence of the SCAR markers.

Difficulty or failure to stabilize the introgressed trait in the recipient genome of *N. tabacum* has been a frequent experience with several other interspecific transfers. Collins and Legg (1969) reported unstable inheritance of black shank resistance from *N. plumbaginifolia* due to disturbances at the later stages of the meiotic divisions (anaphase and tetrads) in the 48-chromosome segmental substitution line. Similarly, Stavely et al. (1973) reported that breeding lines with resistance to *Cercospora nicotianae* and *Meloidogyne javanica* from *N. repanda* could not be stabilized because the resistance-carrying chromosomes tended to be eliminated from the segregating populations. Resistance to TMV introgressed from *N. repanda* also showed irregular inheritance (Gwynn et al., 1986). Resistance of monogenic resistance to wildfire from *N. longiflora* showed normal Mendelian inheritance in some genotypes and erratic inheritance in others (Stokes, 1960; Knoche et al., 1984). Additionally, the resistance to *Thielaviopsis basicola* conferred by the dominant gene from *N. debneyi* does not seem to be equally stable across different flue-cured tobacco genotypes (Nicoletti, 1999, the author of this volume (unpublished observations)).

Disturbed inheritance of resistance to TMV introgressed from *N. glutinosa* was observed in otherwise stable resistant lines of *N. tabacum*. The primary cause of the phenomenon was explained as the loss of the resistance factor as the result of rare reciprocal translocations between the N gene-carrying chromosome and its homoeologue (Chen et al., 2018). The explanation may possibly also apply to the instability of black-root resistance described in the preceding paragraph.

Simple inheritance and dominance are great assets in the early stages of interspecific transfer. In practical breeding experience, however, dominance may prove burdensome later when discriminating between homozygotes and heterozygotes/hemizygotes, and achieving stable inheritance becomes an issue.

4.5.9 Genetic tumors

The formation of tumorous growth in some hybrid combinations, first discovered by Kostoff (1930), is another peculiar manifestation of hybrid instability. It was hypothesized that certain genes combined in a hybrid promote tumor formation (Kehr, 1951, Smith 1968). Näf (1958) divided *Nicotiana* species involved in tumor-forming hybrids into two groups ('plus' and 'minus'), with only 'plus' × 'minus' hybrids developing tumors. According to Näf, *N. forgetiana*, *N. langsdorffii*, *N. sanderae*, *N. alata*, *N. longiflora*, *N. noctiflora*, *N. bonariensis*, *N. plumbaginifolia* belong to the plus group and *N. tabacum*, *N. bigelovii*, *N. suaveolens*, *N. miersii*, *N. paniculata*, *N. debneyi*, *N. glauca*, and *N. rustica* are in the minus group. It is easy to notice that, with the exception of *N. noctiflora*, the "plus species" belong to the section Alatae, and the "minus species" are scattered among different other sections.

Early experiments with the hybrid *N. glauca* × *N. langsdorffii* indicated that tumor formation was controlled at the genomic level and that at least one full haploid set of *N. glauca* chromosomes was required to induce tumor development in 4x (*N. glauca* × *N. langsdorffii*) derivatives with the full diploid complement of *N. langsdorffii* chromosomes (Kehr & Smith, 1954). However, in another experiment, Ahuja (1962) demonstrated that a single chromosome, or even a chromosome fragment from *N. longiflora,* was enough to induce tumorization in the derivatives of the trispecific allopolyploid 3x (*N. debneyi-tabacum* × *N. longiflora*) backcrossed to 4x (*N. debneyi* × *N. tabacum*).

The genetic control of tumor development in tumor-forming hybrids was indirectly indicated in the experiment conducted by Izard (1970), who, by exposing the amphidiploid 4x (*N. glauca* × *N. langsdorffii*) to X-rays, produced a nontumorous mutant lineage of that synthetic species and named it *N. gisquetii*. By studying the tumorization process in the same hybrid, Ichikawa et al. (1990) explained tumor growth as controlled by the *rol* genes that conferred increased sensitivity to endogenous auxins in tumor-prone hybrids.

Sequences of cT-DNA (*rol* genes) were acquired horizontally from *Agrobacterium rhizogenes* by some ancestral forms of *Nicotiana* species in their phylogenetic past and subsequently dispersed sexually across the genus. They have been detected in several present-day species (Matveeva & Lutova, 2014). The direct involvement of the *rol* genes in the tumorization process was postulated by Ichikawa et al. (1990) but was not conclusively confirmed by subsequent studies (Matveeva & Lutova, 2014). Tumorization may be induced through complicated gene interactions affecting hormone levels in plants and their sensitivity to plant hormones, including IAA in *N. rustica* × *N. tabacum* (Kehr & Smith, 1954) and cytokinins (Matveeva & Lutova, 2014). Tumors in hybrids involving *N. tabacum* were reported for *N. tabacum* × *N. alata* (Kostoff, 1943), *N. tabacum* × *N. sanderae* (Kostoff, 1943; Kehr & Smith, 1954; Burk, 1972), a sesquidiploid hybrid *N. tabacum* × *N. forgetiana* (TTF) and *N. tabacum* × *N. langsdorffii* (Burk, 1972). The seedlings of *N. obtusifolia* × *N. tabacum* also developed tumorous malformations (Liu & Marubashi, 2014). That latter hybrid did not involve an alatoid species, so the observed tumors could have belonged to a different category. Tumors were also observed in several trispecific combinations of *N. debneyi* × *N. tabacum* as female parents with members of the section Alatae (*N. alata, N. longiflora, N. langsdorffii, N. plumbaginifolia* and *N. sanderae*) as males (Kehr, 1951; Ahuja, 1962).

In the majority of hybrids in which they occur, genetic tumors usually appear in late stages of plant life, practically after cessation of active growth and flowering and, save for two instances, they have not as yet been reported as interfering with interspecific transfer and introgression processes *per se*. An exception to this rule was reported by Burk (1972). The plants of *N. tabacum* × *N. langsdorffii* started developing tumors from the seedling stage onward, which made them unamenable to breeding manipulations, including fertility restoring treatments. In another case of this kind, very early and massive tumor development by the seedlings of *N. obtusifolia* × *N. tabacum* was the major component of the seedling death syndrome in that hybrid (Liu & Marubashi, 2014). Neither parent of that hybrid belonged to the plus group, so the underlying causes of its abnormal growth may have been different.

However, a phenomenon related to genetic tumor formation may be implicated in the unique deleterious linkage associated with resistance to tomato spotted wilt virus (TSWV) from *N. alata*. The author of this book produced the hybrid *N. tabacum* × *N. setchellii* with the aim of studying the expression of root-sprouting ability, a characteristic feature of *N. setchellii*, in the hybrid with cultivated tobacco. The peculiarity of this hybrid was that its *N. tabacum* parent was the intervarietal F_1 hybrid heterozygous for the alien insertion from *N. alata* that conferred resistance to TSWV (see Sect. 4.6.3). The germinating seeds of the F_1 2x (*N. tabacum* × *N. setchellii*) were treated with colchicine. Among the resultant progeny, there was one female sterile amphidiploid and ca. 50 apparently nonconverted sterile amphihaploids. Both the amphidiploid and the amphihaploids were kept in the greenhouse long after the flowering stage. As the plants grew old, many of them developed heavy leaf malformations and tumorous growths on both stem and leaves that bore striking resemblance to those characteristic of other interspecific combinations in *Nicotiana*, including the paradigmatic hybrid *N. glauca* × *N. langsdorffii*. For further discussion of this particular phenomenon, see Sect. 4.6.3.

4.6 Effects Related to Introgression in *N. tabacum*

4.6.1 *Changes and Modifications in the Expression of Introgressed Trait*

A phenomenon that frequently troubles introgressive tobacco breeders is that the expression of the desired trait, the degree of resistance to a pathogen being the most notable case in point, is not fully recovered once interspecific transfer has been completed and regular inheritance is established.

A well-known case of reduced expression in *N. tabacum* compared to that in the native species is resistance to blue mold from *N. debneyi*. As many as four levels of expression were defined by Clayton (1968) depending on each individual introgression, indicating the involvement of several genes from both the donor and the recipient genomes. Plant age-dependent expression was the principal component of that variation. In lineages with the highest expression (level 1), full resistance was present in all development stages whereas in level 2 homozygotes for the resistance factor were fully resistant whereas heterozygotes were susceptible as seedlings and acquired full resistance as adults. In level 3 the plants were susceptible as seedlings and moderately resistant as adults. In level 4 the age-dependent resistance of adults was lower than in level III and the seedlings were susceptible (Clayton, 1968, Rufty, 1989).

The failure to recover the full resistance of the donor species once it had been transferred to the genome of *N. tabacum* was reported for resistance to PVY from *N. africana* (Lewis, 2007; Doroszewska, 2010; Korbecka-Glinka et al., 2018). The level of *N. africana*-derived tolerance of PVY varied among the tolerant genotypes (Lewis, 2007).

Expression of PVY Resistance and Tolerance Factors in the Unstable Amphidiploid *N. raimondii* × *N. tabacum* The amphidiploid 4x (*N. raimondii* × *N. tabacum*) (F_1 cv., Zamojska 4' × cv. 'Lechia') was produced and studied by Berbeć (1988) as a potential starting material to transfer the hypothetic resistance gene/s for resistance to PVY from *N. raimondii* to tobacco. *N. raimondii* consistently tested immune to various PVY necrotic strains (Sievert, 1972b; Głażewska, 1977; Burk et al., 1982; Doroszewska & Depta, 2011), but the resistance factor/factors were not expressed in hybrids with *N. tabacum* (Burk et al., 1982).

The cvs Zamojska 4 and Lechia were bearers of the symptomless host response to the virus deployed in several Polish tobacco varieties of the time (Mazur, 1975). Both cultivars contributed the symptomless response gene to the intervarietal *N. tabacum* F_1 hybrid that was used to cross with *N. raimondii*. Thus, the *N. tabacum* parent of the amphidiploid did not segregate for the symptomless response locus. The amphidiploid plant responded with systemic necrosis to inoculation with PVY) However, the initial amphidiploid plant was unstable and, when selfed, segregated into true amphidiploids, sesquidiploids, near sesquidiploids and *N. tabacum*-like genotypes and phenotypes. Meiotic configurations in those segregants suggested a largely nonrandom elimination of *N. raimondii* chromosomes from the amphidiploid in its successive selfed generations. Parallel with one-directional segregation and reversion to the *N. tabacum* phenotypes, there was a progressive increase in symptomless vs. necrotic individuals in the amphidiploid's selfed offspring challenged with the necrotic strain PVY^{NZ} (Table 4.8). Mazur (1975) reported strong susceptibility to PVY in F_1 hybrids between resistant and tolerant cultivars, indicating an epistatic interaction between the recessive symptomless response gene in cv. 'Peyod' and the recessive resistance gene of unknown provenance in cv. 'Virginia Krakowska', classified by Mazur as "immune". According to a more recent study by Michel et al. (2018), PVY-tolerant varieties, including Zamojska 4 and Lechia, carry a mutated *NtTPN1* gene that inactivates the inefficient hypersensitive defense response against PVY infection that becomes expressed in susceptible varieties as systemic veinal necrosis. The mechanisms of resistance to PVY in both *N. raimondii* and Virginia Krakowska remain unknown. The majority of the genes that conferred a localized necrosis response of the wild species to various pathogens and which were subsequently transferred to the cultivated tobacco have been consistently reported as dominant rather than recessive. The

Table 4.8 Expression of necrotic symptoms in selfed ten C_2 and eighteen C_3 progenies of the colchicine induced true 72-chromosome amphidiploid 4x (*N. raimondii*) × *N. tabacum* (F_1 cv. Zamojska 4 × LB-838) challenged with the necrotic PVY strain Y^{NZ} (after Berbeć, 1988)

	Amphidiploid					Subamphidiploid					Sesquidiploid					Subsesquidiploid and *N. tabacum*-like phenotypes				
	Number of plants		%	Range of N plants (%)		Number of plants		%	Range of N plants (%)		Number of plants		%	Range of N plants (%)		Number of plants		%	Range of N plants (%)	
Generation	N	T	N	from	to	N	T	N	from	to	N	T	N	from	to	N	T	N	from	to
C2	83	3	96	80	100	20	13	65	0	100	32	50	39	9	100					
C3	84	2	98	86	100	115	16	88	67	100	42	58	41	0	75	14	97	13	0	21

Explanations: Male *N. tabacum* parent (Zamojska 4 × LB-838 syn. 'Lechia') was an intervarietal hybrid homozygous for symptomless host response (tolerance) to the PVY strain Y^{NZ} (Głażewska, 1977). Both Zamojska 4 and Lechia carried the *NtTPN1* gene that confers the tolerance of a wide range of PVY strains by abolishing the systemic necrotic reaction to infection by the virus (Michel et al., 2018)

N necrotic plant, *T* PVY tolerant plant

resistance of *N. raimondii* is very unlikely to be of the hypersensitive type since the studies that listed that species as resistant to PVY (Sievert, 1972b; Głażewska, 1977; Burk et al., 1982; Doroszewska & Depta, 2011) did not report localized necroses after artificially challenging the species with the virus. *N. africana* is as yet the only source of recessive resistance to PVY that has been successfully transferred to *N. tabacum*, although the full recovery of resistant reaction by the wild species in a new genetic milieu was not possible (see the preceding paragraph). Because of the lack of knowledge on the nature of PVY resistance in *N. raimondii,* it is difficult to explain how that specific resistance factor interacts with the mutated *NtTPN1* gene in a PVY-tolerant variety to produce strong susceptibility to the virus. The interaction operates through an as yet unknown epistatic mechanism as a result of which the two resistance factors cancel each other, and their joint presence is signaled as strong susceptibility. It seems that this genetic system could be of some value as a component of future projects to transfer PVY resistance from *N. raimondii* to *N. tabacum.* A *sui generis* reversed precedent to this situation was reported by Wernsman and Rufty (1987) who demonstrated that the pleiotropic association between the resistance to root knot nematodes and the susceptibility to the MSNR strain of potato Virus Y could be used in segregating populations to pick up root knot-resistant genotypes based on their necrotic reaction to the virus.

4.6.2 *Types of Side Effects Related to Interspecific Introgression in* Nicotiana

There are three major classes of side effects that may be produced by an alien gene transferred to the genetic background of a cultivated variety:

(a) Chromosome segment introgressed in the genome of *N. tabacum*, carries a load of alien chromatin along with the desired gene and this is known as linkage drag;
(b) Expression of the alien gene may be affected by the genetic background in which it has been introgressed and the effect is the manifestation of epistasis;
(c) Introgressed gene may affect more than one trait, and this is called pleiotropy;
(d) Position effect that makes the performance of the introgressed trait vary depending on the site of insertion.

On a case-to-case basis, the four types may be difficult to distinguish from one another. Pleiotropic effects were demonstrated to be of little, if of any, importance in the case of TMV resistance conferred by the N gene from *N. glutinosa*. By comparing the performance of two isogenic lines, one containing the N gene introgressed by conventional interspecific transfer and the other transformed with the same N gene in its 'pure' cloned form, the agronomically negative effects associated with resistance to TMV were found to be practically absent in the transgenic line (Lewis et al., 2007a). A singular case of pleiotropy may be involved in the close linkage between resistance to race 0 of the black shank pathogen (*Phytophtora nicotianae*) and the ability to suppress tobacco cyst nematode (*Globodera tabacum*) populations by tobacco genotypes that carry the *Php* gene for resistance to black shank *from N. plumbaginifolia* (Johnson et al., 2009; Parkunan et al., 2009 – see also Sect. 4.6).

The position effect also seemed to have little impact in another case study of resistance to TMV introgressed from *N. glutinosa*. An array of breeding lines of tobacco carrying the N gene on different chromosomes did not differ significantly for agronomic performance (Lewis & Rose, 2010).

The incomplete recovery of the desired trait in the majority of interspecific transfers is usually best explained by the oligogenic inheritance of those traits (see Sect. 4.2).

4.6.3 Linkage Drag

Introduction of an alien chromatin segment containing the introgressed gene into the recipient chromosome is usually associated with many unwelcome events. Apart from the direct effect of the genetic material flanking the gene of interest, other factors may also be involved, such as pleiotropism or position effects. Crossover or breakage-and-reunion translocations involved in introgression can result in the elimination of biologically or agronomically vital genes in the translocation region (Chaplin & Mann, 1978).

Linkage drag seems to be the major negative effect associated with introgression from other *Nicotiana* species. The amount of usually undesirable genetically active material linked to the introgressed gene and incorporated into the recipient chromosome depends largely on the size of the inserted fragment. Genetic resistance is a great asset in commercial cultivars and in many instances makes tobacco production worthwhile. In tobacco, it involves the control of widespread and economically important tobacco pathogens, causal agents of diseases such as tobacco mosaic, tobacco spotted wilt, blue mold, black root rot, powdery mildew to name but a few. Hence, agronomic penalties caused by the presence of linked alien chromatin in resistant varieties are a serious practical issue.

Phenotype-modifying effects of the introgressed alien gene in different genetic environments vary with individual cases. In two independent studies of *N. debneyi*-derived resistance to black root rot (Thielaviopsis basicola), one by Legg et al. (1981) in Burley and the other by Haji et al. (2003) in flue-cured tobacco, the negative impact on yield was similar. Epistatic interactions with genetic background were found for resistance to wildfire race 0 from *N. longiflora* (Legg et al., 1982; Nielsen et al., 1985) and resistance to TMV (Nielsen et al., 1985). Generally, alien introgressions performed better in air-cured dark tobacco and Burley genotypes than in flue-cured tobacco (Chaplin & Burk, 1971).

Resistance to one pathogenic factor may also be associated with susceptibility to another. Varieties resistant to root knot disease of tobacco caused by races 1 and 3 of the nematode *Meloidogyne incognita* have long been known to respond with severe veinal necrosis caused by a particular strain of PVY (M^SN^R), while root knot susceptible varieties were only mildly affected by that strain (Rufty et al., 1983). The gene that confers resistance to root knots is now deemed as derived from *N. tomentosa*, was first designated the Rk gene and later renamed Rk1 (Pollok et al., 2016, see also Sect. 4.6.4).

Rao and Stokes (1963) reported that TMV-resistant varieties were more liable to develop calcium deficiency symptoms than TMV-susceptible varieties. According to these authors, the H chromosome of *N. glutinosa* that carried the hypersensitive response to TMV was also the most likely bearer of the genes that influence the expression of calcium deficiency symptoms. An increased sensitivity to calcium deficiency seemingly related to TMV resistance was also observed by the author of this review (unpublished data). Grown on unamended sphagnum peat moss, a medium frequently low in Ca, the entire populations of juvenile tobacco plants of TMV-resistant Burley cv. 'TN 90' developed growth disturbances characteristic of calcium malnutrition, whereas hybrid populations derived from mating 'TN-90' to TMV susceptible lines cosegregated for TMV resistance and apparent calcium deficiency symptoms.

Many of these genetic background-dependent responses may be attributable, rather than to the resistance gene *per se*, to the genetic material surrounding the introgressed gene with its different dominant, additive or epistatic effects. This was experimentally confirmed by Lewis et al. (2007a) for the N gene conferring resistance to TMV. TMV-resistant isolines of the flue-cured cv. K326 transformed with the cloned N gene from *N. glutinosa* showed none of the adverse effects of their counterparts into which the N gene was transferred through conventional interspecific hybridization and backcrossing. As already mentioned in the preceding subsection, those results exclude pleiotropism as the cause of the impaired agronomic performance associated with N gene-based TMV resistance and increase the likelihood that those ill effects may be reduced through further backcrossing and selection. These findings may also apply to the resistance to TSWV described in the next paragraph. Both resistances apparently share a similar interspecific lethality-based mechanism that involves a temperature-dependent hypersensitive response to infection.

Linkage Drag Associated with Resistance to TSWV from *N. alata* is a peculiar instance of adverse effects related to interspecific introgression. The resistance gene, referred to as *RTSW-al* (Trojak-Goluch et al., 2016a, 2018) or, perhaps erroneously, as RSTV-al (Trojak-Goluch et al., 2016b), was transferred from *N. alata* with the possible involvement of the third species *N. otophora* (Gajos, 1979, 1981, 1984). As mentioned previously, the resistance factor *per se* used to behave in a rather unpredictable manner, but persistent selection for homozygous resistant genotypes apparently has largely stabilized its mode of inheritance and regularity of expression (Moon & Nicholson, 2007; Laskowska & Berbeć, 2010; Trojak-Goluch et al. 2011, 2016b). The most common side effects that are associated with the *RTSW-al* gene include deformed leaf blade and leaf venation (Korbecka-Glinka et al., 2018, 2021; Laskowska & Berbeć, 2010) but also tumorous outgrowths, although the latter usually appear in aged plants (Laskowska & Berbeć, 2010; Moon & Nicholson,

2007, observations by the author of this volume). Another peculiar quality of the malformations linked to TRSW resistance is that they are most pronounced in heterozygous condition, diminish or disappear altogether once the homozygosity for *RTSW-al* is restored, and recur in the heterozygous offspring of TSWV-resistant × nonresistant lines (Laskowska & Berbeć, 2010; Trojak-Goluch et al. 2011, 2016a).

Based on its external morphological manifestations alone, the linkage drag associated with the *RTSW-al* gene bears a striking resemblance to genetic tumors developed by some interspecific hybrid combinations in *Nicotiana*, the hybrid *N. glauca* × *N. langsdorffii* being the prime example. This phenomenon is briefly discussed in Sect. 4.5.9. According to the earlier described 'plus × minus' hypothesis (Näf, 1958) *N. tabacum* belongs to the "minus group" and *N. alata* to the "plus group". Hence, the hybrid *N. tabacum* × *N. alata* is a tumor-forming combination. However, the actual data on tumor formation relating to this hybrid are inconsistent. Näf himself reported the appearance of tumorous outgrowths on all individuals and on plant parts of *N. tabacum* × *N. alata* (Näf, 1958). In contrast to Näf's account, Kostoff described tumors on the hybrid *N. tabacum* × *N. alata* as occurring "rarely" (Kostoff, 1943). Takenaka and Yoneda (1963) also observed some tumors on plants from crossing tetraploid *N. tabacum* with *N. alata*. Other investigators who have reported on *N. tabacum* × *N. alata* hybrids failed to observe such malformations (Nagao, 1979; Gajos, 1981; Berbeć, 1987; Berbeć & Laskowska, 1997). Kehr (1951) and Ahuja (1962) reported tumors on trispecific hybrids that involved the genomes of *N. tabacum* and *N. alata (N. debneyi-tabacum-alata*) and *N. tabacum-glauca-alata*), but tumor formation was limited to the roots of those hybrid plants. These reports suggest that the interaction between factors from *N. alata* and *N. tabacum* may (but not necessarily always do) produce massive teratological changes that have been associated with resistance to TSWV.

Some additional clues in support of the involvement of genetic tumours in the effects linked to the *RTSW-al* gene come from the observations of of tumorous changes developed by the hybrid *N. tabacum* × *N. setchellii* made by the author of this book and described in the last paragraph of Sect. 4.5.9. The *N. tabacum* parent of the hybrid bred true for the hypersensitive response to TSWV and, taking into account its pedigree, it may be assumed that it carried the RTSW-al gene inherited from Polalta. *N. setchellii* and *N. otophora* do not produce tumors either when crossed with each other or with *N. tabacum* so the three species may be included in Näf's "minus group". However, the TSWV-resistant line of *N. tabacum* effectively behaved like a "plus species", its hybrid with the "minus" *N. setchellii* developing classical symptoms of genetic tumors. Regretfully, neither *N. setchellii* nor *N. otophora* has so far been hybridized with *N. alata*, so nothing is known about the tumor-forming potential of those combinations.

With what is known about the transfer and behaviour of the RTSW-al insertion one may hypothesize that the introgression process was accompanied by rearrangements and recombinations within the genomes of *N. otophora*, *N. tabacum* and *N. alata* involved in the transfer. Those rearangengements led to the formation of a block of tightly linked genes that included, along with the *RTSW-al* gene, some gene insertions or deletions that controlled morphological malformations in the manner very peculiar to manifestations of genetic tumors, including enhanced expression in hererozygotes. Thus, it seems plausible that the linkage drag associated with the *RTSW-al* gene shares the common mechanism with the formation of tumorous changes called genetic tumors and observed in many interspecific *Nicotiana* hybrids. The phenomenon of genetic tumors in *Nicotiana* is not necessarily confined to interactions of intact alien genomes (amphihaploids, amphidiploids or of higher ploidy levels) but was demonstrated to be under the control of single alien chromosomes or chromosome fragments that could be transferred and become expressed in the genotypic milieu of another species (Ahuja, 1962; Bayer & Ahuja, 1968).

Some new light on the nature of the alien insertion that includes the *RTSW-al* gene was shed by the studies reported by Korbecka-Glinka et al. (2018, 2021). In the first of the reports (Korbecka-Glinka et al., 2018), the authors found a strong association between RTSW-al and morphological deformations, with approximately half of the plants homozygous or heterozygous for the inserted fragment developing morphological alterations. However, these changes were also observed in ca. 30% of the plants that did not carry the RTSW-al insertion, which might indicate that some of the factors responsible for morphological abnormalities were located outside the introgressed region. In a sequel to the above study (Korbecka-Glinka et al., 2021), the authors reported that the introgressed region was actually composed of three loci, which they named Ala-Ala-Ala. They further demonstrated that recombination within that region is possible. Whereas both the introgressed phenotype Ala-Ala-Ala and the nonintrogressed phenotype Tob-Tob-Tob were practically free of malformations, the homozygous recombinant line Ala-Ala-Tob developed growth irregularities characteristic of the hybrids heterozygous for the alien insertion. Thus, the cause of the malformed phenotypes related to *RTSW-al* may be interpreted as epistatic interactions between the gene/s from *N. alata* and some as yet unidentified recombinant homeologs/s from *N. tabacum,* and the effect seems to be expressed regardless of whether the interacting genes are on the same or on different chromosomes (in coupling or in repulsion). Regretfully, the authors did not report on the performance of Ala-Ala-Ala/Ala-Ala-Tob and Tob-Tob-Tob/Ala-Ala-Tob recombinant heterozygotes.

Another comment that may also be of interest is that abnormal morphology was observed in transgenic cotton that expressed a *Thielaviopsis basicola*-inhibiting protein, NaD1, encoded by the nad1 gene derived from *N. alata* (Pereg, 2013). According to Gajos (1984), both the accession of *N. alata* used in his work and TSWV-resistant derivatives from crossing *N. tabacum* with *N. alata* also expressed resistance to black root rot (see also Sect. 4.6.6).

Table 4.9 lists examples of *Nicotiana* species in which introgression was associated with carryover of undesirable inherited effects.

Table 4.9 Instances of linkage drag effects related to introgression from alien *Nicotiana* species to *N. tabacum* (after Berbeć and Doroszewska (2020) with minor additions and modifications)

Source of introgressed factor	Introgressed factor	Side-effects caused by linkage drag	Reported by:
N. alata	Resistance to TSWV	Morphological malformations and tumors	Gajos (1984), Kennedy and Nielsen (1993), Moon and Nicholson (2007), Laskowska and Berbeć (2010), and Korbecka-Glinka et al. (2021)
N. longiflora	Resistance to *Phytophtora nicotianae* race 0 (black shank) (*Phl* locus)	Reduced yield, reduced cured leaf quality, physiological leaf spotting	Valleau et al. (1960) and Apple (1967)
	Resistance to *Globodera tabacum tabacum tabacum/ solanacearum*[1]	Reduced cured leaf quality	Crowder et al., (2003)
	Resistance to *Pseudomonas syringae* pv. *tabaci* race 0 (wildfire)	No side-effects	Nielsen et al. (1985)
		Reduced leaf number, reduced total alkaloids, reduced yields, other agronomic parameters vary with genetic background	Legg et al. (1982)
	Resistance to *Meloidogyne javanica*	Depressed yield and leaf quality, reduced nicotine and sugars content;	Schweppenhauser (1975)
		Excessively wide internodes and prostrate leaves, susceptibility to angular leaf spot	Mudzengerere (1994), Raeber and Smeeton (1980, 1984), and Shava et al. (2018)
N. plumbaginifolia	Resistance *to Phytophtora nicotianae* race 0 (black shank) (*Php* locus)	Reduced yield, reduced cured leaf quality;	Chaplin (1962) and Lewis (2011)
		Dwarfness, slow maturity	Goins and Apple (1970)
N. glauca	Resistance to Thielaviopsis basicola (black root rot)	Delayed flowering, fewer leaves per plant, reduced leaf body (thickness)	Trojak-Goluch and Berbeć (2009)
N. repanda	Resistance to *Meloidogyne javanica*	Reduced yield, high percentage of low quality leaves	Schweppenhauser (1974)
N. rustica var. *brasilea*	Resistance to *Phytophtora nicotianae* race 0 and race 1 (black shank);	No side-effects	Drake et al. (2015)

(continued)

Table 4.9 (continued)

Source of introgressed factor	Introgressed factor	Side-effects caused by linkage drag	Reported by:
	resistance to *Pseudomonas syringae* pv. *tabaci* race 0 and race 1 (wildfire} and *P. syringae* pv. *angulata* (angular leaf spot (ALS)		
N. africana	Resistance to PVY	Reduced yield	Lewis (2007)
		Delayed flowering	Keum et al. (1994)
N. debneyi	Resistance to *Thielaviopsis basicola* (black root rot)	More ground suckers, delayed flowering, reduced yield, reduced total nitrogen, reduced total alkaloids (in burley tobacco)	Legg et al. (1981)
		Delayed leaf maturity, reduced yield, decreased leaf quality, reduced total nitrogen, reduced total alkaloids (in flue-cured tobacco)	Haji et al. (2003, 2005) and Bai et al. (1995, 1996)
	Resistance to *Peronospora hyoscyami* f.pv. *tabacina* (blue mold)	Increased alkaloid content (Burley tobacco), reduced yield, reduced cured leaf quality (flue-cured tobacco)	Verrier et al. (2016)
N. suaveolens	Male fertility restoring factor	Decreased leaf yield and leaf value, increased nornicotine content	Hosfield and Wernsman (1974)
N. tomentosa	QTL locus affecting leaf number and days to flower	Decrease in percentage total alkaloids and increase in percentage reducing sugars	Eickholt and Lewis (2014)
	Resistance to races 1 and 3 *Meloidogyne incognita*	Associated with severe susceptibility to PVY (strain M^SN^R)	Rufty et al. (1983)
N. glutinosa	Resistance to TMV	Slow growth, reduced leaf size reduced yield, changed green leaf appearance, reduced leaf quality	Chaplin et al., (1961, 1966), Chaplin and Mann (1978), Legg et al. (1979), and Lewis et al. (2007a)
		Yield unaffected, reduced leaf quality	Hitier and Izard (1958) and Johnson and Main (1983)
		Increased sensitivity to calcium deficiency	Rao and Stokes (1963), author of this volume (unpublished)

[1]Monogenic resistance factor in TCN-resistant line PD 4 is most likely derived from *N. longiflora* (Crowder et al., 2003)

4.6.4 The Role of Synthetic Tobaccos in Facilitating Interspecific Gene Transfer and in Alleviating Linkage Drag

In some gene transfers, 'synthetic' rather than 'natural' *N. tabacum* was used with success as an intermediary to eliminate undesirable linkages. Beltsville 771 is a blue mold-resistant line of *N. tabacum* with polygenic resistance factors from *N. debneyi* obtained after many years of assiduous backcrossing and selection (Clayton, 1968). This notwithstanding, Beltsville 771 continued to show irregular inheritance and inferior agronomic performance. A considerable increase in the rate of recovering homozygous resistance to blue mold, simplified patterns of inheritance and improvement in agronomic characters associated with the resistance factors were among the gains when Beltsville 771 was crossed with the so-called 'Kostoff's hybrid' (see also Sect. 2.2.2), at that time believed to be a pure amphidiploid 4x (*N. sylvestris* × *N. tomentosiformis*), but later demonstrated to carry a considerable amount of introgression from *N. tabacum* (Sheen, 1972; Lim et al., 2006).

Present-day commercial cultivars and breeding lines of *N. tabacum* have undergone a considerable downsizing of their functional genomes and narrowing of their genetic base through the evolutionary process (Renny Byfield et al., 2011; Bombarely et al., 2012) and as a consequence of purposeful selection (Hancock & Lewis, 2017). In contrast, 'synthetic tobaccos' (4x (*N. sylvestris* × *N. tomentosiformis*), 4x (*N. sylvestris* × *N. otophora*), especially those synthesized *de novo* (Skalicka et al., 2005), retain the full genetic potential of their ancestors and preserve all homoeologous genes in duplicate. This makes them more receptive to gene exchanges and better buffered against the ill effects of chromosomal rearrangements and loss of biologically essential chromosome regions involved in interspecific translocations. Owing to these qualities, 'synthetic tobaccos' offer a convenient tool to facilitate gene flow between *Nicotiana tabacum* and other *Nicotiana* species (Chaplin & Mann, 1978; Hancock & Lewis, 2017).

A singular, if not controversial, issue was the use of Kostoff's hybrid by Clayton et al. (1958) and later by Schweppenhauser (1975) to solve the problem of resistance to root knot nematodes found within *Nicotiana tabacum* and notorious for its association with small leaf size and prolific suckering (Clayton et al., 1958). Employing Kostoff's hybrid was effective beyond expectations. Not only was leaf size increased, but the resistance changed its mode of inheritance from polygenic to monogenic (Stavely, 1979). In line with the argument discussed in the previous paragraph, it was surmised that the cross with Kostoff's hybrid helped eliminate minor resistance genes, leaving only a major dominant factor. However, Slana et al. (1977) and Stavely et al. (1977) offered an alternative hypothesis. According to them, Kostoff's hybrid was the ultimate source of resistance to root knots in the usable flue-cured lines developed by Clayton and his followers. Stavely et al. (1977) hypothesized that Kostoff's hybrid must have had either *N. tomentosa* or a resistant selection *of N. tomentosiformis* as the tomentosoid parent. The weak point of that hypothesis was that Kostoff's hybrid in possession of the investigators and the stocks used in other studies (Clayton et al., 1958) were susceptible to root knot. Slana et al. (1977) tried to account for the discrepancy by assuming that the original Kostoff's seed material was a heterogeneous and heterozygous population segregating for resistance to root knot. They argued that by a happy coincidence, a resistant individual had been picked up for crosses in Clayton's original project. In the maintenance process, other users of the original segregating population of Kostoff's hybrid stabilized their seed stocks as susceptible to root knot. However, Rufty et al. (1983) were doubtful about the involvement of *N. tomentosa* in this type of resistance on account of the very strong resistance of that species to strain M^SN^R of potato virus Y. This controversy notwithstanding, the ultimate provenance of the Rk1 gene from *N. tomentosa* seems to be now generally accepted (Ng'ambi et al., 1999; Julio et al., 2015; Pollok et al., 2016; Adamo et al., 2021). Some of these authors quote Yi et al. (1998) in support of this claim even though no direct reference to *N. tomentosa* is made in the latter paper.

Very similar to Clayton's (1968) is Schweppenhauser's (1975) account on how the deployment of alloploid *N. tomentosiformis* × *N. sylvestris* helped free his root knot resistant material of deleterious linkages and change the resistance genetics from semidominant to monogenic dominant. Schweppenhauser, however, did not give any details concerning his alloploid except that it was probably the pollen parent in the initial cross. Based on that clue, the use of Kostoff's hybrid or its descendant cannot be excluded, and the argument presented in the preceding paragraph might also apply to Schwepenhauser's case.

4.6.5 Marker-Assisted Recovery and Selection of Introgressed Genes

Breeders strive for the introgressed chromatin fragment to be as small as possible while at the same time retaining the essential gene. The simplest approach is to resort to repeated assiduous backcrossing hoping for a rare recombination to occur within the translocated region. In most cases, that fortuitous approach has very slim chances of succeeding. Opportunities for recombination are thought to be very limited within and around the introgressed region. Johnson et al. (2002a) found that recombination was highly suppressed in the region around the *Ph* gene allelic to another gene for resistance to black shank (*Php*) that had been known to be introgressed from *Nicotiana plumbaginifolia* (Johnson et al., 2002b). However, recombination within the introgressed region is possible, as was recently demonstrated for the insertion from *N. alata*, which carries the gene/s controlling resistance to TSWV in tobacco (Korbecka-Glinka et al., 2021).

Another advocated remedy is to retransfer the gene of interest from its original source species in the expectation of more beneficial and smaller translocations occurring during the initial stages of the repeated process (Chaplin & Mann, 1978; Bai et al., 1996).

Bai et al. (1995) attempted to develop RAPD markers of resistance to black root rot disease. The resistance was originally transferred from *N. debneyi* by Clayton (1969) to the Burley variety 'Burley 49'. It was subsequently bred by the authors of the report (Bai et al., 1995) into the background of the flue cured variety 'Delgold'. As a result, a black root rot resistant line near isogenic (NIL) to Delgold was developed. Two RAPD markers polymorphic between Delgold and NIL were found to be strongly linked with the resistance gene, one in coupling and the other in repulsion. The authors proposed a scheme in which the two markers could be used in segregating populations to detect plants homozygous for the resistance gene thereby eliminating costly challenging with pathogen and progeny testing.

In another experiment, Bai et al. (1996) tried to repeat the transfer of the gene for black root rot resistance from *N. debneyi* by creating a somatic hybrid *N. tabacum* + *debneyi* and screening sexually obtained selfed and backcrossed progenies for resistant individuals. Resistant 50-chromosome alien addition lines thus obtained were intended to be treated with irradiation to generate stable translocations between the *N. debneyi* resistance-carrying chromosome and tobacco chromosomes. These translocations were planned to be searched for the shortest chromatin segments by RAPD marker-assisted backcrossing scheme based on selecting the resistant individuals with the lowest number of markers associated with the *N. debneyi* chromosome. To the authors' knowledge, further results have not been published.

A selection scheme for small-size translocations in the progenies of a 50-chromosome addition line resistant to PVY derived from the hybrid *N. tabacum* × *N. africana* assisted by tissue culture and DNA sequence markers was devised by Lewis (2005). He compared backcross progenies of alien addition plants exposed to tissue culture with progenies of *in vivo* grown plants. The stimulation effect of *in vitro* culture on the rate of translocations between *africana* chromosomes and the *tabacum* genome was manifested by six out of seven alien substitution or/and translocation events being identified in backcross progenies from the tissue culture regenerants. Among the resistant backcross segregants, a 48-chromosome individual was found that was associated with only 6 out of 51 RAPD markers specific for the intact *N. africana* chromosome, indicating a relatively small size of the translocated segment.

Four AFLP markers linked to the region coding for the high leaf number trait introgressed from *N. tomentosa* to the line 'Red Russian' of *N. tabacum* were developed and used in the study on the potential of closely related diploid relatives to increase the DNA polymorphism of the cultivated species that could be used in mapping of genes controlling quantitative traits (Lewis et al., 2007b). In a follow-up study, those markers were used for the easy identification of the high leaf number trait in investigating its effect on the yield and quality of flue-cured tobacco (Eickholt & Lewis, 2014).

Several other DNA markers linked to traits introgressed into *N. tabacum* from different *Nicotiana* species have been developed (Lewis, 2011, 2020a). The lists may be supplemented with SNP (single nucleotide polymorphism) probes for black root rot resistance from *N. debneyi* (Qin et al., 2016) and SSR (simple sequence repeat) probes linked to the gene conferring resistance to blue mold in a race of

N. langsdorffii (Zhang et al., 2012), reportedly a new potential alternative to the Suaveolentes species as a source of resistance to that disease (Zhang & Zaitlin, 2008).

4.6.6 'Genetic Drag' Associated with Beneficial Effects

Some results presented by Lewis (2007) on the effects of the gene conferring tolerance of PVY transferred from *N. africana* indicated that the alien chromatin introgressed together with the resistance gene may actually contribute to improved cured leaf quality of resistant genotypes. According to the author, this is a tentative suggestion that requires further study.

Other instances of the association of introgressed traits with positive effects, in some cases probably unintended, are related to the added benefit of resistance to another disease. There are several flue-cured varieties that were originally bred for resistance to *Phytophthora nicotianae* (black shank) and carried the resistance gene derived from *N. plumbaginifolia* (*Php* gene). These varieties were subsequently also found to exhibit a high degree of tolerance toward *Globodera tabacum solanacearum*, one of the causative agents of cyst nematode disease (Johnson et al., 2009). The two resistances were closely linked and apparently carried on the same chromosome segment with suppressed recombination capabilities. The linkage was so tight that the authors were unable to explain whether the correlations between the resistances to tobacco cyst nematodes and black shank were due to a closely linked gene cluster or to a pleiotropic effect of a single gene (Johnson et al., 2009; Parkunan et al., 2009). From other studies it appears that the very same chromatin segment also contains another resistance locus, *Phl*, originally transferred from *N. longiflora*, nonallelic but positioned very closely to the *Php* gene (Johnson et al., 2002b).

Another example is related to resistance to *Pseudomonas syringae* (wildfire). The resistance to wildfire effective against race 0 of the pathogen originally transferred from *N. longiflora* (Clayton, 1947) was later found to be associated with resistance to *Globodera* nematodes, although the linkage was not very tight (Spasoff et al., 1971; Hayes et al., 1997).

Resistance to wildfire effective against races 0 and 1 of *P. syringae* pv. *tabaci* and against *P. syringae* pv. *angulata* (angular leaf spot) was also transferred from *N. rustica* var. *pumila* (Stavely & Skoog, 1976, 1978) and from *N. rustica* var. *brasilea* (Woodend & Mudzengerere, 1992). In the latter case, the '*Wz*' gene, named after the breeding line in which it had been incorporated, was found to segregate independently of the factor introgressed from *N. longiflora* (Bukuta, 2002). Drake and Lewis (2013) demonstrated that the introgressed chromatin region surrounding '*Wz*' also contained a closely linked resistance factor to black shank, effective against two main races (0 and 1) of *Ph. nicotianae.* In their opinion, the added benefit of black shank resistance had been accomplished inadvertently while transferring the resistance to wildfire. However, from the previously cited account by

Bukuta (2002), it seems that both introgression efforts, i.e., transfers of resistance to wildfire and to black shanks, were conducted consciously, although it is not clear which of them came first or whether they were conducted alongside each other. In any case, it appears from Bukuta's report that the breeder/s were aware of the linkage between the two resistance factors.

In yet another instance, Gajos (1984) found that TSWV-resistant derivatives from crossing *N. tabacum* with *N. alata* also resisted black root rot. The observed added benefit was, in the researcher's account, attributable to unconscious double selection for two closely linked resistance factors derived from an accession of *N. alata* that showed full resistance to both *Thielaviopsis basicola* and TSWV. Generally, *N. alata* is not listed among the sources of resistance to that disease (Burk & Heggestad, 1966; Doroszewska & Przybyś, 2007). This notwithstanding, different accessions of that allogamous and polymorphic species varied substantially in their response to inoculation with *Thielaviopsis basicola* (Doroszewska & Przybyś, 2007), and some other investigators actually reported the species as resistant to BRR (Takenaka, 1960, 1963; Stoyanova, 1979). Interestingly, the defensin-encoding gene nad1, cloned from *N. alata*, was shown to confer protection against *Th. basicola* to some crops other than tobacco (Pereg, 2013).

4.6.7 Linkage Drag in the Heterozygous Condition

Save for the unusual case of the linkage drag associated with resistance to TSWV, many of the adverse effects of introgression can be eliminated or alleviated when the introgressed chromatin is in the heterozygous rather than homozygous condition. Vital genes lost in the translocation region of the rearranged chromosome can be replaced by those present in its unchanged counterpart. On the other hand, since many alien genes inserted together with a transferred gene of interest exhibit additive action, their effects will be diminished in heterozygotes (Lewis, 2011). This was confirmed in F_1 hybrids between resistant and susceptible parents in flue-cured tobacco for resistance to black shank (*Phytophtora parasitica* var. *nicotianae*) from *N. longiflora* (Wernsman & Rufty, 1987), resistance to TMV from *N. glutinosa* (Chaplin et al., 1966; Lewis & Rose, 2010; Lewis, 2011), resistance to PVY from *N. africana* (Lewis, 2007), and black root rot (*Thielaviopsis basicola*) resistance from *N. debneyi* (Haji et al., 2005, 2006). Likewise, deleterious effects of the *N. suaveolens* male fertility restoring factor on the productivity of a restored stamened alloplasmic cms suaveolens line were less penalizing in heterozygous vs. homozygous condition (Hosfield & Wernsman, 1974).

Indeed, commercial utilization of resistance to diseases and possibly of other genes from interspecific sources is largely possible because in F_1 hybrids and, recently, in hybrids involving more than two breeding lines (Berbeć, 2017), the trade-offs of interspecific introgression are considerably mitigated. In an *N. longiflora* chromosome substitution line resistant to black shank race 0, the linkage drag effect included severe leaf spotting caused by a gene present on the substituted chromosome. Due to the recessive character of the leaf-spotting factor, it was present only in true-breeding substitution lines and completely eliminated in F_1 hybrids (Hendrix & Apple, 1967).

Black root rot resistance from *N. debneyi* and from *N. glauca* in flue-cured cultivars developed in Poland is a particularly good example of where persistent selection for yield and quality combined with deployment of the resistance factor in the heterozygous condition allowed a number of resistant hybrid cultivars to become successful replacements of the nonresistant cv. 'Wiślica', from which all of those resistant forms were derived and to which some of them remained very closely related. The results of a relevant field experiment were related by Trojak-Goluch et al. (2017) and are summarized in Table 4.10 of this chapter. The differences in favor of the resistant hybrids, especially those for yield at site one of the trial, although numerically substantial, could not be validated statistically. This is often the case with field trials when within-block and block-to-block environmental variation is overwhelming. At site 2, where the random component was much lower, the superiority of black root rot-resistant varieties over Wiślica in terms of yield was easily demonstrated by HSD multiple comparisons, which was also true of

Table 4.10 Field performance of black root rot (BRR) resistant cultivars of flue-cured tobacco developed at IUNG-PIB, Puławy, Poland against the backdrop of 'Wiślica', the non-resistant Polish-bred cultivar, a typical representative of the "Polish Virginia" market type (adapted after Trojak-Goluch et al., 2017)

	Site 1[5]			Site 2[6]		
Cultivar	Cured leaf yield ($t*ha^{-1}$)	Grades 1–3	Nicotine content	Cured leaf yield ($t*ha^{-1}$)	Grades 1–3	Nicotine content
VRG 2[1]	3.03	98.2	2.21	2.18	64.5	2.21
VRG 4[2]	2.98	92.8	2.41	1.96	65.2	2.19
VRG 5 TL[3]	3.29	95.5	1.52	2.32	70.8	1.47
VRG 10 TL[3]	3.15	93.8	1.85	2.69	70.5	1.74
Wigola[4]	3.04	97.7	1.99	2.34	71.6	1.95
Wiślica	2.06	77.6	2.46	1.96	58.8	2.35
Tukey's HSD	ns[7]		0.072	0.274		0.434

[1]Single-cross hybrid between Wiślica and a black root rot resistant line derived from Wiślica (source of BRR resistance: *N. debneyi*);
[2]Single-cross hybrid between two black root rot resistant lines derived from Wiślica (source of BRR resistance: *N. debneyi*);
[3]Three-way hybrid involving two BRR resistant lines derived from Wiślica and a non-resistant line (source of BRR resistance: *N. debneyi*);
[4]Single-cross hybrid between a black root rot resistant isoline of Wiślica and Wiślica (source of BRR resistance: *N. glauca*);
[5]At south-eastern Poland, typical flue-cured tobacco region; medium heavy silty soil;
[6]At north-eastern Poland, northernmost flue-cured tobacco region, light loamy sand; [7]non significant by Tukey's HSD at 0.05 probability level. Both sites were free of discernible black root rot pressure

Wigola, the near-isogenic resistant counterpart of Wiślica. At both sites, leaf quality measured by the percentage of the first three out of six grades used in the local grading system was also in favor of the black root rot-resistant hybrids. Wiślica followed by its black-root rot resistant derivatives (VRG 2 and VRG 4) were the entries highest in nicotine. Surprisingly, Wigola, which was genetically closest to the relatively high-nicotine Wiślica, showed considerably less nicotine than either its nonresistant counterpart or its resistant relatives. In the test under discussion, Wigola was the only variety in which the resistance was derived from *N. glauca* rather than from *N. debneyi* (Trojak-Goluch & Berbeć, 2009; Trojak-Goluch et al., 2017).

Heterozygous hybrid cultivars of tobacco, such as those described in the preceding paragraph, are technically and economically feasible because they deploy another singular trait accessible through interspecific transfer – cytoplasmic male sterility (see Chap. 5).

4.6.8 Designer or Shuttle Chromosome

Suppressed recombination and inheritance *en bloc* that plagues interspecific introgression was intended to be turned to advantage by the concept of constructing an artificial chromosome on which gene constructs could be gradually added to one another to create a whole package of agronomically useful genes. Such a package would segregate as a single unit and could be shuttled from one breeding line to another in an intact form and thus facilitate the development of improved cultivars. The idea developed by Campbell et al. (1994) and Lewis and Wernsman (2001) took as a starting point a 50-chromosome breeding line of *N. tabacum* with a pair of added chromosomes from *N. africana*. The original *N. africana* chromosome was planned to be redesigned into an artificially made vehicle for gene shuttling by transforming it with multiple transgenes coding for agronomically important traits. The integrity of such a vehicle was thought to be safe because the addition line was assumed to be somatically and meiotically stable and recombination between the *africana* chromosome and the tobacco genome had not been observed to occur. The work on the construction of the designer chromosome was initiated by creating a linkage block that involved the resistance to PVY and TEV native to the *africana* chromosome, the cloned N gene originally from *N. glutinosa* that imparts resistance to TMV and the *dhfr* transgene conferring resistance to the antibiotic methotrexate. Linking potyvirus (PVY and TEV) resistance genes with the cloned N transgene proved to be feasible. However, an occasional loss of the N gene, possibly through recombination with the *N. tabacum* genome, was an unexpected setback. Actually, some recombination involving the *N. africana* chromosome and *N. tabacum* chromosomes in the 50-chromosome addition line did occur in experiments that were not directly related to the construction of the gene shuttle (Lewis, 2005). The authors of that novel vehicle for introgression (Lewis & Wernsman, 2001) argued in favor of continuing the experiments with the designer chromosome, indicating that the expected benefits would ultimately outweigh the initial difficulties and efforts to overcome these difficulties. There has been no follow-up reported to date, however. This is possibly because some incentive on the part of potential beneficiaries of the idea may be lacking. Tobacco leaf merchants and manufacturers strongly object to genetically modified tobacco, and genetic engineering is an essential part of constructing the designer chromosome.

4.7 Summary of Gene Transfer from *Nicotiana* Species to Cultivated Tobacco

Table 4.11 lists *Nicotiana* species from which introgression to tobacco was reported. All accounts of interspecific introgression found in the literature were included, completed and discontinues, and ranging from very detailed and well documented to those that were reduced to fragmentary information available in abstracts or even in brief citations. Full descriptions were not always available, as many of those reports were published in journals that were very hard to access or/and in exotic languages and scripts. For the same reasons, the author may have and almost certainly has missed some important details and maybe also the whole projects, possibly involving other species and other introgressed traits.

Table 4.11 *Nicotiana* species as sources of traits introgressed in *Nicotiana tabacum* (after Berbeć & Doroszewska, 2020); re-designed with additions and modifications

Species source of introgression	Introgressed trait	Authors/ reported by:
N. alata	Resistance to TSWV (tomato spotted wilt virus)	Stoyanova (1979)[1], Gajos (1981[2], 1984)[2], Atanassov et al. (1991)[3], and Patrascu et al. (1999)[3]
	Resistance to *Thielaviopsis basicola* (black root rot)	Gajos (1984)[4]
	Resistance to *Erysiphe cichoracearum*	Stoyanova (1979)[5]
N. longiflora	Resistance to *Phytophtora nicotianae* race 0 (black shank)	Valleau et al. (1960), Collins and Legg (1969), Oka and Ninomi (1961), and Dang et al. (2019)
	Resistance to *Pseudomonas syringae* pv. *tabaci* race 0 (wildfire)	Clayton (1947) and Heggestad et al. (1960)
	Resistance to *Meloidogyne javanica* (root knot nematodes)	Raeber and Schweppenhauser (1964), Raeber and Smeeton (1973b), Schweppenhauser (1975), Mackenzie et al. (1986), Ternouth et al. (1986), and Venkatesvarlu et al. (1998)
	Resistance to *Globodera tabacum solanacearum/tabacum* (cyst nematodes)	Spasoff et al. (1971), Komm and Terril (1982), Hayes et al. (1997), and Crowder et al. (2003)[6]

(continued)

Table 4.11 (continued)

Species source of introgression	Introgressed trait	Authors/ reported by:
N. plumbaginifolia	Resistance to *Phytophtora nicotianae* Race 0	Chaplin (1954), Cameron (1958), Oka and Ninomi (1961)[7], Chaplin (1962), Apple (1962), and Dang et al. (2019)[7]
	Tolerance of *Globodera tabacum* (cyst nematodes)	Johnson et al. (2009)[8]
	Resistance to *Meloidogyne incognita* (root knot nematodes)	Clayton et al. (1958)
	Chlorophyll production gene Ws (pbg)	Moav (1958) and Niwa (1969)
	Purple flower gene	Ar-Rushdi (1957)
N. sanderae	Resistance to TMV (tobacco mosaic)	Yancheva (1989), Palakarcheva (1984)[9], Dorossiev et al. (1990)[9], and Palakarcheva and Krusteva (1995)[9]
	Resistance to *Erysiphe cichoracearum* (powdery mildew)	Palakarcheva et al. (1990)[10] and Dorossiev et al. (1990)[10]
	Resistance to TSWV	Palakarcheva (1984)[9]
N. glauca	Resistance to *Thielaviopsis basicola* (black root rot)	Berbeć (1963b, 1966, 1977)[11], Trojak-Goluch and Berbeć (2009), Czubacka (2022) and Trojak-Goluch et al. (2017)[12]
	Resistance to PVY	Berbeć (1966, 1977)[13] and Stoyanova (1972)[13]
	Tolerance of TMV	Berbeć (1966)[13]
N. noctiflora	Resistance to *Peronospora hyoscyami* f.pv. *tabacina*	Palakarcheva and Dorossiev (1983)[14] and Palakarcheva (1984)
N. benavidesii	Tolerance of PVY (Potato virus Y)	Berbeć and Głażewska (1988)[15] and Czubacka (2022)
N. knightiana	Resistance to *Peronospora hyoscyami* f.pv. *tabacina*	Corbaz (1962)[16] and Nikova and Shabanov (1988)[17]
N. nesophila	Resistance to *Colletotrichum tabacum* (anthracnose)	Sievert (1972a)[18]
	Resistance to *Phytophtora nicotianae* Race 0 and 1	Reed and Collins (1980)[19]
N. repanda	Resistance to *Meloidogyne javanica/incognita/arenaria* (root knot nematodes)	Raeber and Schweppenhauser (1964), Raeber and Smeeton (1973a), Stavely et al. (1973), Schweppenhauser (1974), Gwynn et al. (1986)[20],Mackenzie et al. (1986), Ternouth et al. (1986), and Mudzengerere (1994)
	Resistance to *Globodera tabacum* (cyst nematodes)	Gwynn et al. (1986)[21] and LaMondia (2010)
	Resistance to *Pseudomonas syringae* pv. *tabaci* (wildfire – race not specified)	Gwynn et al. (1986)[22]

(continued)

Table 4.11 (continued)

Species source of introgression	Introgressed trait	Authors/ reported by:
	Resistance to TMV	Gwynn et al. (1986)[23] and Bates (1990)[24]
	Resistance to *Cercospora* sp. *nicotianae* (frogeye)	Stavely et al. (1973) and Wan et al. (1971)[25]
N. stocktonii	Resistance to *Phytophtora nicotianae* Race 0 and 1	Reed and Collins (1980)[26]
N. rustica var. pumila	Resistance to race 0 and race 1 (wildfire} and *Pseudomonas syringae* pv. *angulata* (angular leaf spot (ALS)	Stavely and Skoog (1976, 1978)
N. rustica sp.	Resistance to TMV	Nifong (2008), Holmes (1937b)[27]
	Resistance to *Pseudomonas. syringae* pv. *angulata* (angular leaf spot (ALS)	Mackenzie et al. (1986)[28]
N. rustica var. brasilea	Resistance to *Phytophtora nicotianae* race 0 and race 1 (black shank)	Bukuta (2002)[29]
	Resistance to *Pseudomonas syringae* pv. *tabaci* race 0 and race 1 (wildfire} and *P. syringae* pv. *angulata* (angular leaf spot (ALS)	Woodend and Mudzengerere (1992)[30] and Bukuta (2002)[30]
	Alkaloid/nicotine content	Chaplin (1977[31], 1978)[31], Chaplin and Sisson (1984)[31], and Chaplin (1987)[31]
N. rustica var. chlorotica	Resistance to *Peronospora hyoscyamii* pv. *tabacina*	Pandeya et al. (1986)[32]
N. rustica var. 'Babor'	Alkaloid content	Pandeya and White (1984)[33]
N. africana	Resistance to PVY (potato virus Y)	Keum et al. (1994)[34], Doroszewska and Berbeć (1999)[35], Doroszewska (2010)[35] and Czubacka (2022)
	Resistance to PVY (potato virus Y) and TEV (tobacco etch)	Lewis (2005)[36] and Lewis et al. (2007a[36], b)[36]
N. amplexicaulis	Resistance to *Meloidogyne javanica* (root knot nematodes)	Venkateswarlu et al. (1998)[37]
N. benthamiana	Resistance to *Spodoptera litura* (tobacco caterpillar)	Ramavarma et al. (1980)[38]
	Resistance to aphids	Krusteva et al. (2003b) and Murthy et al. (2014)
	Resistance to *Erysiphe/ Golovinomyces cichoracearum* (powdery mildew)	Dorossiev et al. (1990)[39]
N. debneyi	Resistance to *Peronospora hyoscyami* f.pv. *tabacina* (blue mold)	Clayton (1958)[40], Ternovsky (1964), Bailov et al. (1966), Smith (1968)[41], Berbeć (1977), Palakarcheva and Bailov (1976)[42], Palakarcheva (1981)[42],

(continued)

Table 4.11 (continued)

Species source of introgression	Introgressed trait	Authors/ reported by:
		Palakarcheva and Krusteva (1978)[42], Baranova et al. (2015)[42], Lea (1963)[43], Wark (1963[43], 1970)[43], Clayton (1968)[43], Marani et al. (1971)[44], and Berbeć (1963a[44], 1964)[44]
	Resistance to *Thielaviopsis basicola* (black root rot)	Clayton (1958[45], 1969)[45], Berbeć (1963a)[46], Ternovsky (1964), Brandle et al. (1992)[47], Bai et al. (1996)[48], and Kenward et al. (1999)[48]
	Resistance to *Erysiphe/ Golovinomyces cichoracearum* (powdery mildew)	Ternovsky (1964), Palakarcheva and Krusteva (1978)[49], and Palakarcheva (1976, 1981)[49]
		Smeeton and Ternouth (1990)[50]
N. excelsior	Resistance to *Peronospora hyoscyami* f.pv. *tabacina*	Wark (after Lucas, 1975), Gillham et al. (1977), and Dorossiev et al. (1990)[51]
	Resistance to aphids	Murthy et al. (2014)
N. exigua	Resistance to *Peronospora hyoscyami* f.pv. *tabacina*	Wark (1975), Gillham et al. (1977), Manolov et al. (1978a, b), and Manolov (1980)
N. goodspeedii	Resistance to *Peronospora hyoscyami* f.pv. *tabacina*	Wark (1963, 1970)[52], Wuttke (1969)[52], Palakarcheva (1976, 1981)[53], Dorossiev et al. (1990)[54], and Baranova et al. (2015)
	Resistance to *Erysiphe/ Golovinomyces cichoracearum* (powdery mildew)	Palakarcheva and Bailov (1976), Palakarcheva (1981), and Dorossiev et al. (1990)[55]
	Resistance to TMV	Palakarcheva (1976, 1984)[56]
	Resistance to wildfire	Palakarcheva (1976)
N. gossei	Resistance to *Spodoptera litura* (tobacco caterpillar)	Rao et al. (1980)[57]
	Resistance to aphids	Rao et al. (1980)[58], Georgeva (1996)[59], Dmitrov and Krusteva (1998)[60], Krusteva et al. (2003b)[61], and Murthy et al. (2014)[62]
N. maritima	Resistance to *Peronospora hyoscyami* f.pv. *tabacina*	Palakarcheva and Dorossiev (1983)[63]
N. megalosiphon	Resistance to *Meloidogyne* (root knot nematodes)	Clayton et al. (1958)
	Resistance to *Peronospora hyoscyami* f.pv. *tabacina*	Palakarcheva (1976)[64] and Manolov (1981, 1983)
	Resistance to *Erysiphe/ Golovinomyces cichoracearum* (powdery mildew)	Manolov (1983)

(continued)

Table 4.11 (continued)

Species source of introgression	Introgressed trait	Authors/ reported by:
N. suaveolens	Resistance to *Peronospora hyoscyami* f.pv. *tabacina*	Izard et al. (1964)[65], Izard and Schiltz (1964)[65], Wark (after Lucas, 1975)[66]
	Resistance to *Alternaria alternata*	Stavely, 1979[67]
	Restoration of male fertility to cms line	Schweppenhauser and Mann (1968) and Hosfield and Wernsman (1974)
N. umbratica	Tolerance of leaf curl virus (LCV)	Murthy et al. (2014)[68]
	Resistance to aphids	Murthy et al. (2014)[68]
N. velutina	Resistance to *Peronospora hyoscyami* f.pv. *tabacina*	Wark (1963)[69]
		Corbaz (1962), Gillham et al. (1977), and Powell (1979)
N. kawakamii	Resistance to PVY	Ohashi (1985)[70]
N. otophora	Megachromosome-inducing factor	Burns and Gerstel (1969); Collins et al. (1970)
	Unstable carmine color factor	Gerstel and Burns (1966, 1967, 1968), Burns and Gerstel (1967)
N. setchellii	Resistance to *Erysiphe/ Golovinomyces cichoracearum* (powdery mildew)	Shabanov et al. (1974)[71]
N. tomentosa	Resistance to *Meloidogyne incognita* races 1 and 3 (root knot nematodes)	Slana et al. (1977), Stavely et al. (1977), Stavely (1979), Yi et al. (1998), Legg and Smeeton (1999), Slana and Stavely (1978)[72], Rufty et al. (1983)[72], and Pollok et al. (2016)[72]
	QTL locus affecting leaf number and days to flower	Clausen and Cameron (1944), Lewis et al. (2007b), and Eickholt and Lewis (2014)
	Wh-p "pale white flower" and Pp "purple plant" qualitative factors	Clausen and Cameron (1944)
N. tomentosiformis	Resistance to *Erysiphe/ Golovinomyces cichoracearum* (powdery mildew)	Ternovsky (1941), Ohashi (1985), Ternouth et al. (1986), Smeeton and Ternouth (1990)[73], and Tatemichi (1990)[74]
	Resistance to *Meloidogyne javanica* (root knot nematodes)	Mackenzie et al. (1986) and Ternouth et al. (1986)
	Resistance to tobacco vein mottling virus (TVMV)	Legg and Smeeton (1999)
	Resistance to PVY	Legg and Smeeton (1999)[75]
N. glutinosa	Resistance to TMV (tobacco mosaic)	Holmes (1938)[76], Ternovsky (1941), Gerstel (1943), Kostoff and Georgieva (1944), Gerstel (1945b, 1946), Hitier and Izard (1958), and Oka (1961)

(continued)

Table 4.11 (continued)

Species source of introgression	Introgressed trait	Authors/ reported by:
	Resistance to *Erysiphe/ Golovinomyces cichoracearum* (powdery mildew)	Wark (after Lucas, 1975) Smeeton and Ternouth (1990) and Vinogradov and Larkina (2013)[77] Baranova et al. (2015)[77]
	Reduced flower size	Gerstel (1945b)[78]

Footnotes are listed separately for each species – source of an introgressed trait – in the order of their appearance in the table

N. alata:

[1]TSWV resistant *N. tabacum*- resembling plants were identified in the post-sesquidiploid 1st breakdown generation, no follow-up to that report; [2]transfer of resistance performed through the mediation of the amphidiploid hybrid *N. tabacum* × *N. otophora* (Gajos, 1981); [3]via asymmetric protoplast fusion (Atanassov et al., 1991, Patrascu et al., 1999), no follow-up to either of these reports; [4]allegedly, introgressed from *N. alata* together with resistance to TSWV as a result tight linkage between the two resistance factors; no confirmation of this statement in subsequent studies; [5]powdery mildew-resistant, *N. tabacum*-resembling plants were identified in the post-sesquidiploid breakdown generation

N. longiflora:

[6]Found to be loosely linked to resistance to wildfire (Crowder et al., 2003, although Komm and Terril (1982) reported a tight linkage between the two factors without excluding a pleiotropic effect of the same gene

N. plumbaginifolia:

[7]Transferred as a single chromosome of *N. plumbaginifolia* added to the 48-chromosome complement of *N. tabacum*

[8]Found to be unconsciously introgressed from *N. plumbaginifolia* as closely linked with resistance to *Ph. Nicotianae*

N. sanderae:

[9]Reported without details, no follow-up to these reports; [10]reported without details, no follow-up to these reports

N. glauca:

[11]Both reports lack details, no follow-up; [12]commercial cultivar of flue-cured tobacco developed; [13]reported without details, no follow-up

N. noctiflora:

[14]No details available

N. benavidesii:

[15]Symptomless carrier type

N. knightiana:

[16]Reported without details, no follow-up; [17]transfer of resistance attempted

N. nesophila:

[18]Transfer of resistance attempted; [19]transfer of dominant resistance factor/s accomplished up to BC_3 generation, no follow-up reported

N. repanda

[20,21,22]Transfer of resistance not completed; [23]transfer of resistance not completed, unstable heterozygous substitution; [24] asymmetric protoplast fusion followed by sexual backcross to *N. tabacum*; resistant *N. tabacum*-like phenotypes recovered; [25]carried to advanced generations but not completed, no follow-up reported

N. stocktonii:

[26]Transfer of dominant resistance factor/s accomplished up to BC_3 generation, no follow-up

N. rustica sp.:

(continued)

[27]transfer of plant age-dependent hypersensitive response via a third species (*N. paniculata*); [28]introgressed fragment of the *N. rustica* chromosome linked with resistance to wildfire and ALS
N. rustica var. brasilea:
[29]Linked to resistance factor to black shank (Drake & Lewis, 2013); [30] mentioned as linked to resistance to black shank (Bukuta, 2002); [30] [31]introgression of alkaloid-controlling factors to low-alkaloid flue cured line
N. rustica var. chlorotica:
[32]Resistance identified in progenies from the somatic *N. rustica*+ *N. tabacum* hybrid. Since *N. rustica* is not known to carry resistance to blue mold it was explained as possible complementation or interactions involving nuclear and/or cytoplasmic genomes of parental species. The results put in doubt by Rufty (1989)
N. rustica var. 'Babor':
[33]Simultaneous increase in leaf yield and total alkaloid content (cv. Delgold) attributed to introgression of alkaloid-controlling factors from *N. rustica*
N. africana
[34]Recessive substitution, tolerance (symptomless host response), covers a wide spectrum of PVY strains; [35]inherited in monogenic recessive fashion (Korbecka-Glinka et al., 2018); [36]partially dominant segmental substitution
N. amplexicaulis
[37]Introgression reported as in progress, no follow-up
N. benthamiana
[38]Visual evidence for segmental substitution as a heteromorphic bivalent in meiosis; [39] reported introgressions used the trispecific hybrid 4x (*N. goodspeedii* × *N. tabacum*) × *N. benthamiana* in the initial stage of the transfer. Both *N. goodspeedii* and *N. benthamiana* were reported as resistant to powdery mildew but of the two only *N. goodspeedii* is known to be resistant to blue mold (see this table)
N. debneyi
[40]Oligogenic resistance transferred by means of several translocations involving different *N. debneyi* chromosomes; [41]dominant, oligogenic (three-factorial) resistance; [42]unspecified mode of inheritance; [43]monogenic resistance conditioned by a single major gene; [44]resistance transferred to and expressed in flue-cured tobacco (no details and no follow-up) [45]resistance transferred from *N. tabacum* × *N. debneyi* hybrid; [46]resistance transferred to and expressed in flue-cured tobacco (no details and no follow-up); [47]resistance transferred from a somatic *debneyi+tabacum* hybrid;) [48]Tnd-1 retrotransposon identified as either participating in or closely linked to BRR resistance; [49]dominant resistance factor (s); [50]single dominant resistance factor
N. excelsior
[51]Reported introgressions used the trispecific hybrid 4x (*N. goodspeedii* × *N. tabacum*) × *N. excelsior* in the initial stage of the transfer.
N. goodspeedii and *N. excelsior* were both reported as resistant to blue mold and powdery mildew (see this table)
N. goodspeedii
[52]Dominant monogenic or oligogenic patterns of resistance depending on lineage, the validity of that introgression put in doubt by Milla et al. (2005) who provided evidence that the resistance may actually come from N. debneyi, the detailed character of Wark's account of the transfer notwithstanding (Wark, 1970; Rufty, 1989); [53]dominant resistance factor (s); [54]reported introgressions used the trispecific hybrid 4x (*N. goodspeedii* × *N. tabacum*) × *N. benthamiana* in the initial stage of the transfer. Both *N. goodspeedii* and *N. benthamiana* were reported as resistant to powdery mildew but of the two only *N. goodspeedii* is known to be resistant to blue mold (see this table); [55]see footnote[53]; [56]reported without details, no follow-up
N. gossei
[57]Resistant alien substitutions, segmental substitutions, additions and segmental additions were identified in early breakdown generations, no follow-up; [58]resistant alien substitutions and additions were identified in early breakdown generations; [59]resistance bred into *N. tabacum*, commercial type

(continued)

not specified; [60]monogenic dominant inheritance reported in aphid-resistant *N. tabacum lines*, no details of transfer; [61]no details of transfer reported; [62]resistant lines reported without details of transfer
N. maritima
[63]No details
N. megalosiphon
[64]Attempt to transfer resistance factor (after Tatemichi, 1990), no details
N. suaveolens
[65]Resistant tobacco lines derived from the amphidiploid 4x (*N. suaveolens* × *N. tabacum*); [66] no details available; [67]resistant tobacco lines developed from the amphidiploid 4x (*N. suaveolens* × *N. tabacum*), no further details
N. umbratica
[68]Resistant lines reported without details of transfer
N. velutina
[69]Oligogenic pattern of inheritance; resistance in cv. 'Ovens 62' may actually come from *N. debneyi* (Milla et al., 2005)
N. kawakamii
[70]Reported without details
N. setchellii
[71]Reported without details
N. tomentosa
[72]Dominant, hypersensitive, temperature-dependent factor *Rk1*
N. tomentosiformis
[73]Single dominant factor; [74]two dominant factors introgressed via *N. sylvestris* as the bridging species; [75]resistance of va type allelic to that in 'Virgin Mutante A' (Koelle, 1961; Ano et al., 1995)
N. glutinosa
[76]Dominant, hypersensitive, temperature-dependent factor; [76]single dominant resistance factor transferred from the amphidiploid *N. digluta* by Ternovsky in 1930s–1940s and present in several Russian varieties; [77]transferred as a pair of chromosomes to a 26-chromosome pair double addition line

References

Adamo, N., Johnson, C. S., Reed, T. D., & Eisenback, J. D. (2021). Reproduction of *Meloidogyne arenaria* race 2 on flue-cured tobacco with putative resistance derived from *Nicotiana repanda*. *Journal of Nematology, 53*, e2021–e2064. https://doi.org/10.21307/jofnem-2021-064

Ahuja, M. R. (1962). A cytogenetic Study of heritable tumors in *Nicotiana* species hybrids. *Genetics, 47*, 865–880.

Ahuja, M. R. (1968). An hypothesis and evidence concerning the genetic components controlling tumor formation in *Nicotiana*. *Molecular & General Genetics, 103*, 176–184.

Ahuja, M. R. (1974). Control of tumor-associated peroxidase in genetic tumor system in *Nicotiana*. *Experientia, 30*, 1007–1008.

Ahuja, M. R. (1998). Genetic tumors in *Nicotiana* and other plants. *The Quarterly Review of Biology, 73*, 439–462.

Ano, G., Blancard, D., & Cailleteau, B. (1995). Mise au point sur la résistance récessive aux souches nécrotiques du virus Y de la pomme de terre (PVY) présente chez *Nicotiana tabacum*. *Annales Tabac Section, 2*(27), 35–42.

Anon. (1999). *Role of alien species in tobacco improvement – Reaction of N. tabacum x N. gossei derivatives to tobacco aphid (Myzus nicotianae)*. Rajahmundry.

Appa Rao, K., & Krishna Murthy, K. V. (1963). Studies on multiple polyploids in *Nicotiana*. *Genetica, 34*, 66–78.

Apparao, K., & Gopinath, D. M. (1968). Colchicine-induced mutations in a trispecific *Nicotiana* hybrid. *Canadian Journal of Genetics and Cytology, 10*, 782–787.

Apple, J. L. (1962). Transfer of resistance to black shank (*Phytophthora parasitica* var. *Nicotianae*) from *Nicotiana*. *Phytopathology, 52*, 1.

Apple, J. L. (1967). Occurrence of race 1 *Phytophtora parasitica* var. *nicotianae* in North Carolina and its implications. *Tobacco Science, 11*, 79–83.

Ar-Rushdi, A. H. (1955). Inheritance in *Nicotiana tabacum*. XXVI. Sterility genes from *tomentosae* species. *Journal of Genetics, 54*, 9–22.

Ar-Rushdi, A. H. (1957). The cytogenetics of variegation in a species hybrid in *Nicotiana*. *Genetics, 42*, 312–325.

Atanassov, A., Dimantov, D., Atanassov, I., Dragoeva, A., Vassileva, Z., Vitanov, V., Jankulova, M., & Djilianov, D. (1991). Tranfer of resistance to tomato spotted wilt virus/TSWV/from wild *Nicotiana* species to *N. tabacum* via somatic hybridisation. *Physiologia Plantarum, 82*, A23.

Bai, D., Reeleder, R., & Brandle, J. E. (1995). Indentification of two RAPD markers tightly linked with the *Nicotiana debneyi* gene for resistance to black root rot. *Theoretical and Applied Genetics, 91*, 1184–1189.

Bai, D. P., Reeleder, R., & Brandle, J. E. (1996). Production and characterization of tobacco addition lines carrying *Nicotiana debneyi* chromosomes with a gene for resistance to black root rot. *Crop Science, 36*, 852–857.

Bailov, D., Palakarcheva, M., & Daskalov, S. (1966). Genetic investigations into interspecific hybrids. In *Proceedings of the 4th international tobacco science congress, Athens* (pp. 656–666).

Baldwin, I. T. (2001). An ecologically motivated analysis of plant-herbivore interactions in native tobacco. *Plant Physiology, 127*, 1449–1458.

Baranova, E. G., Ivanitskii, K. I., & Suchkov, V. I. (2015). Formation of the gene pool of sources of cytoplasmatic male sterility of tobacco. apriori-journal.ru/seria2/2-2015 APRIORI. Серия: Естественные и технические науки. 2015. № 2. С. 9.).

Bates, G. W. (1990). Asymmetric hybridization between *Nicotiana tabacum* and *N. repanda* by donor recipient protoplast fusion: transfer of TMV resistance. *Theoretical and Applied Genetics, 80*, 481–487.

Bayer, M. H., & Ahuja, M. R. (1968). Tumor formation in *Nicotiana*: Auxin levels and auxin inhibitors in normal and tumor-prone genotypes. *Planta, 79*, 292–298.

Berbeć, J. (1963a). Zagadnienie hodowli odpornościowej tytoniu w pracach badawczych IUNG w Puławach (Breeding tobacco for disease resistance in research conducted by IUNG at Puławy). *Ochr Rośl, 3*, 12–17.

Berbeć, J. (1963b). On the possibility to improve tobacco production in Poland by exploiting the advancements in breeding for disease resistance (in Polish). *Wiad Tyt, 9*(78), 129–133.

Berbeć, J. (1964). Zakres prac nad wyhodowaniem krajowych odmian tytoniu odpornych na mączniak rzekomy, prowadzonych w Puławach (Experience with developing domestic tobacco varieties resistant to blue mold at Puławy) (in Polish). *Wiad Tyt, 11*, 170–172.

Berbeć, J. (1966). Badania nad krzyżówką Nicotiana glauca x N. tabacum (Investigations on the cross *Nicotiana glauca* × *N. tabacum*) (in Polish). *Information PAN*, 7-213, 340.

Berbeć, J. (1977). Sześćdziesiąt lat prac hodowlanych i uprawowych nad tytoniem w instytucie w Puławach (Sixty years of research on breeding and production of tobacco at the Institute at Puławy) (in Polish). *Wiad Tyt (8)*, 6–11.

Berbeć, A. (1980). Chromosome substitution and gene mutation concurring to produce a white-flower derivative of the interspecific hybrid *Nicotiana tabacum* L. × *N. benavidesii* Goodspeed. *Genetics Pol, 21*, 283–289.

Berbeć, A. (1987). Cytogenetical study on *Nicotiana tabacum* L. cv. Nadwiślański Mały (2x and 4x) × *Nicotiana alata* Link et Otto hybrids. *Genetics Pol, 28*, 251–261.

Berbeć, A. (1988). Morphology, cytogenetics and resistance to potato virus Y of the amphidiploid *Nicotiana raimondii* Macbride × *N. tabacum* L. *Genetics Pol, 29*, 41–52.
Berbeć, A. (2017). Three-way crosses vs. single crosses in tobacco: first agronomic assessment. *Crop Science, 57*, 1363–1372. https://doi.org/10.2135/cropsci2016.07.0624
Berbeć, A., & Doroszewska, T. (1981). Investigations of the interspecific hybrid *Nicotiana amplexicaulis* Burbidge × *Nicotiana tabacum*. *Genetics Pol, 22*, 197–207.
Berbeć, A., & Doroszewska, T. (1992). Alloplasmic forms of cultivated tobacco (*Nicotiana tabacum* L.) with alien cytoplasm introgressed from the species *Nicotiana amplexicaulis, N. knightiana* and *N. raimondii* (in Polish). *Pamiętnik Puławski, 100*, 141–150.
Berbeć, A., & Doroszewska, T. (2020). The use of *Nicotiana* species in tobacco improvement. In N. Ivanov, N. Sierro, & M. Peitsch (Eds.), *The tobacco plant genome. Compendium of plant genomes*. Springer.
Berbeć, A., & Głażewska, Z. (1988). Transfer of resistance to Potato Virus Y from *Nicotiana benavidesii* Goodspeed to *N. tabacum* L. *Genetics Pol, 29*, 323–333.
Berbeć, A., & Laskowska, D. (1997). Compatibility and tentoxin sensitivity in the off-spring of near-amphihaploid hybrid *Nicotiana tabacum* × *N. alata*. *Journal of Applied Genetics, 38*, 143–150.
Berbeć, J., Berbeć, A., Glazewska, Z., Opoka, B., Kobus, I., & Czop, T. (1982). Investigations on obtaining new varieties and breeding lines of *Nicotiana tabacum* resistant to virus leaf rib necrosis (Potato Virus Y) and virus degeneration of plants by Lycopersicum virus 3. USDA Grant no. FG-Po-341 (JB-6). *Final Report, Institute of Soils Science and Plant Cultivation, Pulawy, Poland* (unpublished).
Bolsunov, I. (1970). New *Nicotiana* amphidiploid as a valuable starting material for the control of virulent blue mould lines. *International Tobacco Scientific Congress, Hamburg 1970 Proceedings 1971*, 160.
Bombarely, A., Edwards, K. D., Sanchez-Tamburrino, J., & Mueller, L. A. (2012). Deciphering the complex leaf transcriptome of the allotetraploid species *Nicotiana tabacum*: A phylogenomic perspective. *Genomics, 13*, 406.
Brandle, J. E., Court, W. A., & Gleddie, S. (1992). The application of somatic hybridization to a tobacco breeding program. *Tobacco Chemistry Research Conference, 46*, 30, abstr. 17.
Brandle, J. E., Stobbs, L. W., & Gleddie, S. (1995). Resistance to a necrotic strain of potato Y among *Nicotiana* species, somatic hybrids, and tobacco cultivars. *Plant Disease, 79*, 152–154.
Brandle, J. E., Rogers, W. D., & Ankersmit, J. C. D. (1997). AC Gayed flue cured tobacco. *Canadian Journal of Plant Science, 77*, 157–158.
Brandwagt, B. F., Kneppers, T. J. A., Van Der Weerden, G., Nijkamp, H. J., & Hille, J. (2001). Most AAL toxin-sensitive *Nicotiana* species are resistant to the tomato fungal pathogen *Alternaria alternata* f. sp. lycopersici. *Molecular Plant-Microbe Interactions, 14*, 460–470.
Brieger, F. G., & Forster, R. (1942). Tumores em certos híbridos do gênero *Nicotiana*. *Bragantia*, 2259–2274.
Bukuta, L. (2002). Determination of allelism between different sources of black shank race 0 (*Phytophthora Nicotianae*) resistance and linkage studies between black shank race 0 and angular leaf spot race 1 *Pseudomonas syringae* pv. *tabaci* Tox -) resistance in tobacco (*Nicotiana tabacum*). *CORESTA Study Grant Report, CORESTA Congress, New Orleans.*
Burk, L. G. (1960). Male-sterile flower anomalies in interspecific tobacco hybrids. *The Journal of Heredity, 51*, 27–31.
Burk, L. G. (1967). An interspecific bridge-cross – *Nicotiana repanda* through *N. sylvestris* to *N. tabacum*. *The Journal of Heredity, 58*, 215–218.
Burk, L. G. (1972). Viable hybrids from monosomics *N. tabacum* by *N. langsdorffii*. *Tobacco Science, 16*, 43–45.
Burk, L. G. (1973). Partial self-fertility in a theoretical amphidiploid progenitor of *Nicotiana tabacum*. *The Journal of Heredity, 64*, 348–350.
Burk, L. G., & Dean, C. E. (1975). Hybrid fertility and aphid resistance in the cross *Nicotiana tabacum* × *N. gossei*. *Euphytica, 24*, 59–63.

Burk, L. G., & Heggestad, H. E. (1966). The genus *Nicotiana* as a source of resistance to diseases of cultivated tobacco. *Economic Botany, 20*, 76–88.

Burk, L. G., & Neas, M. O. (1964). 4n(*N. tabacum* × *N. nudicaulis*), a colchicine induced fertile hybrid. *Tobacco Science, 8*, 65–66.

Burk, L. G., Gooding, G. V., & Chaplin, J. F. (1982). Reaction of *Nicotiana* species and cultivars or breeding lines of *Nicotiana tabacum* to three strains of Potato Virus Y. *Tobacco Science, 26*, 85–88.

Burns, J. A., & Gerstel, D. U. (1967). Flower color variegation and instability of a block of heterochromatin in *Nicotiana. Genetics, 57*, 155–167.

Burns, J. A., & Gerstel, D. U. (1969). Consequences of spontaneous breakage of heterochromatic chromosome segments in *Nicotiana* hybrids. *Genetics, 63*, 427–439.

Busconi, M., Reggi, S., Lorenzoni, C., & Fogher, C. (2010). Interspecific crosses in the genus *Nicotiana*. In *Proceedings of 54th Italian Society Agricultural Genetics Annual Congrres, Post 7.02*

Butenko, R. G., & Luneva, M. Z. (1966). Application of the sterile culture technique for the culturing of distant hybrids of *Nicotiana* (in Russian). *Fizjologija Rastenij, 13*, 733–736.

Calitz, P. C., & Milne, D. L. (1962). Reaction of *Nicotiana* species and species crosses to the root knot nematode, Meloidogyne javanica. *South African Journal of Plant Science, 5*, 123–126.

Cameron, D. R. (1958). Alien substitution of a locus affecting immunity to black shank in *Nicotiana tabacum. Proc × Intl Congr Genet Univ Toronto, 2*, 41.

Cameron, D. R., & Moav, R. M. (1957). Inheritance in *Nicotiana tabacum* XXVII. Pollen killer, an alien genetic locus inducing abortion of microspores not carrying it. *Genetics, 42*, 326–335.

Campbell, K. G., Wernsmann, E. A., Fitzmaurice, W. P., & Burns, J. A. (1994). Construction of a designer chromosome in tobacco. *Theoretical and Applied Genetics, 87*, 837–842.

Chaplin, J. F. (1954). Use of tetraploid *Nicotiana tabacum* in the initial cross for transferring black shank resistance factors from *N. plumbaginifolia* to *N. tabacum. Proceedings of Association of Southern Agricultural Workers, 56.*

Chaplin, J. F. (1962). Transfer of black shank resistance from *Nicotiana plumbaginifolia* to flue-cured *N. tabacum. Tobacco Science, 6*, 184–189.

Chaplin, J. F. (1977). Breeding for varying levels of nicotine in tobacco. In *Proceedings of American Chemical Society Symposium – Recent Advances in the Chemical Composition of Tobacco and Tobacco Smoke 173rd American Chemical Society Meeting, New Orleans*, pp. 328–339.

Chaplin, J. F. (1978). Genetic manipulation for tailoring tobacco plant to meet the requirements of the grower manufacturer and consumer. *CORESTA International Tobacco Science Symposium Information Bulletin Sofia*, 17–31.

Chaplin, J. F. (1987). Die Zuchtung von Tabak im Hinblick auf verschieden hohe Alkaloidgehalte. *Beitrag Tabakforschg Intl, 14*, 1–9.

Chaplin, J. F., & Burk, L. G. (1971). Interspecific hybridization and gene transfer in *Nicotiana*: Problems and possible solutions. *Fifth International Tobacco Science Congress, Hamburg, Germany Proceedings, 59*, 67.

Chaplin, J. F., & Mann, T. J. (1961). Interspecific hybridization, gene transfer and chromosomal substitution in *Nicotiana. North Carolina Agricultural Experimental Station Technical Bulletin, 128.*

Chaplin, J. F., & Mann, T. J. (1978). Evaluation of tobacco mosaic resistance factor transferred from burley to flue-cured tobacco. *The Journal of Heredity, 69*, 175–178.

Chaplin, J. F., & Sisson, V. A. (1984). Interspecific hybridization between *N. rustica* and *N. tabacum* for different alkaloid levels. *CORESTA Congress Bulletin Specifications, Vienna, 93*, A17, ISSN.0525-6240.

Chaplin, J. F., Mann, T. J., & Apple, J. L. (1961). Some effects of the *Nicotiana glutinosa* type of mosaic resistance on agronomic characters of flue-cured tobacco. *Tobacco Science, 5*, 80–83.

Chaplin, J. F., Matzinger, D. F., & Mann, T. J. (1966). Influence of the homozygous and heterozygous mosaic-resistance factor on quantitative character of flue-cured tobacco. *Tobacco Science, 10*, 81–84.

Chari, M. S., & Patel, N. G. (1972). Relative susceptibility of different species of *Nicotiana* to *Spodoptera litura*(Fabricius). *Indian Journal Of Entomology, 34*, 249–250.

Chase, M. W., Knapp, S., Cox, A. V., Clarkson, J., et al. (2003). Molecular systematics, GISH and the origin of hybrid taxa in *Nicotiana* (Solanaceae). *Annals of Botany, 92*, 107–127.

Chen, S., Ren, F., Zhang, L., Liu, Y., et al. (2018). Unstable allotetraploid tobacco genome due to frequent homeologous recombination, segmental deletion, and chromosome loss. *Molecular Plant, 11*, 914–927.

Chu, H. Y. (1954). *The mechanism of transfer of a gene from Nicotiana glutinosa to N.tabacum.* PhD thesis. University of California.

Chung, S. C., Yoshida, K. T., & Takeda, G. (1996). Reproductive characteristics of amphidiploids derived from an interspecific hybrid between *Nicotiana* trigonophylla Du*N.* And *N. tabacum* L. *Japan Journal of Breeding, 46*, 29–33.

Chyi, Y. S., & Sanford, J. C. (1985). "Egg transformation" induced by irradiated pollen in *Nicotiana*: A re-examination. *Theoretical and Applied Genetics, 70*, 433–439.

Clausen, R. E. (1952). The cytogenetics of introgression (abstract). *Science, 115*, 481.

Clausen, R. E., & Cameron, D. R. (1944). Inheritance in *Nicotiana tabacum.* XVIII. Monosomic analysis. *Genetics, 29*, 447–477.

Clausen, R. E., & Cameron, D. R. (1957). Inheritance in *Nicotiana tabacum.* XXVIII. The cytogenetics of introgression. *Proceedings of the National Academy of Sciences of the United States of America, 43*, 908–913.

Clayton, E. E. (1947). A wildfire resistant tobacco. *The Journal of Heredity, 38*, 35–40.

Clayton, E. E. (1954). Identifying disease resistance suited to interspecific transfer. *The Journal of Heredity, 45*, 273–277.

Clayton, E. E. (1958). The genetics and breeding progress in tobacco during the last 50 years. *Agronomy Journal, 50*, 352–356.

Clayton, E. E. (1968). The transfer of blue mold resistance to tobacco from *Nicotiana debneyi.* Part IV. Breeding progress 1957–1967. *Tobacco Science, 12*, 112–124.

Clayton, E. E. (1969). The study of resistance to the black root disease of tobacco. *Tobacco Science, 13*, 30–37.

Clayton, E. E., Graham, T. W., Todd, F. A., Gaines, J. G., & Clark, F. A. (1958). Resistance to the root knot disease of tobacco. *Tobacco Science, 2*, 53–63.

Collins, G. B., & Legg, P. D. (1969). Cytogenetics of a burley tobacco breeding line derived from a cross with *Nicotiana longiflora* Cav. *Canadian Journal of Genetics and Cytology, 11*, 382–388.

Collins, G. B., Legg, P. D., & Anderson, M. K. (1970). Cytophotometric determination of DNA content in *Nicotiana megachromosomes*. *Canadian Journal of Genetics and Cytology, 12*, 769–778.

Corbaz, R. (1962). Research on the control of blue mould of tobacco. *CORESTA Congress Bulletin Information Guangzhou*, 9–19, AP-2.

Crowder, B. J., Wilkinson, C. A., Johnson, C. S., & Eisenbach, J. D. (2003). Inheritance of resistance to tobacco cyst nematode in flue-cured tobacco. *Crop Science, 43*, 1305–1312.

Czubacka, A. (2022). The use of the Polish germplasm collection of *Nicotiana tabacum* in research and tobacco breeding for disease resistance. *Agriculture, 12*. https://doi.org/10.3390/agriculture12121994

Dang, J., Wang, J., Yang, Y., Shang, W., Guo, K., & Liang, G. (2019). Resistance of *Nicotiana tabacum* to Phytophthora parasitica var. nicotianae Race 0 Is enhanced by the addition of *N. plumbaginifolia* chromosome 9 with a slight effect on host genomic expression. *Crop Science, 59*, 2667–2678. https://doi.org/10.2135/cropsci2019.01.0005

Davis, E. L., Rich, J. R., & Gwynn, G. R. (1988). Reaction of selected *Nicotiana* spp. × *Nicotiana tabacum* crosses grown in microplots to 3 *Meloidogyne* spp. *Nematropica, 18*, 109–116.

Depta, A., Kursa, K., Doroszewska, T., Laskowska, D., & Trojak-Goluch, A. (2018). Reaction of *Nicotiana* species and cultivars of tobacco to Tobacco mosaic virus and detection of the N gene that confers hypersensitive resistance. *Czech Journal of Genetics and Plant Breeding, 54*, 143–146. https://doi.org/10.17221/81/2017-CJGPB

DeVerna, J. W., Myers, J. R., & Collins, G. B. (1987). Bypassing prefertilization barriers to hybridization in *Nicotiana* using in vitro pollination and fertilization. *Theoretical and Applied Genetics, 73*, 665–671.

Dmitrov, B., & Krusteva, D. (1998). Resistance of tobacco to tobacco aphids (*Myzus nicotianae* Blackman) originating from *N. gossei. CORESTA Congress Bulletin Specifications, Brighton, 1998, 152, PPOST14.*

Dorossiev, L., Palakarcheva, M., & Yancheva, A. (1990). Application of *in vitro* methods in the development of disease resistant Oriental tobacco hybrids and lines. *Genetics and Breeding, 23*, 306–315.

Doroszewska, T. (2010). Transfer of tolerance to different Potato virus Y (PVY) isolates from *Nicotiana africana* Merxm. to *Nicotiana tabacum* L. *Plant Breeding, 129*, 76–81.

Doroszewska, T., & Berbeć, A. (1999). Transfer of resistance to Potato Virus Y to tobacco cultivars. *CORESTA Meet Agro-Phyto Groups, Suzhou, China AP4.*

Doroszewska, T., & Berbeć, A. (2000). Cytogenetical investigations of polyploid interspecific hybrids of *Nicotiana africana* with different cultivars of *N. tabacum. Journal of Genetics and Breeding, 54*, 77–82.

Doroszewska, T., & Depta, A. (2011). Resistance of wild *Nicotiana* species to different PVY isolates. *Phytopathologia, 59*, 9–24.

Doroszewska, T., & Przybyś, M. (2007). Charakterystyka odporności gatunków *Nicotiana* na czarną zgniliznę korzeni *Thielaviopsis basicola* [BERK. and BROOME] FERR (in Polish). *Zeszyty Problemowe Postepów Nauk Rolniczych, 517*, 253–266.

Drake, K., & Lewis, R. S. (2013). An Introgressed *Nicotiana rustica* genomic region confers resistance to Phytophthora *Nicotiana*e in cultivated tobacco. *Crop Science, 53*, 1366–1374.

Drake, K. E., Moore, J. M., Bertrand, P., Fortnum, B., Peterson, P., & Lewis, R. S. (2015). Black shank resistance and agronomic Performance of fue-Cured tobacco lines and hybrids carrying the Introgressed *Nicotiana rustica* region, Wz. *Crop Science, 55*, 79–86.

Eickholt, D. P., & Lewis, R. S. (2014). Effect of an introgressed *Nicotiana tomentosa* leaf number QTL on yield and quality characteristics in flue-Cured tobacco. *Crop Science, 54*, 1–9. https://doi.org/10.2135/cropsci2013.07.0464

Engvild, K. C. (1985). Pollen irradiation and possible gene transfer in *Nicotiana* species. *Theoretical and Applied Genetics, 69*, 457–461.

Evans, D. A., Flick, C. E., & Jensen, R. A. (1981). Disease resistance: Incorporation into sexually incompatible somatic hybrids of the genus *Nicotiana. Science, 213*, 907–909.

Fardy, A., & Hitier, H. (1945). Espèces tétraploides et hybrides interspécifiques amphidiploides et triples diploides ed *Nicotiana*, obtenus par l'action de la colchicine. *Mémorial du Serv d'Expl Industr des Tabacs et des Allumettes (SEITA) Series B*, 1–117

Furusato. (1960). Breeding of amphidiploids from a hybrid between *Nicotiana rustica* and *N. tabacum. National Institute of Genetics Annual Report, 11*, 61.

Gajos, Z. (1975). Rozmnażanie roślin międzygatunkowego mieszańca *Nicotiana tabacum* L. x *Nicotiana alata* Link z kalusa wytwarzanego in vivo (Propagation of the interspecific hybrid *Nicotiana tabacum* L. x *Nicotiana alata* Link. from callus produced *in vivo*) (in Polish). *Biul Inf Centr Lab Przem Tyton, 1–2*, 15–20.

Gajos, Z. (1976). Odporność międzygatunkowych mieszańców Nicotiana tabacum x N. glauca na zakażenie wirusem Lycopersicum virus 3 Smith (On the resistance of interspecific hybrids *N. tabacum* x *N. alata* to infection by tomato spotted wilt virus) (*Lycopersicum virus* 3 Smith) (in Polish). *Zeszyty Problemowe Postepów Nauk Rolniczych, 182*, 83–88.

Gajos, Z. (1979). Próby wykorzystania mieszańców *Nicotiana tabacum* L. x *Nicotiana otophora* Gris. w hodowli tytoniu odpornego na *Peronospora tabacina* Adam (PT-2) i na inne choroby. (Attempt to use hybrids of *Nicotiana tabacum* L. × *N. otophora* Gris for breeding tobacco resistant to Peronospora tabacina Adam (PT-2) and to other diseases) (in Polish). *Biul Inf Centr Lab Przem Tyton, 1–2*, 11–23.

Gajos, Z. (1981). Przeniesienie odporności na wirus brązowej plamistości pomidora (Tomato spotted wilt virus) z Nicotiana alata Link et Otto do tytoniu szlachetnego przez skrzyżowanie obu gatunków (Transfer of resistance to tomatto spotted wilt virus from *N. Nicotiana alata* Link et Otto to tobacco by crossing the two species) (in Polish). *Biul Inf Centr Lab Przem Tyton, 1–2*, 3–24.

Gajos, Z. (1984). Genetyczna odporność międzygatunkowych mieszańców *Nicotiana tabacum* L. × *Nicotiana alata* link. et Otto na czarną zgniliznę korzeni (Thielaviopsis basicola Ferr). (Genetic resistance to black root rot (*Thielaviopsis basicola* Ferr.) of the interspecific hybrids of *Nicotiana tabacum* L. × *Nicotiana alata* link. et Otto (in Polish). *Biul Inf Centr Lab Przem Tyton, 1–4*, 3–14.

Georgeva, Y. (1996). Relation between phenolic compounds in tobacco leaves and the resistance to aphids. *Proceedings of Balkan Tobacco Conference Biotechnology & Biotechnological Equipment, Suppl 10*, 56.

Gerstel, D. U. (1943). Inheritance in *Nicotiana tabacum*. XVII. Cytogenetical analysis of *glutinosa*-type resistance to mosaic disease. *Genetics, 28*, 533–556.

Gerstel, D. U. (1945a). Inheritance in *Nicotiana tabacum*. XIX. Identification of the *tabacum* chromosome replaced by one from *N. glutinosa* in mosaic-resistant Holmes Samsoun tobacco. *Genetics, 30*, 448–454.

Gerstel, D. U. (1945b). Inheritance in *Nicotiana tabacum*. XX. The addition of *Nicotiana glutinosa* chromosomes to tobacco. *The Journal of Heredity, 36*, 197–206.

Gerstel, D. U. (1946). Inheritance in *Nicotiana tabacum* XXI. The mechanism of chromosome substitution. *Genetics, 31*, 421–427.

Gerstel, D. U. (1960). Segregation in new allopolyploids of *Nicotiana*. Comparison of 6x (*N. tabacum* × *N. tomentosiformis*) and 6x (*N. tabacum* × *N. otophora*). *Genetics, 45*, 1713–1734.

Gerstel, D. U. (1961). Chromosomal control of preferential pairing in *Nicotiana*. *Science, 133*, 579–580.

Gerstel, D. U., & Burk, L. G. (1960). Controlled introgression in *Nicotiana*. A cytological study. *Tobacco Science, 4*, 147–150.

Gerstel, D. U., & Burns, J. A. (1966). Flower variegation in hybrids between *Nicotiana tabacum* and N. *otophora*. *Genetics, 53*, 551–567.

Gerstel, D. U., & Burns, J. A. (1967). Phenotypic and chromosomal abnormalities associated with the introduction of heterochromatin from *Nicotiana otophora* into *N. tabacum*. *Genetics, 56*, 483–502.

Gerstel, D. U., & Burns, J. A. (1968). Flower variegation in *Nicotiana* hybrids: Reproductive instability. *Tobacco Science, 12*, 23–24.

Gerstel, D., & Burns, J. (1983). Anther restoration in male sterile tobacco by nucleolus organizing chromosomes: A restorer from *Nicotiana bigelovii*. *Tobacco Science, 27*, 151–153.

Gerstel, D. U., & Parry, D. C. (1973). Production and behavior of nullisomic S in *Nicotiana tabacum*. *Tobacco Science, 17*, 78–79.

Gerstel, D. U., Burns, J. A., & Burk, L. G. (1978). Cytoplasmic male sterility in *Nicotiana*, restoration of fertility and the nucleolus. *Genetics, 89*, 157–169.

Giddings, G. D., & Rees, H. (1992). A Nicotiana gametosomatic hybrid and its progenies. A *Nicotiana* gametosomatic hybrid and its progenies. *Journal of Experimental Botany, 43*, 419–425.

Gillham, F. E. M., Wark, D. C., & Harrigan, E. K. S. (1977). Disease resistant flue-cured tobacco breeding lines for north Queensland. I. Resistance to blue mould, Peronospora tabacina. *Australian Journal of Experimental Agriculture and Animal Husbandry, 17*, 652–558.

Głażewska, Z. (1977). Resistance of wild *Nicotiana* species and cultivars of *N. tabacum* and *N. rustica* to necrotic strains of PVY (in Polish). *Materiały XVII Sesji Naukowej IOR*, 277–287.

Goins, R. B., & Apple, J. L. (1970). Inheritance and phenotypic expression of a dominant factor for black shank resistance from *Nicotiana plumbaginifolia* in a *Nicotiana tabacum* milieu. *Tobacco Science, 14*, 7–11.

Goodspeed, T. H. (1954). *The Genus Nicotiana*. Chronica Botanica Co.

Gopinath, D. M., Krishnamurthy, K. V., & Krishnamurthy, A. S. (1965). Cytological studies in interspecific hybrids of N*icotiana* involving a new Australian species, *Nicotiana amplexicaulis*. *Canadian Journal of Genetics and Cytology, 7*, 328–340.

Greenleaf, W. H. (1941). Sterile amphidiploids: Their possible relation to the origin of *Nicotiana tabacum*. *The American Naturalist, 75*, 394–399.

Gupton, C. L., Gooding, G. V., & Corbin, T. C. (1982). Tolerance to potato virus Y and tobacco vein mottling virus in black shank resistant burley tobacco (*Nicotiana tabacum* L.). *Tobacco Science, 26*, 89–93.

Gwynn, G. R., Barker, K. R., Reilly, J. J., & Komm, D. A. (1986). Genetic resistance to tobacco mosaic virus, cyst nematodes, root-knot nematodes, and wildfire from *Nicotiana repanda* incorporated into *Nicotiana tabacum*. *Plant Disease, 70*, 958–962.

Haji, H. M., Brammall, R. A., & VanHooren, D. L. (2003). Effect of the *Nicotiana debneyi* black root rot resistance gene on the yield and quality characteristics of flue-cured tobacco in Ontario. *Canadian Journal of Plant Science, 83*, 939–942.

Haji, H., Mishra, S., & DeVos, M. (2005). Canadian tobacco hybrids: Where do we stand? *Mao Agriculture Consulting*. http://satg.org/maoconsulting/Canadian%20Tobacco%20Hybrids.htm

Haji, H. M., Misha, S., & Devos, M. (2006). Host plant resistance management strategies for control of black root rot resistance. *CORESTA Meet Agro-Phyto Groups, Paris, PPost 08*.

Hancock, W. G., & Lewis, R. S. (2017). Heterosis, transmission genetics, and selection for increased growth rate in a *N. tabacum* × synthetic tobacco cross. *Molecular Breeding, 37*, 53. https://doi.org/10.1007/s11032-017-0654-4

Hancock, W. G., Kuraparthy, V., Kernodie, S. P., & Lewis, R. S. (2015). Identification of maternal haploids of *Nicotiana tabacum* aided by transgenic expression of green fluorescent protein: evidence for chromosome elimination in the N-*tabacum* × N-*africana* interspecific cross. *Molecular Breeding, 35*, 179.

Hayes, A. J., Wilkinson, C. A., & Johnson, C. S. (1997). Evaluation of tobacco accessions for resistance to tobacco cyst nematode and wildfire. *Crop Science, 37*, 586–591.

Heggestad, H. E., Clayton, E. E., Neas, M. O., & Skoog, H. A. (1960). Development of Burley 21, the first wildfire-resistant tobacco variety, including results of variety trials. *TN Agricultural Experiment Station Bulletin, 321*.

Hendrix, J. W., & Apple, J. L. (1967). Stem resistance to *Phytophtora parasitica* var. *nicotianae* in tobacco derived from *Nicotiana longiflora* and *N. plumbaginifolia*. *Tobacco Science, 11*, 148–150.

Hillman, W. S., & Smith, H. H. (1965). Induced flowering in a vegetative tobacco hybrid: Response of a previously non-flowering *Nicotiana* clone to gibberellic acid, low temperature and short days. *The Journal of Heredity, 56*, 3–6.

Hitier, H., & Izard, C. (1950). Observations on the karyological behaviour of some *Nicotiana* hybrids. Observations sur le comportement caryologique de quelques hybrides de *Nicotiana*. *CR Academy Science Paris, 230*, 1305–1310.

Hitier, H., & Izard, C. (1958). Essais de lutte contre la mosaique du tabac par le chimiotherapie et par l'hybridation et la selection genetique. *Annals of Institute Experiement Tabac, Bergerac, 3*, 85–96.

Holmes, F. O. (1936). Interspecific transfer of a gene governing type of response to tobacco mosaic infection. *Phytopathology, 26*, 1007–1014.

Holmes, F. O. (1937a). Genes affecting response of *Nicotiana tabacum* hybrids to tobacco mosaic virus. *Science, 85*, 104–105.

Holmes, F. O. (1937b). Hereditary factors affecting tobacco-mosaic disease in solanaceous plants. *Phytopathology, 27*, 131–132.

Holmes, F. O. (1938). Inheritance of resistance to tobacco-mosaic disease in tobacco. *Phytopathology, 28*, 553–561.

Hosfield, G. L., & Wernsman, E. A. (1974). Effect of alien cytoplasm and fertility restoring factor on growth, agronomic characters and chemical constituents in a male sterile variety of flue-cured tobacco. *Crop Science, 14*, 575–577.

Iancheva, T. (1989). Inheritance of the resistance to tomato spotted wilt virus, *Nicotiana* virus 1 Smith, and *Erisyphe cichoracearum* d.c originating from the wild species *Nicotiana sanderae* in interspecific crosses with *Nicotiana tabacum*. *Genetics Sel, 22*, 412–419.

Ichikawa, T., Ozeki, Y., & Syono, K. (1990). Evidence for the expression of the rol genes of *Nicotiana glauca* in genetic tumors of *N. glauca* × *N. langsdorffii*. *Molecular & General Genetics, 220*, 177–180.

Iizuka, T., Kuboyama, T., Marubashi, W., Oda, M., & Tezuka, T. (2012). *Nicotiana debneyi* has a single dominant gene causing hybrid lethality in crosses with *N. tabacum*. *Euphytica, 186*, 321–328.

Ivancheva-Gabrovska, T. (1978). Response of *Nicotiana* species to tomato spotted wilt virus and *Thrips tabaci* Lind. *1st National Symposium Plant Immunity Report*, pp. 67–78.

Ivancheva-Gabrovska, T., & Kutova, I. (1981). Resistance of *Nicotiana species* to black root rot (*Thielaviopsis basicola*). *Rasten Nauk, 18*, 128–134. [*CORESTA Information Bulletin, (2)*, 97].

Ivancheva-Gabrovska, T., & Manolov, A. (1982). Investigations on the resistance of interspecific hybrids *N. tabacum* x *N. alata* and *N. tabacum* x *N. sanderae* to tomatto spotted wilt virus (in Bulgarian). *2nd National Symposium Plant Immunity Rep Plovdiv, 2*, 65–78.

Iwai, S., Kishi, C., Nakata, K., & Kubo, S. (1985). Production of a hybrid of *N. repanda* × *N. tabacum* by ovule culture. *Plant Science, 41*, 175–178.

Izard, C. (1952). Recherches sur les tumeurs hereditaires des hybrides interspecifiques de Nicotiana. *CR Academy Science Paris, 234*, 1305–1306.

Izard, C. (1970). A propos del'hybride tumoral *N. glauca* × *N. langsdorffii*, du mutant non tumoral et d'une variante obtenue avec l'espece *N. glauca* var. *laterita*. *CR Academy Science Paris, 270*, 2296–2298.

Izard, C., & Schiltz, P. (1964). Recherche de varietes resistantes a *Peronospora tabacina* a partir de l'amphidiploide *N. suaveolens* × *N. tabacum* var. *chinensis*. *SEITA Annals, 1*, 105–113.

Izard, C., Lacharpague, J., & Schiltz, P. (1964). Recherche de varietes resistantes a *Peronospora tabacina* a partir de l'amphidiploide *N. suaveolens* x *N. tabacum* var. *chinensis*. *Etudes CORESTA. Vienne. CORESTA Information Bulletin, 1028.*

Jack, A. M. (2010). Breeding for resistance to *Meloidogyne javanica* in Zimbabwe. 44th Tobacco Workers' Conference.

Johnson, C. S., & Main, C. E. (1983). Yield/quality trade-offs of tobacco mosaic virus – Resistant tobacco cultivars in relation to disease management. *Plant Disease, 67*, 886–890.

Johnson, E. S., Wolff, M. F., & Wernsman, E. A. (2002a). Marker-assisted selection for resistance to black shank disease in tobacco. *Plant Disease, 86*, 1303–1309.

Johnson, E. S., Wolff, M. F., Wernsman, E. A., Atchley, W. E., & Shrew, H. D. (2002b). Origin of the black shank resistance gene, Ph, in tobacco cultivar Coker 371-Gold. *Plant Disease, 86*, 1080–1084.

Johnson, E. S., Wernsman, E. A., & LaMondia, J. A. (2009). Effect of a chromosome fragment marked by the *Php* gene for resistance to *Phytophtora nicotianae* on reproduction of tobacco cyst nematodes. *Plant Disease, 93*, 309–315.

Julio, E., Verrier, J. L., & Dorlhac de Borne, F. (2006). Development of SCAR markers linked to three disease resistances based on AFLP within Nicotiana tabacum L. *Theoretical and Applied Genetics, 112*, 335–346.

Julio, E., Cotucheau, J., Decorps, C., Laurent, T., & Dorlhac de Borne, F. (2015). Characterisation of the Rk1 resistance to Meloidogyne in tobacco by RNA-Seq analysis. *CORESTA Agro/Phyto Meeting, Izmir. AP 29.*

Kehr, A. E. (1950). Genetic tumors in *Nicotiana*. *The American Naturalist, 85*, 51–64.

Kehr, A. E. (1951). Genetic Tumors in *Nicotiana*. *The American Naturalist, 99*, 73–79.

Kehr, A. E., & Smith, H. H. (1952). Multiple genome relationships in *Nicotiana*. *Cornell University Agricultural Experimental Station Memoir, 311*, 3–19.

Kehr, A. E., & Smith, H. H. (1954). Genetic tumors in *Nicotiana* hybrids. *Brookhaven Symposia in Biology, 6*, 55–78.

Kennedy, B. S., & Nielsen, M. T. (1993). Characterization of tomato spotted wilt virus (TSWV) resistance in the tobacco cultivar 'Polalta' (Abstr.). *Phytopathology, 83*, 1420.

Kenward, K. D., Bai, D., Ban, M. R., & Brandle, J. E. (1999). Isolation and characterization of Tnd-1, a retrotransposon marker linked to black root rot resistance in tobacco. *Theoretical and Applied Genetics, 98*, 387–395.

Keum, W. S., Chung, Y. H., Choi, S. J., Myung, J. C., & Lee, S. C. (1991). Inheritance of resistance to Potato Virus Y – Necrosis strain of *Nicotiana africana*. *Journal of the Korean Society of Tobacco Science, 13*, 48–51.

Keum, W. S., Chung, Y. H., Choi, S. J., Myung, J. C., & Lee, S. C. (1994). Transfer of Potato Virus Y (necrotic strain) resistance from *Nicotiana africana* to *Nicotiana tabacum*. *Journal of Korean Society Tobacco Science, 16*, 102–107.

Kitamura, S., Tanaka, A., & Inoue, M. (2005). Genomic relationships among *Nicotiana* species with different ploidy levels revealed by 5S rDNA spacer sequences and FISH/GISH. *Genes & Genetic Systems, 80*, 251–260.

Knoche, K. K., Clayton, M. K., & Fulton, R. W. (1984). Comparison of resistance in tobacco to *Pseudomonas syringae* pv. races 0 and 1 by infectivity titrations and bacterial multiplication. *Phytopathology, 77*, 1364–1368.

Kobus, I. (1973). Preliminary study of the breeding value of an interspecific tobacco hybrid (in Polish). *Pamiętnik Puławski, 58*, 261–276.

Koelle, G. (1961). Genetische Analyse einer Y-virus (Rippenbraun) resistenten Mutante der Tabaksorte Virgin A. *Der Züchter, 31*, 71–72.

Komm, D. A., & Terrill, T. R. (1982). A rapid screening technique for the determination of resistance in *Nicotiana tabacum* to a tobacco cyst nematode. *Globodera solanacearum. CORESTA Information Bulletin*, 64–65.

Korbecka-Glinka, G., Czubacka, A., Przybys, M., & Doroszewska, T. (2017). Resistance vs. tolerance to *Potato virus* Y in tobacco – Comparing effectiveness using virus isolates from Central Europe. *Breeding Science, 67*, 459–465. https://doi.org/10.1270/jsbbs.17019

Korbecka-Glinka, G., Trojak-Goluch, A., Doroszewska, T., & Goepfert, S. (2018). The influence of *Nicotiana alata*-derived introgression on plant malformations of tobacco breeding lines resistant to tomato spotted wilt virus. *CORESTA Congress, Kunming. Agro/Phyto Groups, AP 49.*

Korbecka-Glinka, G., Trojak-Goluch, A., Bakaher, N., & Goepfert, S. (2021). Agronomic performance of Polalta-derived breeding lines resistant to tomato spotted wilt virus. *CORESTA Agro/Phyto Meet, AP 11.*

Kostoff, D. (1930). Tumours and other malformations on certain *Nicotiana* hybrids. *Zbl Bakt II Abt, 81*, 244–260.

Kostoff, D. (1931). A fertile triple hybrid, *Nicotiana tabacum* × (*Nicotiana sylvestris* × *Nicotiana rusbyi*). Preliminary report. *American Journal of Botany, 18*, 112–113.

Kostoff, D. (1933). Cytogenetic studies of the triple fertile hybrid *N. tabacum* × (*N. sylvestris* × *N. rusbyi*) – *N. triplex. Bulletin of Applied Botany, Genetics and Plant Breeding, S6*, 2(5), 167–205.

Kostoff, D. (1935). Heritable tumors in plants experimentally produced. *Genetica, 17*, 367–376.

Kostoff, D. (1936). The origin of cultivated tobacco. *Current Science, 4*, 872.

Kostoff, D. (1937). Studies on polyploid hybrid plants. XVI. *Nicotiana rustica* x *Nicotiana tabacum* amphidiploid. *CR Academy Science URSS, 11*, 453–455.

Kostoff, D. (1943). *Cytogenetics of the genus Nicotiana*. State Printing House.

Kostoff, D., & Georgieva, R. (1944). Resistance to tobacco mosaic virus. II. Inheritance of necrotic reactions and plant breeding value of of the strains *N. tabacum* var. *virii*. In *Centr Agr Exp Inst.*

Kostoff, D., & Radjably, I. (1935). Cytogenetic studies of certain composite hybrids in *Nicotiana*. *Bulletin of Institute of General USSR, 10*, 29–63.

Kramer, M. G., & Reed, S. M. (1988). An evaluation of maternal nullihaploidy in the production of nullisomics of *Nicotiana tabacum* L. II. A pollen irradiation and ovule culture approach. *The Journal of Heredity, 79*, 469–472.

Krishnamurthy, K. V., Murty, G. S., & Appa Rao, K. (1960). Cytogenetics of the trispecific hybrid *Nicotiana tabacum* × (*N. glutinosa* × *N. trigonophylla*) and its reciprocal. *Euphytica, 9*, 111–121.

Krusteva, D., Dimitrov, B., & Nikova, V. (2003a). *Nicotiana gossei* Domin as a source of aphid resistance in *Nicotiana tabacum* (in Bulgarian). *Tiutun (Tobacco), 53*, 217–220.

Krusteva, D., Nikova, V., Vladova, R., Dimitrov, B., Petkova, A., & Iancheva, A. (2003b). *Nicotiana benthamiana* Domin. As a source of aphid resistance in *N. tabacum* L. *CORESTA Agro-Phyto Meet Bucharest APOST11.*

LaMondia. (2010). Management of the tobacco cyst nematode on cigar wrapper tobaccos in Connecticut. *44th tobacco workers' conference.*

Larkina, N. I. (2015). *Scientific principles of interspecific hybridization using maternal Nicotiana tabacum* (Monograph). All-Russian Institute of Tobacco and Makhorka. ISBN 978-5-93491-681-8.

Laskowska, D., & Berbeć, A. (2003). Preliminary study of the newly discovered tobacco species *Nicotiana wuttkei* Clarkson and Symon. *Genetic Resources and Crop Evolution, 50*, 835–839.

Laskowska, D., & Berbeć, A. (2005). Cytology and fertility of viable hybrids of *Nicotiana tabacum* L. cv. TB-566 with *N. alata* Link et Otto. *Journal of Applied Genetics, 46*, 11–18.

Laskowska, D., & Berbeć, A. (2010). TSWV resistance in DH lines of tobacco (Nicotiana tabacum L.) obtained from a hybrid between Polalta and Wiślica. *Plant Breeding, 129*, 731–733.

Laskowska, D., Doroszewska, T., Depta, A., Kursa, K., Olszak-Przybys, H., & Czubacka, A. (2013). A survey of *Nicotiana* germplasm for resistance to Tomato Spotted Wilt Virus (TSWV). *Euphytica, 193*, 207–219.

Laskowska, D., Berbeć, A., Van Laere, K., Kirov, I., Czubacka, A., & Trojak-Goluch, A. (2015). Cytology and fertility of amphidiploid hybrids between *Nicotiana wuttkei* Clarkson et Symon and *N. tabacum* L. *Euphytica, 206*, 597–608. https://doi.org/10.1007/s10681-015-1459-3

Lea, H. W. (1963). The transfer of resistance against blue mold (*Peronospora tabacina* Adam) from *Nicotiana debneyi* to cultivated tobacco. *CORESTA Information Bulletin, 3*, 13–15.

Lee, C. B., Page, L. E., McClure, B. A., & Holtsford, T. P. (2008). Post-pollination hybridization barriers in *Nicotiana* section *Alatae*. *Sexual Plant Reproduction, 21*, 183–195.

Legg, P. D., & Smeeton, B. W. (1999). II. Breeding and genetics. In L. Davis & M. T. Nielsen (Eds.), *Tobacco: Production, chemistry and technology*. Blackwell Science Ltd.

Legg, P. D., Collins, G. B., & Litton, C. C. (1979). Effects of the N mosaic-resistance factor on agronomic and chemical traits in Burley tobacco. *Crop Science, 19*, 455–457.

Legg, P. D., Litton, C. C., & Collins, G. B. (1981). Effects of *N. debneyi* black root rot resistance factor on agronomic and chemical traits of burley tobacco. *Theoretical and Applied Genetics, 60*, 365–368.

Legg, P. D., Litton, C. C., & Collins, G. B. (1982). Effects of *Nicotiana longiflora* Cav. resistance to race 0 Phytophthora parasitica var. *Nicotianae* on agronomic and chemical traits in burley tobacco. *Crop Science, 22*, 35–38.

Lehmann, H. (1936). On interspecific cross sterility in the genus *Nicotiana* (in German). *Plant Breed*, Abst 7 (1936), abst 1023.

Lewis, R. S. (2002). Tissue culture and molecular marker-assisted introgression of PVY resistance from *N. africana* to *N. tabacum. CORESTA Congress, New Orleans, Agro-Phyto Groups, AP28.*

Lewis, R. S. (2005). Transfer of resistance to Potato Virus Y (PVY) from *Nicotiana africana* to *Nicotiana tabacum*: Possible influence of tissue culture on the rate of introgression. *Theoretical and Applied Genetics, 110*, 678–687.

Lewis, R. S. (2007). Evaluation of *Nicotiana tabacum* genotypes possessing *Nicotiana africana*-derived genetic tolerance to Potato Virus Y. *Crop Science, 47*, 1975–1984.

Lewis, R. S. (2011). *Nicotiana*. In C. Kole (Ed.), *Wild crop relatives: Genomic and breeding resources, plantation and ornamental crops*. Springer. https://doi.org/10.1007/978-3-642-21201-7_10

Lewis, R. S. (2020a). Nicotiana tabacum L.: Tobacco. In J. Novak & W. D. Blüthner (Eds.), *Medicinal, aromatic and stimulant plants. Handbook of plant breeding* (Vol. 12). Springer. https://doi.org/10.1007/978-3-030-38792-1_9

Lewis, R. S. (2020b). Tobacco. In C. Kole (Ed.), *Wild crop relatives: Genomic and breeding resources, plantation and ornamental crops* (Vol. 2011, pp. 185–206). Springer.

Lewis, R. S., & Rose, C. (2010). Agronomic performance of tobacco mosaic virus-resistant tobacco lines and hybrids possessing the resistance gene N introgressed on different chromosomes. *Crop Science, 50*, 1339–1347.

Lewis, R. S., & Wernsman, E. A. (2001). Efforts to initiate construction of a disease resistance package on a designer chromosome in tobacco. *Crop Science, 41*, 1420–1427.

Lewis, R. S., Linger, L. R., Wolff, M. F., & Wernsman, E. A. (2007a). The negative influence of N-mediated TMV resistance on yield in tobacco: Linkage drag versus pleiotropy. *Theoretical and Applied Genetics, 115*, 169–178.

Lewis, R. S., Milla, S. R., & Kernodle, S. P. (2007b). Analysis of an introgressed *Nicotiana tomentosa* genomic region affecting leaf number and correlated traits in *Nicotiana tabacum. Theoretical and Applied Genetics, 114*, 841–854.

Li, B. C., Bass, W. T., & Cornelius, P. L. (2006). Resistance to tobacco black shank in Nicotiana species. *Crop Science, 46*, 554–560.

Lilienfeld, F. A. (1952). Cytogenetical problems in the genus *Nicotiana. National Institute of Genetics Annual Report, 3*, 38–39.

Lilienfeld, F. A. (1953). Cytogenetical problems in *Nicotiana.* b. Continued investigation of synthesized *N. tabacum. National Institute of Genetics Annual Report, 4*, 33.

Lim, K. Y., Souckova-Skalicka, K., Sarasan, V., Clarkson, J. J., Chase, M. W., Kovarik, A., & Leitch, A. R. (2006). A genetic appraisal of a new synthetic *Nicotiana tabacum* (Solanaceae) and the Kostoff synthetic tobacco. *American Journal of Botany, 93*, 875–883.

Liu, H., & Marubashi, W. (2014). Species origin of genomic factors in *Nicotiana nudicaulis* Watson controlling hybrid lethality in interspecific hybrids between *N. nudicaulis* Watson and *N. tabacum* L. *PLoS ONE, 9*, e97004. https://doi.org/10.1371/journal.pone.0097004

Lucas, G. B. (1975). *Diseases of tobacco*. Biological Consulting Associates.

Lucas, G. B., Gooding, G. V., Sasser, J. N., & Gerstel, D. U. (1980). Reaction of *Nicotiana africana* to black shank, Granville wilt, root knot, tobacco mosaic virus and potato virus Y. *Tobacco Science, 24*, 141–142.

Lucov, Z., Cohen, S., & Moav, R. (1970). Effects of low temperature on the somatic instability of an alien chromosome in *Nicotiana tabacum. Heredity, 25*, 431–439.

Luneva, M. Z., & Poddubnaya Arnoldi, V. A. (1979), Morphological and cytoembryological study of F_1 diploids and amphidiploids Nicotiana *didebta* X *N. glauca. Probl otdalen gibridizatsii*, 235–246

Mackenzie, J., Smeeton, B. W., Jack, A. M., & Ternouth, R. A. F. (1986). Review on breeding for resistance to root knot, Meloidogyne javanica in flue-cured tobacco in Zimbabwe. *Bulletin Specifications of CORESTA Symposium Taormina, 64*, abstr P05.

Mallach, G. S. (1943). Inheritance in *Nicotiana tabacum*. XVI. Structural differences among the chromosomes of a selected group of varieties. *Genetics, 28*, 525–532.

Manolov, A. (1980). Transfer of *Peronospora* resistance from *Nicotiana exigua* to oriental tobacco (in Bulgarian). *Bulg Tiutun, 25*, 11–18.

Manolov, A. (1981). Transfer of complex resistance to blue mold and powdery mildew. *Bulg Tiutun, 12*, 9–17.

Manolov, A., Leshcheva, D., & Baldjzhiyeva, E. (1978a). Use of interspecific hybridization in tobacco for the development of initial material resistant to diseases. *1st National Symposium on Plant Immunity, 2*, 117–122.

Manolov, A., Slavkov, M., Leshcheva, D., & Baldjzhiyeva, E. (1978b). Use of interspecific hybridization in tobacco for creating an initial material with valuable economic qualities. *Bulletin Specifications of CORESTA Symposium Sophia, 87*, P13.

Marani, A., Fishler, G., & Amirar, A. (1971). The inheritance of resistance to blue mold (*Peronospora tabacina* Adam.) in two cultivars of tobacco (*Nicotiana tabacum* L). *Euphytica, 21*, 97–105.

Mattingly, C. F., & Collins, G. B. (1974). The use of anther derived haploids in *Nicotiana.* III. Isolation of nullisomics from nullisomic lines. *Chromosoma, 46*, 29–36.

Matveeva, T. V., & Lutova, L. A. (2014). Horizontal gene transfer from *Agrobacterium* to plants. *Frontiers in Plant Science, 5*, article 326.

Matzinger, D. F., & Wernsman, E. A. (1967). Genetic diversity and heterosis in *Nicotiana* I. Interspecific crosses. *Theoretical Applied Genetics (Der Zuchter), 37*, 188–191.

Mazur, M. (1975). Study on the inheritance of resistance to the necrotic strain of PVY (Y-N) and on the agronomic performance of several hybrids and cultivars of flue-cured tobacco infected by PVY (in Polish). *Biul Inf Centr Lab Przem Tyton*, (3–4), 13–33.

McCorkle, K. L., Drake-Stowe, K., Lewis, R. S., & Shew, D. (2018). Characterization of *Phytophthora nicotianae* resistance conferred by the introgressed *Nicotiana rustica* region, Wz, in flue-cured tobacco. *Plant Disease, 102*, 309–317.

McCray, F. A. (1932). Compatibility of certain *Nicotiana* species. *Genetics, 17*, 621–636.

McRae, N. A. (1941). Genetic analysis of *N. triplex* segregation and their relationship to existing *N. tabacum* types. *Technical Bulletin of Department of Agriculture Canada, 33*, 1–36.

Michel, V., Julio, E., Candresse, T., Cotucheau, J., Decorps, C., Volpatti, R., Moury, B., Glais, L., et al. (2018). NtTPN1: aRPP8-like R gene required for Potato virusY-induced veinal necrosis in tobacco. *The Plant Journal, 95*, 700–714. https://doi.org/10.1111/tpj.13980

Milla, S. R., Levin, J. S., Lewis, R. S., & Rufty, R. C. (2005). RAPD and SCAR markers linked to an introgressed gene conditioning resistance to *Peronospora tabacina* D.B. Adam. *Tobacco Crop Science, 45*, 2346–2354.

Moav, R. (1958). Inheritance in Nicotiana tabacum XXIX: relationship of residual-chromosome homology to interspecific gene transfer. *The American Naturalist, 92*, 267–278.

Moav, R., & Cameron, D. R. (1960). Genetic instability in *Nicotiana* hybrids. I. The expression of instability in *N. tabacum* × *N. plumbaginifolia. American Journal of Botany, 47*, 87–93.

Moav, R., & Cameron, D. R. (1961). Chromosome complement dosage in relation to seed development of species hybrids in *Nicotiana. Botanical Gazette, 123*, 70–77.

Modilevsky, J. (1939). Cytogenetic investigations of the genus Nicotiana. VII . Crossing amphidiploid *N. disualovii* with some species of genus *Nicotiana. Journal of Institute Botanical Academy of Science Ukraine, 21–22*(29–30), 107–137.

Moon, H., & Nicholson, J. S. (2007). AFLP and SCAR markers linked to tomato spotted wilt virus resistance in tobacco. *Crop Science, 47*, 1887–1894.

Moore, G. A., & Collins, G. B. (1982). Isolation of nullihaploids from diverse genotypes of *Nicotiana tabacum. The Journal of Heredity, 73*, 192–196.

Morgan, O. D. (1964). Experimental infection of *Nicotiana* species and interspecific crosses with *Cercospora nicotianae* and *Cercospora kikuchii. Plant Disease Report, 48*, 693–695.

Mudzengerere, E. T. (1994). Breeding for resistance to Meloidogyne javanica in Burley tobacco in Zimbabwe. *Bulletin Specifications of CORESTA Congress Harare, 114*, P30.

Murthy, T. G. K., Sredhaar, U., & Siva Raju, K. (2014). Incorporation of aphid resistance from *N. gossei. N. repanda*, N x. *umbratica-nesophila* and N × *benthamiana–repanda. CTRT annual report 2013-2014*, Rajamundry, India, pp. 17–20.

Murty, B. R., & Swaminathan, M. S. (1957). Cytogenetic studies in derivatives of *Nicotiana rustica* × *N. tabacum. Euphytica, 6*, 227–236.

Näf, U. (1958). Studies on tumour formation in *Nicotiana* hybrids I. The classification of parents into two etiologically significant groups. *Growth, 22*, 167–180.

Nagao, T. (1979). Somatic hybridization by fusion of protoplasts. II. Combinations of *Nicotiana tabacum* with *N. glutinosa* and of *N. tabacum* with *N. alata. Japanese Journal of Crop Science, 48*, 385–392.

Neal, J. W., Pittarelli, G. W., & Gott, K. M. (1987). *Nicotiana* species with high resistance to greenhouse whitefly *Trialeurodes vaporariorum. Tobacco Science, 31*, 61–62.

Ng'ambi, T. B. S., Rufty, R. C., Barker, K. R., & Melton, T. A. (1999). Identification of sources of resistance to four species of root-knot nematodes in tobacco. *Journal of Nematology, 31*, 272–282.

Nicoletti, R. (1999). La resistenza genetica al marciume nero delle radici del tabacco. *Il Tabacco, 7*, 15–19.

Nielsen, M. T., Legg, P. D., & Litton, C. C. (1985). Effects of two itntrogressed disease resistance factors on agronomic characteristics and certain chemical components in burley tobacco. *Crop Science, 25*, 698–701.

Nifong, J. M. (2008). *Assessment of the diversity and disease resistance properties within a collection of Nicotiana rustica*. MS thesis, NCSU Raleigh, NC.

Nikova, V. M., & Shabanov, D. S. (1988). *Nicotiana amplexicaulis* Burbidge as a source of cytoplasm male sterility in tobacco (*Nicotiana tabacum* L.). *Comptes Rendus de l 'Academie Bulgare des Sciences, 41*, 83–86.

Nikova, V., & Vladova, R. (2002). Wild *Nicotiana* species as a source of cytoplasmic male sterility in *Nicotiana tabacum*. *Beitr Tab Int/Contrib, 20*, 301–311.

Nikova, V., & Zagorska, N. (1987). Nicotiana *maritima* as a source of tobacco male sterility. *Genetics Selection, 20*, 224–231 [*CORESTA Information Bulletin, 1987*, 224–231].

Nikova, V. M., Zagorska, N. A., & Pundeva, R. S. (1991). Development of four tobacco male sterile sources using *in vitro* techniques. *Plant Cell, Tissue and Organ Culture, 27*, 289–295.

Nikova, V., Vladova, R., Pundeva, R., & Shabanov, D. (1997). Cytoplasmic male sterility in *Nicotiana tabacum* L. obtained through interspecific hybridization. *Euphytica, 94*, 375–378.

Nikova, V., Krusteva, D., Dimitrov, B., Pandeva, R., & Vladova, R. (2003). Investigation of the interspecific hybrid *Nicotiana tabacum* L. × *N. sanderae* Hort. obtained via in vitro techniques. *CORESTA Agro/Phyto Meet, Bucharest, APOST12*.

Niwa, M. (1969). Radiation induced interspecific transfer of ws(pbg) gene from *Nicotiana plumbaginifolia* to *N. tabacum*: II. Estimation of frequency of the gene transfer induced by irradiation at the gametogenetic stage. *Japan Journal of Breeding, 19*, 84–88.

Ohashi, Y. (1985). Breeding studies of wild species related to tobacco, with special reference to disease resistance. *Bulletin of Iwata Tobacco Experimental Station, 17*, 1–62.

Oka, H. (1961). A breeding study on interspecic transfer of disease resistance in tobacco. *Hatano Tobacco Experimental Station Bulletin, 49*.

Oka, K., & Ninomi, K. (1961). Interspecifie transfer of black shank resistance in tobacco: cytogenetical observation on pollen sterility in a black shank resistant allotrisomic plant with *N. plumbaginifolia* chromosome. *Japan Journal of Breeding, 11*, 245.

Olmo, H. P. (1935). Genetical studies of monosomic types of *Nicotiana tabacum*. *Genetics, 20*, 286–300.

Opoka, B. (1969). Odporność dzikich gatunków *Nicotiana* na *Lycopersicon virus* 3 (Brittlebank) Smith (Resistance of wild *Nicotiana* species to *Lycopersicon virus* 3 (Brittlebank) Smith) (in Polish). *Hod Rosl Aklim Nas, 13*, 83–87.

Oupadissakon, S., & Wernsman, E. A. (1977). Agronomic performance and nature of gene effects in progenitor-species derived genotypes of tobacco. *Crop Science, 17*, 843–847.

Palakarcheva, M. (1976). Disease resistant tobacco hybrids and varieties as a result of interspecific hybridization (in Bulgarian). *Rasten Nauk, 13*, 35–42.

Palakarcheva, M. (1981). Genetic peculiarities of tobacco breeding for disease resistance (in Bulgarian). *Genetics Sel, 14*, 277–284.

Palakarcheva, M. (1984). Interspecific hybridization in the genus *Nicotiana*. *Bulletin of Special CORESTA Congress Vienne, 142*, PP31.

Palakarcheva, M., & Bailov, D. (1976). Disease resistant hybrids and varieties of tobacco obtained from interspecific hybridization. *Rasten Nauk, 13*, 35–42. (*Review of Plant Pathology, 56*, 3705).

Palakarcheva, M., & Dorossiev, L. (1983). Amphidiploids *N. maritima* X *N. tabacum* (4n=80) and *N. noctiflora* X *N. tabacum* (4n=72), a useful source of disease resistance. *Rasten Nauk, 20*(3), 68–72.

Palakarcheva, M., & Krusteva, D. (1978). Study on the inheritance of powdery mildew (*Erysiphe cichoracearum* D.C.) resistance in intercultivar tobacco hybrids. *Genetics Sel, 11*, 126–134.

Palakarcheva, M., & Krusteva, D. (1995). Creation of isogenic tobacco lines, possessing gene N from *Nicotiana sanderae* W for resistance to tomato and tobacco strain of TMV (Nicotiana virus 1 Smith). *Comptes Rendus de l 'Academie Bulgare des Sciences, 48*, 65–68.

Palakarcheva, M., Yantcheva, A., & Delon, R. (1990). Study of cytoplasmic sterile male hybrids (CMS)-cytoplasm derived from *Nicotiana goodspeedii. Bulletin Specification of CORESTA Symposium, Kallithea, 141*, A32. ISSN.0525-6240.

Pandey, K. K. (1974). Overcoming interspecific pollen incompatibility through the use of ionising radiation. *Heredity, 33*, 279–284.

Pandey, K. K. (1975). Sexual transfer of specific genes without gametic fusion. *Nature, 256*, 310–313.

Pandey, K. K. (1978). Gametic gene transfer in *Nicotiana* by means of irradiated pollen. *Genetica, 49*, 53–69.

Pandey, K. K. (1980a). Parthenogenetic diploidy and egg transformation induced by irradiated pollen in *Nicotiana. New Zealand Journal of Botany, 18*, 203–207. https://doi.org/10.1080/0028825X.1980.10426918

Pandey, K. K. (1980b). Further evidence for egg transformation in *Nicotiana. Heredity, 45*, 15–29.

Pandey, K. K. (1981). Egg transformation through the use of irradiated pollen: poor man's genetic engineering. *Incompatibility Newsletters, 13*, 119–123.

Pandeya, R. S., & White, W. H. (1981). Nordel, a new bright tobacco cultivar. *Canadian Journal of Plant Science, 61*, 795–797.

Pandeya, R. S., & White, F. H. (1984). Delgold: A new flue-cured tobacco. *Canadian Journal of Plant Science, 64*, 233–236.

Pandeya, R. S., Douglas, G. C., Keller, W. A., Setterfield, G., & Patrick, Z. A. (1986). Somatic hybridization between *Nicotiana rustica* and *N. tabacum*: Development of tobacco breeding strains with disease resistance and elevated nicotine content. *Z Pflanzerzüchtg, 96*, 346–352.

Pandeya, R. S., Brandle, J. E., Ankersmit, J. C. D., Rogers, V. D., & Keller, W. A. (1991). Delfield flue-cured tobacco. *Canadian Journal of Plant Science, 71*, 575–577.

Parkunan, V., Johnson, C. S., & Eisenback, J. B. (2009). Effects of Php gene-associated versus induced resistance to tobacco cyst nematode in flue-cured tobacco. *Journal of Nematology, 41*, 261–266.

Parr, J. C., & Thurston, R. (1968). Toxicity of *Nicotiana* and *Petunia* species to the larvae of the tobacco hornworm. *Journal of Economic Entomology, 61*, 1525–1531.

Patel, K. A., & Gerstel, D. U. (1961). Additional information on the mechanism of chromosome substitution in *Nicotiana. Tobacco Science, 5*, 18–20.

Patrascu, M., Paunescu, D., & Ciuperca, A. (1999). Transfer of the resistance to TSWV from *Nicotiana alata* to some tobacco cultivars, using the *in vitro* culture technique. In *CORESTA Agro/Phyto Meeting Suzhou, China, POST8.*

Pereg, L. L. (2013). Black root rot of cotton in Australia: The host, the pathogen and disease management. *Crop & Pasture Science, 64*, 1112–1126. https://doi.org/10.1071/CP13231

Pittarelli, G. W., & Sisson, V. A. (1989). Registration of cytoplasmic male sterile tobacco germplasm Bel MS-2. *Crop Science, 29*, 836–837.

Pittarelli, G. W., & Stavely, J. R. (1975). Direct hybridization of *Nicotiana repanda* × *N. tabacum. The Journal of Heredity, 66*, 281–284.

Pollok, J. R., Johnson, C. S., Eisenback, J. D., & Reed, T. T. (2016). Reproduction of *Meloidogyne incognita* Race 3 on flue-cured tobacco homozygous for Rk1 and/or Rk2 resistance genes. *Journal of Nematology, 48*, 79–86.

Powell, J. R. (1979). Ovens 62. *Journal of Australian Agricutlural Science, 46*, 144.

Qin, Q., Li, Y., Martinez, N., Miller, R., Ding, N., Wu, X., Li, D., & Yang, S. (2016). Development of user-friendly markers for disease resistance to black root rot through genotyping by sequencing. *TSRC, Tobacco Science Research Conference 70*, abstr. 42.

Raeber, J. G., & Schweppenhauser, M. A. (1964). Breeding for resistance to root-knot nematode (*M. javanica*). *Tobacco Research Board Annual Report*, 1963–1964.
Raeber, J. G., & Smeeton, B. (1973a). Variety trials. *Tobacco Research Board Annual Report*, 1972–1973.
Raeber, J. G., & Smeeton, B. (1973b). Agronomic trial of standard broad leaf lines. *Tobacco Research Board Annual Report*, 1972–1973.
Raeber, J. G., & Smeeton, B. (1980). Breeding for resistance to root-knot. *Tobacco Research Board Annual Report*, 1979–1980.
Raeber, J. G., & Smeeton, B. (1984). Observation trial of RL-L-L lines (Root-knot and white mould resistant lines). *Tobacco Research Board Annual Report*, 1983–1984.
Ramavarma, K. T., Appa Rao, K., & Narayanan, A. I. (1976). Studies on the nullisomic progenies of a haploid plant in *Nicotiana tabacum*. *Tobacco Science, 20*, 187–188.
Ramavarma, K. T., Apparao, K., & Narrayanan, A. I. (1977). Cytogenetics of a trispecific hybrid (*N. benthamiana* × *N. glutinosa*) × *N. tabacum* and its alloploids in genus *Nicotiana*. *Cytologia, 42*, 703–709.
Ramavarma, K. T., Appa Rao, K., & Narayanan, A. I. (1978). Male sterile flower anomalies in an inter specific *Nicotiana* hybrid. *Tobacco Research, 4*, 29–33.
Ramavarma, K. T., Apparao, K., Sitharamaiah, S., & Narayanan, A. I. (1980). Interspecific transfer of resistance to tobacco caterpillar (Spodoptera litura F.) from *Nicotiana benthamiana* to *Nicotiana tabacum*. *Cytologia, 45*, 103–111.
Ramavarma, K. T., Apparao, K., Narayanan, A. I., Sitaramaiah, S., & Joshi, B. G. (1991). Evolving tobacco varieties resistant to the leaf-eating caterpillar *Spodoptera litura* F. through interspecific hybridization. *Tobacco Research, 17*, 29–32.
Rao, P. N., & Stokes, G. W. (1963). Role of chromosome H in the development of calcium deficiency symptoms in burley tobacco. *Crop Science, 3*, 261–264.
Rao, K. A., Ramavarma, K. T., & Joshi, B. G. (1980). Interspecific hybridization and breeding for pest resistance in tobacco. *Tobacco Science, 24*, 46–48.
Reed, S., & Collins, G. B. (1980). Chromosome pairing relationship and black shank resistance in three *Nicotiana* interspecific hybrids. *The Journal of Heredity, 71*, 423–426.
Renny Byfield, S., Chester, M., Kovarik, A., Le Comber, S., Grandbastien, M.-A., Deloger, M., Nichols, R. A., Macas, J., Novak, P., Chase, M. W., & Leitch, A. R. (2011). Next generation sequencing reveals genome downsizing in allotetraploid *Nicotiana tabacum*, predominantly through the elimination of paternally derived repetitive DNAs. *Molecular Biology and Evolution, 28*, 2843–2854.
Rufty, R. C. (1989). Genetics of host resistance to tobacco blue mold. In *Blue mold of tobacco* (pp. 141–164). APS Press.
Rufty, R. C., Wernsman, E. A., & Powell, N. T. (1983). A genetic analysis of the association between resistance to *Meloidogyne incognita* and a necrotic response to infection by a strain of potato virus Y in tobacco. *Phytopathology, 73*, 1413–1418.
Sabour, M., Simmonds, J., & Setterfield, G. (1986). Variation in nicotine content of cultured cell lines of *Nicotiana* species and their somatic and sexual hybrids. *Plant Breeding, 97*, 324–333.
Scholthof, K.-B. (2017). Spicing up the N gene: F. O. Holmes and tobacco mosaic virus resistance in *Capsicum* and *Nicotiana Plants*. *Phytopathology, 107*, 148–157.
Schweppenhauser, M. A. (1968). Recent advances in breeding tobacco resistant to *Meloidogyne javanica*. *CORESTA Information Bulletin, 1*, 9–20.
Schweppenhauser, M. A. (1974). Interspecific bridge transfer in *Nicotiana* of resistance to *Meloidogyne javanica*. *South African Journal of Science, 70*, 312–314.
Schweppenhauser, M. A. (1975). Rootknot resistance from *Nicotiana longiflora*. *Tobacco Science, 19*, 26–29.
Schweppenhauser, M. A., & Mann, T. J. (1968). Restoration of staminal fertility in *Nicotiana* by introgression. *Canadian Journal of Genetics and Cytology, 10*, 401–411.

Shabanov, D. S., & Nikova, V. M. (1984). Cytoplasmic male sterility in *Nicotiana tabacum* with cytoplasm from *Nicotiana rustica. Comptes Rendus de l 'Academie Bulgare des Sciences, 37*, 525–528.

Shabanov, D., Popchristev, W., & Tomov, N. (1974). Interspecific hybrid *Nicotiana tabacum* × *N. setchellii* (in Bulgarian). *Nauch Trud, 4*, 12–22.

Shava, J. G., Richardson-Kageler, S., Dari, S., Magama, F., & Rukuni, D. (2018). Incorporation of root-knot nematode (*Meloidogyne javanica* (Treub) Chitwood) resistance in the Zimbabwean elite Virginia tobacco (*Nicotiana tabacum L.*). *Agricultural Reviews, 39*, 333–337.

Sheen, S. J. (1972). Isozyme evidence bearing on the origin of *Nicotiana tabacum. Evolution, 26*, 143–154.

Shinkareva, I. K. (1979). Meiosis in hybrids of tobacco with the wild species *Nicotiana suaveolens* Lehm. *Soviet Genetics, 15*, 588–592 (*Genetika, 5*, 887–889).

Sievert, R. C. (1972a). Resistance to anthracnose in the genus *Nicotiana. Tobacco Science, 16*, 32–34.

Sievert, R. C. (1972b). Sources of resistance to potato virus Y in the genus *Nicotiana. Tobacco Science, 106*, 92–94.

Skalicka, K., Lim, K. Y., Matyasek, R., Matzke, M., Leitch, A. R., & Kovarik, A. (2005). Preferential elimination of repeated DNA sequences from the paternal, *Nicotiana tomentosiformis* genome donor of a synthetic allotetraploid. *The New Phytologist, 166*, 291–303.

Slana, L. J., & Stavely, J. R. (1978). High soil temperatures break resistance to *Nicotiana tomentosa* and NC 95 tobacco to *Meloidogyne incognita* in tobacco. *Phytopathology News, 67*, 537–543.

Slana, L. J., Stavely, J. R., Grosso, J. J., & Golden, A. M. (1977). Probable source of *Meloidogyne incognita* resistance in tobacco as indicated by reactions to five *Meloidogyne* isolates. *Phytopathology, 67*, 537–543.

Slusarkiewicz-Jarzina, A., & Zenkteler, M. (1983). Development of hybrid plants from ovules of *Nicotiana tabacum* pollinated in vitro with pollen grains of *Nicotiana knightiana. Experimentia, 39*, 1399–1400.

Smeeton, B. W., & Ternouth, R. A. F. (1990). Sources of resistance to powdery mildew, wildfire, angular leaf spot, and *Alternaria*. In *Bulletin Spec CORESTA Sympsium, Kallithea*, pp. 179, P22

Smith, H. H. (1950). Differential photoperiod response from an interspecific gene transfer. *The Journal of Heredity, 41*, 199–203.

Smith, H. H. (1958). Genetic plant tumors in *Nicotiana. Annals of the New York Academy of Sciences, 71*, 1163–1178.

Smith, H. H. (1968). Recent cytogenetic studies in the genus *Nicotiana. Advances in Genetics, 14*, 1–43.

Smith, H. H. (1988). The Inheritance of genetic tumors in *Nicotiana* hybrids. *The Journal of Heredity, 79*, 277–283.

Smith, H. H., & Abashian, D. V. (1963). Chromatographic investigations on the alkaloid content of *Nicotiana* species and interspecific combinations. *American Journal of Botany, 50*, 435–447.

Smith, H. H., Stevenson, H. Q., & Kehr, A. E. (1958). Limits and consequences of multiple allopolyploidy in *Nicotiana. The Nucleus (Calcutta), 1*, 205–222.

Spasoff, L., Fox, J. A., & Miller, L. I. (1971). Multigenic inheritance of resistance to Osborne's cyst nematode in tobacco. *Journal of Nematology, 3*, 329–330.

Sproule, A., Donaldson, P., Dijak, M., Bevis, E., Pandeya, R., Keller, W. A., & Gleddie, S. (1991). Fertile somatic hybrids between transgenic *Nicotiana tabacum* and transgenic *N. debneyi* selected by dual antibiotic resistance. *Theoretical and Applied Genetics, 82*, 450–456.

Stavely, J. R. (1979). Disease resistance. In R. D. Durbin (Ed.), *Nicotiana*. Procedures for experimental use. U.S. Department of Agriculture Technical Bulletin 1586 (pp. 111–120).

Stavely, J. R., & Skoog, H. A. (1976). Transfer of resistance to a virulent strain of Pseudomonas tabaci from *Nicotiana rustica* to *Nicotiana tabacum* breeding lines. *Proceedings of American Phytopathology Society, 3*, 231.

Stavely, J. R., & Skoog, H. A. (1978). Stabilization in *Nicotiana tabacum* of a dominant gene from *N. rustica* for resistance to a virulent strain of Pseudomonas tabaci. *Phytopathology News, 12*, 181.

Stavely, J. R., Pittarelli, G. W., & Burk, L. G. (1973). *Nicotiana repanda* as a potential source for disease resistance in *N. tabacum. The Journal of Heredity, 64*, 265–271.

Stavely, J. R., Kung, S. D., & Slana, L. J. (1977). Comparison of polypeptide composition of Fraction I protein of certain Nicotiana species and hybrids related to root knot resistant tobacco. *Tobacco Science, 179*, 31–32.

Stokes, G. W. (1960). Difference in the behaviour of the *Nicotiana longiflora* wildfire resistance locus in tobacco varieties Burley 21 and KY 61. *Phytopathology, 50*, 770–772.

Stoyanova, M. (1972). Investigations of the hybrids between the species *Nicotiana tabacum* and *N. glauca* (in Bulgarian). *B: Otdal Hybridiz Rast C BAN*, 127–136.

Stoyanova, M. P. (1979). Resultats de l'hybridation entre les especies *Nicotiana tabacum* L. Et *Nicotiana alata* Link. *Comptes Rendus de l 'Academie Bulgare des Sciences, 32*, 683–686.

Subhashini, U., Venkateswarlu, T., & Anjani, K. (1986). Embryo rescue in *Nicotiana* hybrids by in vitro culture. *Plant Science, 43*, 219–222.

Swaminathan, M. S., & Murthy, B. R. (1959). Aspects of asynapsis in plants. I. Random and non-random chromosome associations. *Genetics, 44*, 1271–1272.

Szilagyi, L. (1975). Elimination of chromosomes in an alloploid hybrid of *Nicotiana tabacum* × *N. glauca. Acta Botanica Academiae Scientiarum Hungaricae, 21*, 433–441.

Takenaka, Y. (1960). Cytogenetic studies of the genus Nicotiana XIII. *National Institute of Genetics Annual Report, 11*, 63–64.

Takenaka, Y. (1962). Cytogenetic studies in Nicotiana XVIII. Reduction divisions in three interspecific hybrids between autotetraploid *N. tabacum* and three other species. *Japan Journal of Plant Breeding, 12*, 275–277.

Takenaka, Y. (1963). Cytogenetic studies in *Nicotiana*. XV. Reduction divisions in three interspecific hybrids and one amphidiploid. *Japanese Journal of Genetics, 38*, 135–140.

Takenaka, Y., & Yoneda, Y. (1963). Tumorous hybrids in *Nicotiana. National Institute of Genetics Annual Report, 13*, 63–64.

Tatemichi, Y. (Ed.). (1990). *Illustrated book of the genus Nicotiana. plant breeding and genetics research laboratory*. Japan Tobacco Inc..

Ternouth, R. A. P., MacKenzie, J., & Shepherd, J. A. (1986). *Introduction of Meloidogyne javanica resistance into flue-cured tobacco in Zimbabwe*. CORESTA Inf Bull Symp.

Ternovsky, M. F. (1934). Die Fragen der Immunitat bei Vertretern der Gattung *Nicotiana. Der Zuchter, 6*, 140–144.

Ternovsky, M. F. (1941). Methods of breeding tobacco varieties resistant to tobacco mosaic and powdery mildew (in Russian). *Sb Nauch Issled Rabot WITIM, 143*, 126–141.

Ternovsky, M. F. (1962). Summary and prospects of interspecific hybridization in the genus *Nicotiana*. In *Wide hybridization of plants (Otdalennaya gibridizatsiya rastenii) Proceedings of the Conference on Wide Hybridization of Plants and Animals; collection of reports*, pp. 331–345.

Ternovsky, M. F. (1963). Levee de la sterilite des hybrides interspecifiques de Nicotiana. *Proceedings of the Conference of Polyploidy Breed*, 70–80 (CORESTA Inf Bul 1966 (2):60, Abst 2821).

Ternovsky, M. F. (1964). Distant hybridization in developing new tobacco varieties and market types. *Tabak, 2*, 22–25.

Ternovsky, M. F., & Larkina, N. I. (1978). Hybridization of native tobacco cultivars with the wild species *Nicotiana rosulata. Soviet Genetics, 14*, 1038–1045.

Thurston, R. (1961). Resistance in *Nicotiana* to green peach aphid and some other tobacco insect pests. *Journal of Economic Entomology, 54*, 946–949.

Trojak-Goluch, A., & Berbeć, A. (2002). *Potential of Nicotiana glauca as a source of tolerance to black root rot caused by Chalara elegans in tobacco improvement*. CORESTA Congress Agro-Phyto Groups, New Orleans, APPOST20.

Trojak-Goluch, A., & Berbeć, A. (2003). Cytological investigations of the interspecific hybrids of *Nicotiana tabacum* L. × *N. glauca* Grah. *Journal of Applied Genetics, 44*, 45–54.
Trojak-Goluch, A., & Berbeć, A. (2005). Potential of *Nicotiana glauca* (Grah.) as a source of resistance to black root rot *Thielaviopsis basicola* in tobacco improvement. *Plant Breeding, 124*, 507–510.
Trojak-Goluch, A., & Berbeć, A. (2009). Growth, development and chemical characteristics of tobacco lines carrying black root rot resistance derived from *Nicotiana glauca* (Grah.). *Plant Breeding, 130*, 92–95. https://doi.org/10.1111/j.1439-0523.2009.01755.x
Trojak-Goluch, A., Laskowska, D., Agacka, M., Czarnecka, D., Kawka, M., & Czubacka, A. (2011). Effectiveness of combining resistance to *Thielaviopsis basicola* and Tomato spotted wilt virus in haploid tobacco genotypes. *Breeding Science, 61*, 389–393.
Trojak-Goluch, A., Korbecka-Glinka, G., Goepfert, S., Doroszewska, T., & Berbec, A. (2016a). Tobacco breeding for TSWV resistance using RTSW-al factor derived from cultivar Polalta. CORESTA Congress, Berlin, 2016, Agronomy/Phytopathology Groups, APPOST 24.
Trojak-Goluch, A., Laskowska, D., & Kursa, K. (2016b). Morphological and chemical characteristics of doubled haploids of flue-cured tobacco combining resistance to *Thielaviopsis basicola* and TSWV. *Breeding Science, 66*, 293–299.
Trojak-Goluch, A., Berbeć, A., & Doroszewska, T. (2017). Wykorzystanie gatunków z rodzaju Nicotiana w najnowszej krajowej hodowli odpornościowej tytoniu. Praca przeglądowa (The use of *Nicotiana* species in the latest domestic tobacco breeding. A review) (in Polish). *Agron Science, 72*, 47–56.
Valleau, W. D. (1952). Breeding tobacco for disease resistance. *Economic Botany, 6*, 69–102.
Valleau, W. D., Stokes, G. W., & Johnson, E. (1960). Nine years' experience with the *Nicotiana longiflora* factor for resistance to Phytophthora parasitica var *Nicotiana*e in the control of black shank. *Tobacco Science, 4*, 92–94.
Venkateswarlu, T., Nagarajan, K., & Subhashini, U. (1998). Root-knot nematode resistance through interspecific hybridization. *Tobacco Symposium, 7*, B20.
Verrier, J. L., Malpica, A., & L'Humeau, J. (2016). *Breeding for blue-mould resistance in Burley and flue-cured tobacco: An overview of results*. CORESTA Congress Berlin, Agro/Phyto Groups, AP7.
Vinogradov, V. A., & Larkina, N. I. (2013). Типы и механизмы устойчивости к грибным болезням у табака (Types and mechanisms of resistance to fungal diseases of tobacco) (in) Innovative research and development for securing the environmentaly safe production of agricultural crops (in Russian). In *Proceedings of the research and extension conference*. Krasnodar.
Wan, H., Ueng, I. N., Lin, T. T., & Chen, C. H. (1971). Interspecific transfer of frogeye resistance from *N. repanda* to *N. tabacum*. *Taiwan Tob Wine Monop Bur Tob Res Annu Rep*, 34–35.
Wark, D. C. (1963). *Nicotiana* species as sources of resistance to blue mold (Peronospora tabacina Adam) for cultivated tobacco. In *Proceedings of 3rd Tobacco Science Congrress Salisbury, South Rhodesia*. Tobacco Research Board, pp. 252–259.
Wark, D. C. (1970). Development of flue cured tobacco cultivars resistant to common strain of blue mold. *Tobacco Science, 14*, 147–150.
Wark, D. C. (1975). The development of blue mold resistant cultivars in Australia. *CSIRO Division of Plant Industrial Annual Report, 1975*, 31–33.
Wernsman, E. A. (1992). Varied role for the haploid sporophyte in plant improvement. In H. T. Stalker & J. P. Murphy (Eds.), *Plant breeding in the 1990's (proceedings of the symposium in Plant Breeding in the 1990's)* (pp. 461–484). CAB International.
Wernsman, E. A., & Rufty, R. C. (1987). Tobacco. In W. W. Fehr (Ed.), *Principles of cultivar development. Vol. 2. Crop species* (pp. 669–698). Macmillian.
Wernsman, E. A., Matzinger, D. F., & Mann, T. J. (1976). Use of progenitor species germplasm for the Improvement of a cultivated allotetraploid. *Crop Science, 16*, 800–803.

Williams, E., & Pandey, K. K. (1975). Meiotic chromosome pairing in interspecific hybrids of *Nicotiana*. II. South American species hybrids: The influence of genotype on pairing. *New Zealand Journal of Botany, 13*, 611.

Wong, Y. N. (1975). Genome analysis of tobacco interspecific hybrids. *Taiwan Tob Wine Monop Bur Tob Res Inst Res Rep, 2*, 22–31.

Woodend, J. J., & Mudzengerere, E. (1992). Inheritance of resistance to wildfire and angular leaf spot derived from *Nicotiana rustica* var. Brasilea. *Euphytica, 64*, 149–156.

Wuttke, H. H. (1969). Different levels of resistance in blue mold resistant tobacco. *Australian Journal of Agricultural Animal Husbandry, 9*, 545–548.

Yancheva, A. (1989). Inheritance of the resistance to tomato spotted wilt virus *Nicotiana* virus 1 Smith and *Erysiphe cichoracearum* d.c. originating from the wild species *Nicotiana sanderae* in interspecific crosses with *Nicotiana tabacum*. *Genetics Selection Evolution, 22*, 412–419.

Yang, S. J. (1960). Numerical chromosome instability in *Nicotiana* hybrids. I. Interplant variation among offspring of amphiploids. *Genetics, 50*, 745–756.

Yi, H. Y., Rufty, R. C., Wernsman, E. A., & Conkling, M. C. (1998). Mapping the root-knot nematode resistance gene (Rk) in tobacco with RAPD markers. *Plant Disease, 82*, 1319–1322.

Yuan, X., Yan, C., Wu, Z., Ren, F., Zhang, H., Baker, B., Chen, J., & Kuang, H. (2015). Frequent gain and loss of resistance against tobacco mosaic virus in *Nicotiana* species. *Molecular Plant, 8*, 1813–1815. https://doi.org/10.1016/j.molp.2015.09.001

Zhang, S., & Zaitlin, D. (2008). Genetic Resistance to *Peronospora tabacina* in *Nicotiana langsdorffii*, a South American Wild Tobacco. *Phytopathology, 98*, 519–528.

Zhang, S., Gao, M., & Zaitlin, D. (2012). Molecular linkage mapping and marker-trait associations with NlRPT, a downy mildew resistance gene in *Nicotiana langsdorffii*. *Frontiers in Plant Science, 3*, 185. https://doi.org/10.3389/fpls.2012.00185

Chapter 5
Species of *Nicotiana* as the Sources of Cytoplasmic Male Sterility for Cultivated Tobacco

5.1 Introductory Notes

The use of F_1 hybrids and heterosis in tobacco. In tobacco, as in many other crops, cytoplasmic male sterility (CMS) finds utility in the economically viable production of intervarietal hybrids as an alternative to purebred cultivars. The major incentive to develop hybrid cultivars in allogamous crops such as maize was the spectacular increase in vigor and yields as a result of heterosis (Crow, 1998). In tobacco, many authors agree that little commercially exploitable heterosis per se could be expected from intervarietal hybrids (Chaplin, 1966; Jones & Henderson, 1978; Legg, 1991; Billenkamp, 1997). This notwithstanding, in the 1960s, Matzinger and Wernsman (1968) showed evidence that although heterosis in hybrids of inbred lines within the same market type (flue-cured tobacco) was minimal, it was considerably increased above the midparent value when some flue-cured varieties were crossed to those of the oriental market class. Even greater heterosis values were obtained when flue-cured varieties were crossed to the putative progenitor species of tobacco or to its close relatives (Matzinger & Wernsman, 1967), although in the latter case, the heterotic effect had no commercial use due to serious leaf quality issues (Wernsman & Matzinger, 1966). More recent studies revealed that the use of mid-parent heterosis in flue-cured tobacco can be a viable breeding strategy. Dexter-Boone and Lewis (2019) showed heterotic effects for yield to be substantially higher than those in the previous estimates from the 1960s. Furthermore, they argued that heterosis had potential in transgressive breeding of flue-cured tobacco and that heterotic effects could be fixed in new breeding lines. Similar positive results were reported for heterotic effects in dark air-cured tobacco (Pscheidt et al., 2021).

Other agronomic and commercial benefits from cytoplasmic male sterility. In tobacco, a self-pollinating crop, the ease with which some important traits can be combined into one genotype (Verrier et al., 2000; Haji et al., 2006) and the mitigation of deleterious linkages (see Sect. 4.6.7) are regarded among the major merits accruing from the deployment of hybrid cultivars.

A. Berbeć, *A Century of Interspecific Hybridization and Introgression in Tobacco*,
https://doi.org/10.1007/978-3-031-54964-9_5

Once the decision has been made to deploy a hybrid cultivar in preference to an open-pollinated variety or an inbred line, the production of pollen by the flowers of the maternal line must be eliminated for safe and cost-effective hybrid seed production. Cytoplasmic male sterility is practically the only choice that may serve this purpose in tobacco. Since tobacco is a leaf crop, no fertility-restoring lines are required to produce commercial hybrid seed material, as is the case in crops grown for their generative organs.

The protection of hybrid cultivars against unauthorized propagation and other infringements of proprietary rights is another added value of cytoplasmic male sterility that has been gaining importance. Legally protected genes or other proprietary germplasms can be "locked" using the cms system as an additional safeguard (Lewis, 2020; Pscheidt et al., 2021). It must be noted, however, that the very cms system deployed is not physically protected.

For the reasons stated before, hybrid cultivars have made substantial gains in the area planted with tobacco over the past two decades, and cytoplasmic male sterility has become an important factor in tobacco improvement (Lewis, 2020). In Europe, including Germany, France and Poland, newly released varieties are usually male sterile. Outside Europe, the use of cytoplasmic male sterility has also become a standard practice (Zheng et al., 2018; Shava et al., 2020).

5.2 Genetic Background and Phenotypic Manifestations of Cytoplasmic Male Sterility

5.2.1 Genetics of Cytoplasmic Male Sterility in N. tabacum

Manifestations of heritable effects that had their source outside the chromosomes were first observed in the first decades of the twentieth century. At first, they were treated as *sui generis* oddities, but soon extrachromosomal inheritance became a generally accepted fact.

In higher plants, the apparatus of cytoplasmic inheritance is located in chloroplasts and mitochondria. In the majority of crop plants, including tobacco, both chloroplasts and mitochondria are inherited maternally. However, exceptions to that rule and rare cases of transmission of plastids (Horlow et al., 1990; Medgyesy et al., 1985, 1986) and both plastids and mitochondria (Svab & Maliga, 2007) through the pollen have been documented. This rare phenomenon will be recalled in Sect. 5.4.2, where doubts regarding the identity of certain cms lines are discussed.

In tobacco, cytoplasmic male sterility (CMS) seems to be the most extensively studied and practically exploited aspect of extrachromosomal inheritance. CMS is a peculiar case of interaction between nuclear and cytoplasmic DNA (Gerstel et al., 1978; Gerstel, 1980; Hanson & Conde, 1985). If the chromosomal and cytoplasmic genes essential for the normal development of male organs and for regular microsporogenesis are mutually compatible, male fertility is ensured (Gerstel, 1980).

Conversely, if that complementarity is disrupted, e.g., by a mutation, CMS sets in as a result. Initially, there was controversy about the location of CMS-controlling elements in the cytoplasm of tobacco. Chen et al. (1977) and Frankel et al. (1979) favored chloroplast DNA as the site of cms genes. Their view was objected by Belliard et al. (1978), on the grounds that in the cybrids resulting from fusing the protoplasts of *N. tabacum* cms debneyi (alien cytoplasm) with those of mf *N. tabacum* (native cytoplasm) the expression of the cms character was not affected by the type of chloroplast DNA in fusion products. The involvement of mitochondria in the expression of cytoplasmic male sterility in tobacco was demonstrated using asymmetric somatic hybrids of *N. tabacum* with several other *Nicotiana* species: *N. plumbaginifolia* (Nagy et al., 1983), *N. knightiana* (Nagy et al., 1981), *N. quadrivalvis* (Aviv & Galun, 1987; Aviv et al., 1984; Kofer et al., 1991; Raineri et al., 1992), *N. rustica* (Donaldson et al., 1993), *N. repanda* (van ter Hakansson et al., 1988; Hakansson & Glimelius, 1991; Bergman et al., 1995), *N. debneyi* (Asahi et al., 1988; van ter Hakansson et al., 1988; Hakansson & Glimelius, 1991), *N. megalosiphon* (Donaldson et al., 1995), (*N. suaveolens* (van ter Hakansson et al., 1988; Hakansson & Glimelius, 1991; Kofer et al., 1992), *N. glutinosa* (Donaldson et al., 1994), and *N. undulata* (Aviv & Galun, 1986; Kofer et al., 1991; Bergman et al., 1995). The association between mitochondria and cytoplasmic male sterility was also shown by comparing mitochondrial DNA from alloplasmic cms stocks of *N. tabacum* involving the cytoplasm of *N. glauca*, *N. rustica*, *N. repanda* and *N. suaveolens* (Li et al., 2011) and by comparing the mitochondrial DNA of *N. tabacum* cms *repand*a with that of its nuclear isoline (Sun et al. (2005).

Apart from their practical implications, the studies on the transfer and expression of CMS in interspecific somatic tobacco hybrids have significantly contributed to the understanding of general rules governing cytoplasmic inheritance. Chen et al. (1977) and Belliard et al. (1979) demonstrated that the chloroplasts in fused cells quickly segregate, resulting in only one or the other chloroplast genome in fusion products. In chloroplasts, DNA recombination is very rare, which was demonstrated by Donaldson et al. (1993) for the somatic hybrid *N. tabacum* + *N. rustica*. Although mitochondria also tend to be transferred uniparentally (Menczel et al., 1981), the interchanges of mitochondrial DNA are a much more frequent phenomenon (Belliard et al., 1979; Nagy et al., 1981; Aviv & Galun, 1986; Asahi et al., 1988; Kofer et al., 1991; Fitter et al., 2005).

In the current state of knowledge, the lack of complementarity between the nucleus and the cytoplasm may have its ultimate source in rearrangenents within the mitochondrial genome of cytoplasmically male sterile forms. Mitochondria in plant cells tend to fuse extensively, and the recombination of mitochondrial DNA following mitochondrial fusion is well documented (Svab & Maliga, 2007). According to Zheng et al. (2018) and Wang et al. (2020), chimeric open reading frames (ORFs) are formed in recombinant mitochondrial DNA, leading to the production of aberrant RNA transcripts and, consequently, toxic proteins that interfere with normal stamen development. Zheng et al. (2018) performed a study of the mitochondrial DNA basis for CMS in *N. tabacum* with an introgressed cytoplasm of *N. suaveolens*. Six ORFs were unique for that alloplasmic form and had no homology with the mitochondrial genomes of fertile *N. tabacum*. The six ORFs also had no homology with those that amplified in other cms forms in the study, except one that was shared with cms *glutinosa* (Zheng et al. (2018). The above mechanism may account for the substantial differences in the manifestations of CMS in different alloplasmic combinations of *N. tabacum* resulting from the evolution of mitochondrial DNA in *Nicotiana* species – sources of CMS in tobacco (see Sect. 5.2.2). However, it has particular relevance to cytoplasmic male sterility obtained by asymmetric protoplast fusion (see Sect. 5.3.1) and to the appearance of spontaneous and induced cms mutations. In many allogamous species, mutations of mitochondrial DNA are the main source of cytoplasmic male sterility. In autogamous tobacco, potential male sterile mutations are quickly eliminated from the population, and very few have been reported (Berbeć, 1974; Kobus, 1978). All undisputed cytoplasmic male sterile forms in tobacco were produced by moving the whole nuclear genome from one species and reinstalling it in the cytoplasmic milieu of another, thereby disrupting the complementarity between the nucleus and the cytoplasm (Gerstel, 1980), the precise underlying cause notwithstanding.

5.2.2 *Phenotypic Manifestations of Cytoplasmic Male Sterility in* N. tabacum

In the majority of *Nicotiana tabacum* alloplasmics, male sterility is accompanied by malformations of the androecium (Gerstel, 1980). The character and extent of those deformities may vary depending on the source of alien cytoplasm but is also influenced by the nuclear genotype and the environment. Stamens may be missing altogether or be rudimentary e.g. in cms *glauca* (Berbeć, 1972, 2001), cms *excelsior* (Nikova & Vladova, 2002) or cms *wuttkei* (Laskowska & Berbeć, 2007). Feminization of male organs is the most frequent malformation. It may entail feminization of anthers which are tipped with small stigmas e.g. in cms *plumbaginifolia* or cms *quadrivalvis* (Chaplin, 1964; Burk, 1960). The feminization may involve whole stamens which become pistilloid or even carpelloid e.g. cms *goodspeedii* (Tsikov et al., 1977; Berbeć, 2001), cms *gossei* (Gerstel, 1980), cms *suaveolens* (Chaplin, 1959; Berbeć, 2001). In several other plasmatypes of *N. tabacum* e.g. cms *undulata* stamens are transformed into large petaloid structures, sometimes tipped with minuscule stigmas (Chaplin, 1964; Frankel & Galun, 1977). These alterations of male organs are classified as staminal male sterility since they prevent the production of microspores by interfering with normal stamen development. Staminal male sterility is sometimes accompanied by alterations of other flower parts eg. by split, petalous corolla in cms *debneyi* (Chaplin, 1964; Frankel & Galun, 1977) or degenetative changes in the pistil and the ovary affecting female ferility. The latter phenomenon is most notable in cms *suaveolens* (Berbeć, 2001). Considerably shorteded corolla, resulting in a protruding pistil, is characteristic of cms *undulata* (Chaplin, 1964; Frankel & Galun, 1977) but also of several other plasmatypes of *N.*

tabacum representing the same type of flower modifications (Chaplin, 1964; Nikova et al., 1991; Berbeć & Berbeć, 1992).

A few alloplasmics of *Nicotiana tabacum*, notably those involving some species of the section *Paniculata*e such as cms *knightiana* (Berbeć & Doroszewska, 1992), cms *raimondii* (Berbeć, 2001)[1] cms *paniculata* (Nikova & Vladova, 2002) but also *N. rustica* (Hart, 1965) represent the postmeiotic or sporogeneous type of cms in which microspores are produced but are either aborted or develop into dysfunctional or mostly dysfunctional pollen grains. However, the collapsed development of the male gametophyte in the alloplasmics with the postmeiotic type of male sterility did not preclude them from producing haploid plants through anther culture. Anther-derived haploid plants were obtained from cms *debneyi* (Zagorska et al., 1978) and from cms *knightiana* and cms *raimondii* (Berbeć & Laskowska, 1994).

Some alloplasmic lineages with the postmeiotic type of cms (cms *raimondii*, cms *knightiana*) produce apparently normal pollen grains with vestigial germination ability on sucrose agar that makes them partially self-fertile (Berbeć, 1994a, b, 2001). Apparently, this may also be the case with at least some lineages of cms *plumbaginifolia* (Moav et al., 1968; Burk & Durbin, 1978). Sand and Christoff (1972, 1973) found a similar expression of nearly restored male fertility in segregating early backcross populations of the hybrid *N. debneyi* × *N. tabacum*.

In cms *glauca*, expression of the cms trait may vary from staminal (absence of stamens) to postmeiotic: 1 to 3 stamens with normal anthers on shortened filaments, stamens modified into secondary pistils or secondary stigmas (Berbeć, 2001; Nikova & Vladova, 2002). The breakdown of microsporogenesis in the postmeiotic type of cms was observed to occur from early meiosis to the tetrad stage (Nikova & Vladova, 2002). According to observations made by the author of this review, the expression of cms *glauca* may vary among different alloplasmic lineages and is strongly environment-dependent. With decreasing daylength, cms *glauca* flowers tend to develop vestigial stamens, which are at times stigmatoid (Berbeć, 2001). Especially in the winter season under greenhouse conditions, normal-sized stamens with shrunken or even well-developed anthers appear occasionally producing scant pollen (unpublished observations by the author of this volume).

The alloplasmics of *N. tabacum* with the cytoplasm of *N. sylvestris* are fully male fertile, which was an early argument in favor of *N. sylvestris* or its very close extinct relative being the maternal progenitor of cultivated tobacco (Cameron, 1965). Much like the cytoplasm of *N. sylvestris*, that of *N. glutinosa* was also reported, directly or by implication, not to produce male sterility in *N. tabacum* (Ternovsky & Nosova, 1970, 1971; Burk & Durbin, 1978; Gerstel, 1980, also see the discussion of that topic in Sect. 5.4.2).

[1] The cited paper (Berbeć, 2001) lists two plasmatypes as cms *raimondii*: *raimondii* I and *raimondii* II. It is the latter that is true cms *raimondii*. The former is most probably is a lineage of cms *paniculata*. See also "Other doubtful cases of cytoplasmic male sterility in *N. tabacum* germplasm collections", Sect. 5.4.2 of this volume.

Thus, the male sterilizing effect of alien cytoplasm in the nuclear background of *N. tabacum* may cover different manifestations ranging from nil to affected pollen production, affected anthers, affected both anthers and filaments with different degrees of degeneration including the total absence of male organs (Chaplin, 1959; Cameron, 1965; Hart, 1965; Burk, 1967; Berbeć et al., 1990, 1994a, b, c, 2001; Nikova et al., 2001; Nikova & Vladova, 2002).

Reaction of *Nicotiana* species to tentoxin, a cytoplasmic marker. Tentoxin, a cyclic polypeptide produced by the fungus *Alternaria alternata*, causes the disruption of chlorophyll production in the species of *Nicotiana* that carry in their chloroplasts a receptor site for that toxic polypeptide. In sensitive species, tentoxin binds to that receptor site. As a result, photophosphorylation and light-dependent RNA synthesis become inhibited, and chloroplast development is severely disturbed (Burk & Durbin, 1978). Insensitive species do not have the tentoxin receptor, and their chloroplasts develop normally. Sensitivity to tentoxin is maternally inherited, and the alloplasmics developed by replacing the native cytoplasm by that of cultivated tobacco retain their sensitivity or insensitivity to tentoxin throughout the transfer of the alien cytoplasm to the pollen parent (Durbin & Uchytil, 1977b; Burk & Durbin, 1978). Phenotypically, the seedlings of tentoxin-sensitive species and alloplasmics turn yellow upon exposure to the toxin, while the insensitive cytoplasmic types stay green.

For hybrid seed production purposes, tentoxin can be regarded as a relatively reliable tool to distinguish, as early as the seedling stage, between the hybrid phenotype and inadvertent (or purposeful) contamination with any pollen-fertile tobacco seed. This is because the native *N. tabacum* cytoplasm is insensitive to tentoxin, while all alloplasmics that are likely to be used in tobacco hybrid seed production (cms *suaveolens*, cms *tabacum*-mutant, cms *glauca*, cms 'bigelovii', cms *undulata*) are tentoxin-sensitive.

However, it must be noted that tentoxin sensitivity is not a cms marker *per se*. The trait is associated with chloroplast DNA, and cms factors are coded in mitochondrial DNA. In exceptional cases, e.g. in somatic hybrids, chloroplast-coded tentoxin sensitivity and mitochondria-coded male sterility may segregate independently of each other (Aviv & Galun, 1980).

A list of known tentoxin-sensitive and tentoxin-insensitive species is given in Table 5.1.

Table 5.1 Reaction to tentoxin of *Nicotiana* species

Species	Reaction to tentoxin	Reaction to tentoxin reported by
N. alata	+	Durbin and Uchytil (1977a)
N. azambujae	?	
N. bonariensis	+	Durbin and Uchytil (1977a)
N. forgetiana	+	Durbin and Uchytil (1977a)
N. langsdorffii	+	Durbin and Uchytil (1977a)
N. longiflora	+	Durbin and Uchytil (1977a)
N. mutabilis	?	

(continued)

Table 5.1 (continued)

Species	Reaction to tentoxin	Reaction to tentoxin reported by
N. plumbaginifolia	+	Durbin and Uchytil (1977b)
N. sp. 'Rastroensis'	?	
N. sanderae	+	
N. tabacum	—	Burk and Durbin (1978)
N. acaulis	+	
N. ameghinoi	?	
N. glauca	+	Durbin and Uchytil (1977a)
N. noctiflora	+	
N. paa	?	
N. petunioides	+	Durbin and Uchytil (1977a)
N. benavidesii	+	
N. cordifolia	+	Durbin and Uchytil (1977a)
N. cutleri	?	
N. knightiana	—	Burk and Durbin (1978)
N. paniculata	—	Burk and Durbin (1978)
N. raimondii	+	Durbin and Uchytil (1977a)
N. solanifolia	+	Durbin and Uchytil (1977a)
N. acuminata	+	Durbin and Uchytil (1977a)
N. attenuata	+	
N. corymbosa	+$^{?}$	
N. longibracteata	?	
N. linearis	+	
N. miersii	+	Durbin and Uchytil (1977a)
N. pauciflora	+	Durbin and Uchytil (1977a)
N. spegazzinii	+	
N. clevelandii	+	Durbin and Uchytil (1977b)
N. quadrivalvis	+	Durbin and Uchytil (1977a)
N. nesophila	+	Durbin and Uchytil (1977a)
N. nudicaulis	+	Durbin and Uchytil (1977a)
N. repanda	+	Durbin and Uchytil (1977a)
N. stocktonii	+	Durbin and Uchytil (1977a)
N. rustica	—	Burk and Durbin (1978)
N. africana	—	Burk and Durbin (1978)
N. amplexicaulis	+	Berbeć (2001)
N. benthamiana	+	Durbin and Uchytil (1977a)
N. burbidgeae	?	
N. cavicola	+	
N. debneyi	—	Burk and Durbin (1978)
N. eastii	+	Berbeć (2001)
N. excelsior	+	
N. exigua	+	Berbeć (2001)
N. fatuhivensis	?	

(continued)

Table 5.1 (continued)

Species	Reaction to tentoxin	Reaction to tentoxin reported by
N. faucicola	?	
N. fragrans	–	Burk and Durbin (1978)
N. goodspeedii	+	Durbin and Uchytil (1977a)
N. gossei	–	Burk and Durbin (1978)
N. hesperis	+	
N. heterantha	?	
N. ingulba	+	
N. maritima	+	Durbin and Uchytil (1977a)
N. megalosiphon	+	Durbin and Uchytil (1977a)
N. monoschizocarpa	?	
N. occidentalis	+	Durbin and Uchytil (1977a)
N. rosulata	+	
N. rotundifolia	+	
N. simulans	?	Burk and Durbin (1978)
N. stenocarpa	+	
N. suaveolens	+	Durbin and Uchytil (1977b)
N. truncata	?	
N. umbratica	+	
N. velutina	+	Durbin and Uchytil (1977a)
N. wuttkei	?	
N. sp. 'Corunna'	?	
N. sylvestris	–	Burk and Durbin (1978)
N. kawakamii	?	
N. otophora	+	Flick and Evans (1982)
N. setchellii	+	
N. tomentosa	+	
N. tomentosiformis	+	Durbin and Uchytil (1977a)
N. obtusifolia	+	Durbin and Uchytil (1977a)
N. palmeri	+	
N. arentsii	+	
N. glutinosa	+	Durbin and Uchytil (1977a)
N. thyrsiflora	+	
N. undulata	–	Burk and Durbin (1978)
N. wigandioides	+	

Adapted after Durbin and Uchytil (1977a) and Burk and Durbin (1978) with minor additions and modifications

Explanations

+ (plus) denotes leaf chlorosis

– (minus) denotes normal green color of cotyledons after exposing the germinating seeds to tentoxin

? (question mark) information on reaction to tentoxin is lacking and could not be covered by the cited statement of Burk and Durbin (1978) because the species was not known or available for study at that time

$^{?}$(superscript question mark) positive reaction inferred from the statement by Burk and Durbin (1978) that "the remaining species are sensitive"(save for the 10 species that were listed as insensitive)

5.3 Transfer of Alien Cytoplasm and Production of Cytoplasmically Male Sterile Alloplasmics in Tobacco

5.3.1 *Development of* N. tabacum *Alloplasmics Using Different Cytoplasm Donors and Different Cytoplasm Transfer Routes*

The first CMS forms in *Nicotiana* did not involve *N. tabacum*. Among the segregating progeny of the compatible self-fertile intrasectional hybrid *N. sanderae* × *N. langsdorffii* East (1932) selected a cytoplasmically male sterile form of *N. langsdorffii* with the cytoplasm contributed by *N. sanderae*. Christoff (1938) reported stamen abnormalities in a backcross lineage derived from crossing maternal *N. suaveolens* with *N. rustica*. Other instances of cytoplasm-nucleus interactions outside *N. tabacum* resulting in male sterility are *N. repanda* cms *sylvestris* (Burk, 1967), *N. otophora* cms *tomentosiformis* reported by Gerstel and Burns (1974), *N. sylvestris* cms *quadrivalvis* produced by Aviv and Galun (1986) *N. glutinosa* cms *debneyi*, *N. plumbaginifolia* cms *megalosiphon*, *N. sylvestris* cms *suaveolens* (Zelcer et al., 1978; Hanson & Conde, 1985) and *N. sylvestris* cms *undulata* genetically engineered by Maliga and Svab (Thyssen et al., 2012).

The first transfer of an alien male sterile cytoplasm into the nuclear background of *N. tabacum* was performed by Clayton (1950). His method became the paradigm for the majority of subsequent cms transfers and essentially mimicked the classic introgression scheme for chromosome-borne traits. It started from the amphidiploid 4x (*N. debneyi* × *N. tabacum*) and went through the sesquidiploid and breakdown generations, which segregated for male sterile, partly fertile and male fertile plants reflecting the segregation of compatible and incompatible factors from *N. tabacum* and *N. debneyi* in the hybrid nuclei. Once the nuclear genome of *N. debneyi* had been completely eliminated, the cytoplasmically male sterile line *Nicotiana tabacum* cms *debneyi* stabilized. The cms stocks are perpetuated, and their genomic integrity is preserved by backcrossing them to their male fertile maintainer lines (isolines).

Apart from this standard method, there is another method that deploys somatic hybridization (fusion of protoplasts) as the vehicle, at least initially, to transfer alien cytoplasm into the genome of *N. tabacum*. Basically, the method relies on genomic asymmetry, spontaneous or induced. In this approach, protoplasts are used to combine the cytoplasmic factors of one species with the nuclear genome of another. Three variants of this method are used:

(a) Non-prefusion-treated protoplasts from both parental species are fused to produce a somatic hybrid. This would normally yield a mixture of nuclear hybrids and both nuclear and mitochondrial recombinants of varying degrees of asymmetry (Ilcheva et al., 2000). A series of conventional backcrosses to the recipient parent is needed to recover and stabilize cms in the new genetic background (Chen et al., 2012). Therefore, this approach does not improve the economy of cms transfer. It may be advantageous if mitochondrial recombinants, including novel recombinant cms systems, are sought (Kofer et al., 1991). However, the same advantage may become a burden if the transfer of a cms system in its intact form is intended.

(b) Prior to protoplast fusion, the nuclear genome of the cytoplasm donor is inactivated by ionizing radiation to achieve the maximum nuclear asymmetry of fusion products. By this expedient, the cybrids thus obtained would contain the active nuclear genome of the cytoplasm recipient and the cytoplasmic organelles from the donor, the recipient or both. Provided that an effective selectable marker system is employed, this theoretically one-step procedure may substantially speed up the process of cms transfer, but it does not relieve the risk of undesired mitochondrial recombination. Nonetheless, irradiation of cytoplasm donors seems to have worked well with several cms transfers (Aviv et al., 1984, Kumashiro et al., 1988; Kubo et al., 1988).

(c) Naked protoplasts of both the donor and the recipient are prefusion treated: protoplasts of the donor are irradiated to inactivate the nucleus, and those of the recipient are treated with a genome-fragmenting toxin such as iodoacetate, iodoacetamide or rhodamine to inactivate the cytoplasm. In this so-called donor-recipient approach, the fusion of both the nuclear and cytoplasmic genetic structures is highly asymmetric, and most of the selected fusion products contain the cytoplasm of the donor and nucleus of the recipient. The donor-recipient method minimizes the generation of mitochondrial recombinants either by fusion per se or in the subsequent regeneration process, thereby ensuring the integrity of the transferred cms system. According to Matibiri and Mantell (1994), the deployment of dual inactivation fusion followed by postfusion selection based on selectable nuclear and chloroplast markers reduces the time required to accomplish the cms transfer to a mere 12 months, a substantial time gain over the conventional sexual method.

Table 5.2 lists alloplasmic forms of tobacco that have been hitherto reported in the literature and the researchers that were the first to report them, usually no more than two reports for one alloplasmic cms combination. Of 31 species that have been recorded to be involved in alloplasmics with *N. tabacum,* 18 belong to the section Suaveolentes, and the remaining 13 are dispersed across eight other sections. Twelve alloplasmics of *N. tabacum* with different *Nicotiana* species were recorded by Gerstel (1980). The number of alloplasmic combinations has since grown considerably, as has the proportion of those involving Suaveolentes species as cms donors.

There are a few alloplasmic forms included in Table 5.2 that require commenting upon. In these cms lineages, *N. tabacum* rather than an alien species was the cytoplasmic parent of the initial interspecific hybrid. This unusual derivation is indicated by putting the name of *N.tabacum* first followed by the name of the alien species in parentheses. In all of those cases male sterility appeared either in the F_1 or in backcross progenies to the recurrent male *N. tabacum* parent and subsequently became stabilized as cytoplasmic male sterility. To the author's knowledge, the mode of origin of these alloplasmics has never been clarified. The plasmatypes cms

Table 5.2 *Nicotiana* species reported as sources of cytoplasmic male sterility bred into *N. tabacum*

Source of cytoplasmic male sterility	cms type	Authored or reported/studied by:	Flower alterations and other notes
N. alata	ST	Atanassov et al. (1998)[1]	Normal corolla, normal pistil, petaloid stamens tipped with stigmatoid structures;
		Nikova et al. (1999)	Normal corolla, three-loculed or deformed pistil, stamens stigmatoid or absent
N. longiflora	PM	Nikova and Vladova (2002), Nikova et al. (2001)	Shortened corolla, normal pistil, stamens with shortened filaments with shriveled anthers
N. plumbaginifolia	PM, S?)	D. R. Cameron (after Chaplin, 1964); Burk and Durbin (1978), Pino Perez (2012)	Shortened flower tube, pistil protrudes above corolla, stamens with shortened filaments with small or stigmatoid anthers void of pollen; male-fertile lineages reported;
		Bates et al. (1987)	Shortened corolla, protruding pistil, shortened stamens; observations made on regenerated 48 or 49-chromosome asymmetric hybrids
N. tabacum (*alata*)	ST	Nikova et al. (1999)[2]	Possibly a mitochondrial recombinant resulting from cytoplasm transfer through pollen; normally developed or deformed pistil with three-lobed stigma, stamens missing
N. tabacum (*plumbaginifolia*)	ST	Burk (1960)[3] and Chaplin (1964)[3]	Male sterile plant appeared in the BC_2 generation of *N. tabacum* x *N. plumbaginifolia* backcrossed to *N. tabacum*; normal pistil protrudes above the shortened corolla, petaloid stamens, involvement of chloroplast genome from *N. glutinosa* suggested by the data of Chen et al. (1977)
N. tabacum (cms 'tabacum – mutant')	ST	Berbeć (1967, 1974)[4], Berbeć and Berbeć (1976) and Berbeć and Laskowska (2005)	Corolla normal; pistil normal; stamens normally missing, under some ambient conditions display varying degrees of development: from vestigial and deformed to nearly normal
		author of this volume (unpublished)	In very rare cases (greenhouse, winter) normally developed stamens with anthers producing some pollen
N. tabacum (*glauca*)	ST	Stoyanova (1972)[5] and Tsikov and Tsikova (1981)[5]	Possibly resulting from cytoplasm transfer through pollen; normal pistil, flower tube and corolla; stamens mostly missing, sometimes deformed,

(continued)

Table 5.2 (continued)

Source of cytoplasmic male sterility	cms type	Authored or reported/studied by:	Flower alterations and other notes
N. tabacum (*glutinosa*)	ST to PM	Burk (1960)[6], Chaplin (1964)[6] and Nikova and Tsikov (1976)	Possibly a mitochondrial recombinant resulting from cytoplasm transfer through the pollen; floral morphology the same as in cms tabacum (*plumbaginifolia*): normal pistil protrudes above the shortened corolla, petaloid stamens tipped with diminutive stigmatoid structures
N. glauca	ST/PM	Berbeć (1966)[7], Berbeć (1972)[7] and Nikova and Vladova (2002)[7]	Normal corolla, stamens missing or rudimentary
		Sun et al. (1999a)[8] and Chen et al. (2012)[8]	Developed from somatic *hybrid* *N. glauca* + *N. tabacum*
N. knightiana	PM	Kubo (1985), Nikova and Shabanov (1992)	Normal flower morphology, pollen aborted
		Berbeć (1994b) and Berbeć and Doroszewska (1992)[9]	Stainable pollen with severely suppressed germination capacity (below 4% of germinable pollen tubes)
N. paniculata	PM	Kubo (1985), Nikova et al. (1991) and Nikova and Vladova (2002)	Normal flower morphology, anthers with no pollen or collapsed pollen grains
N. raimondii	PM	Berbeć and Doroszewska (1992), Berbeć (1994b) and Berbeć (2001)	Lineages with normal flower morphology (pistil slightly shortened) anthers containing aborted pollen or apparently normal (stainable) pollen grains
N. clevelandii	PreM	Kaul (1988)	Premeiotic breakdown of PMC development, feminization of male organs
N. quadrivalvis	ST	Chaplin (1959), Burk (1960), Nikova and Tsikov (1976), Aviv et al. (1984), Frankel and Galun (1977), Aviv and Galun (1986), Kofer et al. (1990) and Spangenberg et al. (1992), Pino Perez (2012)	Corolla either normal or deeply split to form five narrow petals, pistil normal, stamens with filaments of normal length usually tipped with reduced anthers or anthers transformed into petaloid or feather-like structures

(continued)

Table 5.2 (continued)

Source of cytoplasmic male sterility	cms type	Authored or reported/studied by:	Flower alterations and other notes
N. repanda	ST/ PM	Burk (1967), Stavely et al. (1973), Gerstel et al. (1978) and Reed and Burns (1986), Farbos et al. (2001)	Shortened or normal flower tube, pistil normal or shortened with split stigma, stamens vestigial with reduced filaments and shriveled anthers sometimes capped by a small stigma
		Sun et al. (1999b)[10]	Small-sized, whitish corolla, normal pistil; feminized, stigmatoid anthers
N. rustica	ST/PM	Hart (1965)	Partly male fertile (plants (5% of germinating pollen grains) or male sterile plants with unaltered flower morphology; male fertile alloplasmics;
		Kubo (1985) and Shabanov et al. (1986)	Male sterile and male fertile plants depending on *N. tabacum* variety
		Pittarelli and Sisson (1989)	Fully male sterile with normal corolla and pistil, stamens normal with anthers producing dysfunctional pollen grains;
		Nikova et al. (1997) and Nikova and Vladova (2002)	Fully male sterile, pistil normal or deformed, stamens missing or rudimentary pistiloid/carpeloid
N. africana	PM	Kumashiro et al. (1988) and Nikova and Vladova (2002)	Normal corolla, normal pistil, stamens with normally developed filaments and anthers void of pollen
N. amplexicaulis	ST	Nikova and Shabanov (1988), Nikova et al. (1997), Nikova and Vladova (2002) and Berbeć et al. (1990)	Split corolla, pistil normal or shortened, stamen filaments of normal length, anthers feathery or deformed and void of pollen
N. benthamiana	ST	Ramavarma et al. (1978), Nikova (1984), Nikova et al. (1991), Nikova and Vladova (2002)[11] and Atanassov (1993)[12]	Corolla considerably shortened, pistil of normal length protruding above corolla, anthers transformed into large petaloid structures
N. debneyi	ST	Clayton (1950), Chaplin (1964), Berbeć (1972), Frankel and Galun (1977), Kumashiro and Kubo (1986) and Spangenberg et al. (1992)	Corolla of normal length, sometimes shortened, most often deeply split to the point of being transformed to long petal-like fragments; pistil normal to fasciated and shortened; stamen filaments normal, tipped with small stigmatoids, or transformed into stigmatoid strugtures

(continued)

Table 5.2 (continued)

Source of cytoplasmic male sterility	cms type	Authored or reported/studied by:	Flower alterations and other notes
N. eastii	ST	Berbeć and Berbeć (1992)[13]	Shortened corolla, pistil normal protruding above corolla, anthers transformed into large petaloid structures
N. excelsior	ST	Nikova (1986), Nikova et al. (1997) and Nikova and Vladova (2002)	Corolla from normal to completely split, pistil normal occasionally fasciated, stamens missing
N. exigua	ST	Berbeć (1966) and Berbeć (1972)	Corolla normal, pistil normal or shortened, fasciated or multi-lobed, stamens rudimentary or carpeloid
N. goodspeedii	ST	Palakarcheva (1968), Palakarcheva et al. (1980), Tsikov et al. (1977) and Nikova (1979)	Depending on *N. tabacum* genotype flower tube normal or inflated and shortened; corolla normal or split; pistil normal or shortened; stamens missing or deformed, carpeloid or petaloid; multi-lobed stigmas
N. gossei	ST	Frankel and Gerstel (cited after Gerstel (1980), Hanson and Conde (1985), Tsikov et al. (1974), Gerstel et al. (1980), Tsikov and Tsikova (1986)	Shortened corolla, shortened pistil with multi-lobed stigma, fasciated carpeloid stamens
N. hesperis	?	Kubo (1985)	
N. maritima	ST/PM	Nikova et al. (1991), Nikova and Vladova (2002) and Nikova et al. (1990)	In stabilized cms lineages corolla shortened or/and split, pistil normal protruding above the corolla or shortened, stamen filaments normal tipped with feathery or reduced anthers void of pollen Slightly split corolla, missing stamens
N. megalosiphon	ST	Clayton (1950), Chaplin (1964), Frankel and Galun (1977) and Nikova (1979), Pino Perez (2012)	Corolla normal or shortened, pistil normal protruding above corolla, stamens missing, vestigial or carpeloid; multi-lobed stigmas
N. occidentalis	ST	Ternovsky et al. (1973)	Corolla slightly shortened, pistil normal or fasciated and multilobed protruding slightly above corolla, stamens missing or petaloid
N. rotundifolia	?	Ilcheva et al. (2001)[14]	Flower morphology varied from lack of stamens to feminized anthers, to anthers void of pollen
N. simulans	?	Kubo (1985)[15]	

(continued)

Table 5.2 (continued)

Source of cytoplasmic male sterility	cms type	Authored or reported/studied by:	Flower alterations and other notes
N. suaveolens	ST	Izard and Hitier (1955), Chaplin (1959), Schweppenhauser and Mann (1968), Frankel and Galun (1977), Spangenberg et al. (1992) and Zheng et al. (2018)[16]	Corolla normal, pistil normal or shortened and fasciated with multi-lobed stigma; stamens missing, stigmatoid or carpeloid; extent of flower modifications varies with nuclear genotype and environment
N. velutina	ST	Nikova et al. (1991) and Nikova and Vladova (2002)	Corolla shortened and/or split, pistil normal or multilobed protruding above the corolla, both stamenless flowers and normally developed stamens with anthers void of pollen reported
N. wuttkei	ST	Laskowska and Berbeć (2007)	Shortened and split corolla, slightly shortened pistil with normal stigma, stamens missing
N. undulata	ST	D. R. Cameron (Chaplin, 1965; Burns & Gerstel, 1981; Gerstel and Burns (1983), Chaplin, 1964; Aviv et al., 1984; Frankel & Galun, 1977; Aviv & Galun, 1986), Pino Perez (2012)	Normal pistil protrudes widely above shortened corolla, stamens transformed into large petaloid structures, occasional small stigmata capping the petaloids
N. glutinosa	?	Naumenko (2012), Baranova et al. (2015), Burk and Durbin (1978)[17] and Gerstel (1980)[17]	No details Male fertility and floral morphology not affected

Compiled after Gerstel (1980), Tsikov and Tsikova (1981), and Berbeć and Doroszewska (2020), with additions and modifications

Abbreviations: *PreM* premeiotic (sporogenous tissue is developed, PMC's degenerate before meiotic divisions); *PM* postmeiotic (stamens normally developed), anthers morphologically normal to shrunken with aborted or non-germinating pollen or void of pollen; *ST* staminal type (stamens absent or transformed into petaloid or stigmatoid structures)

[1]Obtained by asymmetric protoplast fusion (*N. tabacum* + *N. alata*), mix of mitochondrial DNA from both parents

[2] Found in the backcross progenies of the cross of maternal *N. tabacum* × *N. alata*

[3]Found in backcross progenies of the cross of maternal *N. tabacum* × *N. plumbaginifolia*

[4]Spontaneous cms mutation discovered in the field of a normal male fertile *N. tabacum* variety; flower morphology very close to that of cms *glauca,* involvement of *N. glauca* cytoplasm suggested by Gerstel (1980), also implied by the data provided by Chen et al. (1977) and Kung (1977); cms *tabacum* and cms *glauca* differ from each other for some aspects of growth and development

(continued)

[5]Found in backcross progenies of the cross of maternal *N. tabacum* × *N. glauca*
[6]Male sterile plant appeared in the F1 generation of *N. tabacum* x *N. glutinosa*, subsequently backcrossed to *N. tabacum*
[7]Developed from sexual hybrids *N. glauca* × *N. tabacum*
[8]Developed from somatic hybrid *N. glauca* + *N. tabacum*
[9]Alloplasmic forms developed by Berbeć (1994a, 2001) showed rudimentary pollen fertility, those obtained by Kubo (1985) and by Nikova and Shabanov (1992) showed full male sterility
[10]Developed from somatic *hybrid N. repanda* + *N. tabacum*
[11]Nikova et al. (1991) transferred cms factors directly from *N. benthamiana*, Ramavarma et al. (1978) used *N. glutinosa* as a bridging species
[12]Transfer of cms reported by asymmetric protoplast fusion
[13]Cms supposedly of androgenetic origin
[14]Male sterile alloplasmics obtained among highly asymmetric regenerants from protoplast fusion, varied flower morphology probably due to due to chromosomal and mitochondrial segregation
[15]Mode of development and floral morphology unknown
[16]Developed by asymmetric somatic hybridization
[17]Absence of male sterilizing effect reported

tabacum (*alata*) (Nikova et al., 1999) and cms *tabacum* (*glauca*) (Stoyanova, 1972; Tsikov & Tsikova, 1981) were discovered later and their mode of origin was very similar to that of Burk's cms stocks. For further comments on these plasmatypes go to Sect. 5.4.1).

Table 5.3 separately lists those cms forms that were synthesized via somatic hybridization. They account for a relatively small proportion of all alloplasmics involving *N. tabacum,* much like the somatic hybrids account for a small proportion of all interspecific combinations in the genus.

5.3.2 Restoration of Male Fertility to cms Alloplasmics

The complementary relationship between the alien cytoplasm and native chromosomes in the alloplasmic lines can be restored by reintroduction of specific genes or chromosomes from the donor species. Sand and Christoff (1972) reported that in the breakdown generations from backcrossing the hybrid *N. debneyi* × *N. tabacum* to *N. tabacum,* the expression of male sterility varied from stigmas instead of anthers to nearly restored fertility with pollen grains that failed to germinate on sucrose agar. Burk (1960) developed a double alien addition line containing two pairs of chromosomes from *N. quadrivalvis* that restored fertility to the alloplasmic cms *quadrivalvis*. Male fertility can also be restored at the mitochondrial level. By fusing alloplasmic lines *N. tabacum* cms *quadrivalvis* and *N. tabacum* cms *undulata* Kofer et al. (1991, 1992) obtained fertile mitochondrial recombinants in both alloplasmics that apparently acquired mt DNA sequences complementary to *N. tabacum* chromosomal genes responsible for the expression of male fertility.

Expression of cytoplasmic male sterility and restoration of male fertility are probably closely related to the presence and activity of nucleolar-organizing chromosomes. These are satellited chromosomes carrying nucleus organizing regions

Table 5.3 Cytoplasmic male sterility transferred to *N. tabacum* by protoplast fusion

Source of cytoplasmic male sterility in *Nicotiana tabacum*	Authors	Remarks
N. alata	Atanassov et al. (1998)	cms factors transferred from gamma-irradiated donor species; recovered cms lineages represented mitochondrial DNA recombinants of both parents
N. plumbaginifolia	Menczel et al. (1986) and Bates et al. (1987)	
N. glauca	Chen et al. (2012)	cms factors transferred by fusion with non-inactivated donor species followed by sexual backcrossing to pollen parent
N. quadrivalvis	Aviv et al. (1984) and Aviv and Galun (1986)	cms factors transferred from X-irradiated donor species; Floral morphology of cms cybrids resembled closely cms *quadrivalvis* from sexual backcrosses
N. repanda	Kumashiro et al. (1989) Bates (1990) Sun et al. (1999b)	cms factors transferred from X-irradiated donor species cms factors transferred from gamma-irradiated donor species cms factors transferred to rhodamine-inactivated recipient species
N. africana	Kumashiro et al. (1988)	cms factors transferred from X-irradiated donor species
N. benthamiana	Atanassov (1993)	cms factors transferred by asymmetric protoplast fusion
N. debneyi	Kumashiro and Kubo (1986)	cms factors transferred from X-irradiated donor species by asymmetric fusion
N. maritima	Nikova et al. (1990)	cms factors transferred from gamma-irradiated donor species
N. rotundifolia	Ilcheva et al. (2001)	*N. tabacum*-like cms segregants selected among symmetric fusion products
N. megalosiphon	Kasza and Kandra (1990)	No details available
N. suaveolens	Kubo (1985) and Kubo et al. (1988) Matibiri and Mantell (1994)	cms factors transferred from X-irradiated donor species cms factors transferred from X-irradiated donor species to iodoacetamide inactivated recipient (double inactivation fusion)
N. undulata	Aviv et al. (1984) and Aviv and Galun (1986)	cms factors transferred from X-irradiated donor species, floral morphology of cms cybrids resembled closely cms *undulata* from sexual backcrosses

(NORs) which exhibit nucleolus-forming activities by becoming physically attached to the nucleolus at meiosis (Gerstel et al., 1978, Burns and Gerstel, 1981). In interspecific combinations the activity of NORs from one species may be suppressed in the presence of NORs from another one and the phenomenon is called amphiplasty (Burns and Gerstel 1981). In this manner, pollen-producing functionality can be restored to a cytoplasmically male sterile alloplasmic by introducing nucleolar organizers from the cytoplasm donor (Gerstel et al., 1978, 1980). The mechanism was experimentally demonstrated by Gerstel et al. (1978) who introgressed a NOR-bearing doubled chromosome fragment, most likely from *N. repanda*, to *N. tabacum* cms *repanda*. In contrast to the regular cms *repanda*, the alien addition alloplasmic line thus developed showed full male fertility. In the restored lineage, it is the alien chromosome fragment from *N. repanda* that organized the nucleolus at the same time inhibiting the nucleolus-forming activity of native *N. tabacum* chromosomes. Basically, the same mechanism of anther restoration was observed in several other alloplasmics with added chromosome fragments from an alien cytoplasm donor. These alloplasmics included cms *debneyi* (Burns et al., 1978, Gerstel et al., 1980), cms *undulata* (Burns and Gerstel, 1981) and cms *quadrivalvis* (Gerstel and Burns, 1983). Unlike the restored *cms repanda*, these latter cms combinations showed only partial or minimal reversion to pollen fertility.

In most cases, restoration of male sterility is cytoplasm-nucleus specific. Grebenkin (1968) mated *N. tabacum* cms *debneyi* to *N. debneyi*, *N. tomentosa*, *N. tomentosiformis*, *N. otophora*, *N. setchellii*, *N. solanifolia*, *N. raimondii*, *N. benavidesii* and *N. glauca* to assess the potential of those species to restore normal flower morphology to the respective F_1 hybrids. All the F_1s, except *N. tabacum* cms *debneyi* x *N. debneyi*, developed flower malformations characteristic of their maternal alloplasmic parent. Normal flower morphology was recovered only in the amphihaploid 2x (*N. debneyi* x *N. tabacum*) that, upon chromosome doubling, could be converted to a fully fertile amphidiploid.

Examples of restored male fertility in different alloplasmics of *N. tabacum* are listed in Table 5.4. In some cases, a male fertile line with restorer gene/s from one species could restore male fertility to more than one type of alloplasmics (Reed & Burns, 1986). A curious case that may fall into the latter category was recently reported by Liao et al. (2017). The report was discussed earlier in this review in connection with diploid materials in the offspring of interspecific matings (see Sect. 3.5.2). The mixed progeny from hybridizing *N. tabacum* cms *glauca* with *N. alata* consisted of sterile phenotypes intermediate between the parental species and self-fertile plants that resembled the phenotype of the cultivated parent (cv. K 326). Genotyping with SSR markers showed that both classes of progeny, i.e., the plants that appeared to be regular F_1 hybrids *N. tabacum* x *N. alata* and the *N. tabacum*-resembling plants, contained some DNA material from *N. alata* and thus were of hybrid rather than of asexual origin. Both classes also shared restored stamen morphology, although in the putative allohaploid plants, pollen grains were aborted, as would be expected of this interspecific hybrid. The researchers further hypothesized that the *N. alata*-derived DNA sequences in the *N. tabacum*-like plants were responsible for restored male fertility in those maternal offspring of the *N. tabacum* × *N. alata* cross, a supposition supported by the restored stamens in what appeared to be the regular *N. tabacum* cms *glauca* x *N. alata* hybrids. As the authors themselves admitted, these unusual findings needed further study.

Gurdon et al. (2016) demonstrated that male-fertile floral morphology and pollen production could be restored to *Nicotiana tabacum* cms *undulata* by grafting. Male fertile branches were regenerated from the fusion zone of the graft junction with

Table 5.4 Male fertility restoration in alloplasmic *N. tabacum* lines involving cms factors from different *Nicotiana* sources

Cytoplasmic male sterility introgressed from	Authors	Notes
N. longiflora	Ahuja (1962)	Tumor-forming and male fertility-restoring factors were demonstrated to share the same region of a *N. longiflora* chromosome
N. glauca	Liao et al. (2017)	Male fertility restored in *N. tabacum*-like phenotypes obtained from maternal *N. tabacum* cms *glauca* plants pollinated with *N. alata* pollen. *N. alata* is likely to carry factors that restore male fertility to cms *glauca* alloplasmics as indicated by restored stamen morphology in the F_1 hybrids *N. tabacum* cms *glauca* x *N. alata*
N. quadrivalvis	Burk (1960)	Male fertility partly restored in a double alien addition line involving 2 pairs of chromosomes from *N. quadrivalvis*
	Aviv and Galun (1986)	Male fertility restored through donor recipient asymmetric fusion and recombination of mitochondrial DNA
	Kofer et al. (1991)	Male fertility restored by protoplast fusion and complementary recombination of mt DNA from *N. undulata* and *N. quadrivalvis*
N. repanda	Burk and Mann (1970), Burk and Durbin (1978), Gerstel et al. (1978), Gerstel (1980) and Reed and Burns (1986)	Male fertility restorer genes introgressed as an alien addition of a doubled chromosome fragment presumably from cytoplasm
	Reed and Burns (1986)	Male fertility completely restored by cms *suaveolens* restorer line
N. rustica	Hart (after Smith, 1968)	Partial male fertility restored by the presence one or more *N. rustica* chromosomes
N. debneyi	Sand and Christoff (1973)	several *N. debneyi* chromosomes restore normal flower morphology
	Burns et al. (1978) and Reed and Burns (1986)	Fully restored flower morphology but aborted inviable pollen
	Reed and Burns (1986)	Male fertility restored by cms quadrivalvis restorer line
N. suaveolens	Schweppenhauser and Mann (1968)	Fertility restorer genes (four genes – Linked or independent) introgressed from cytoplasm donor. Male fertility restored to a varying extent

(continued)

Table 5.4 (continued)

Cytoplasmic male sterility introgressed from	Authors	Notes
	Kofer et al. (1992)	Male fertility restored through donor-recipient asymmetric fusion and recombination of mitochondrial DNA
	Kandra (1984)	Male fertility partially restored by crossing to *N. africana*
	Reed and Burns (1986)	Male fertility completely restored by cms suaveolens restorer line and partly restored by cms repanda restorer line
N. undulata	Burns and Gerstel (1981) and Reed and Burns (1986)	Male fertility partly restored in a 50-chromosome alien addition line carrying fertility restoring genes from *N. undulata*
	Aviv and Galun (1986) and Raineri et al. (1992)	Male fertility restored through donor recipient asymmetric fusion and recombination of mitochondrial DNA
	Kofer et al. (1991)	Male fertility restored by protoplast fusion and complementary recombination with mt DNA from *N. undulata* and *N. quadrivalvis*
	Reed and Burns (1986)	Male fertility completely restored by cms *suaveolens* restorer line
	Gurdon et al. (2016)	Male fertility restored by mitochondrial transfer and recombination through a graft junction with *N. sylvestris*

N. sylvestris, indicating the horizontal transfer of mitochondrial genes from *N. sylvestris* and their integration with the cms *undulata* mitochondrial genome through recombination.

5.4 Origin and Identity of Some Plasmatypes in *N. tabacum*

5.4.1 CMS Systems Inherited Maternally that Originated with the Involvement of an Alien Species as the Pollen Parent

In Sect. 5.3.1 were discussed, among other cms lineages, several alloplasmic forms that arose from interspecific crosses in which *N. tabacum* was the maternal parent that supplied the cytoplasmic factors. They included *tabacum* (*plumbaginifolia*), cms *tabacum* (*glutinosa*), cms *tabacum* (*alata*) and cms *tabacum* (*glauca*). Reporting on his cms *tabacum* (*plumbaginifolia*) and cms *tabacum* (*glutinosa*) Burk (1960) was unable to offer an explanation other than that it was the result of some undefined interactions of plasmagenes with non-specific nuclear components in the first or in the early backcross generations of his hybrids. Gerstel (1980) rejected the possibility that the plasmon causing male sterility in *tabacum* (*glutinosa*) was introduced from paternal *N. glutinosa*. He did so on the grounds that the reciprocal transfer involving *N. glutinosa* as the cytoplasmic nonrecurrent parent had resulted in plants with normal anthers. However, the more recent findings on the rare cases of chloroplast and mitochondrial DNA transfer through the sperm accompanied by mitochondrial recombinations have rendered the pollen transfer hypothesis much more plausible (see the subsequent Sect. 5.4.2).

On page 433 of the monograph "Biochemical Aspects of Crop Improvement", Svistava and Gupta (1991) included a piece of information to the effect that both cms *tabacum* (*glutinosa*) and cms *tabacum* (*plumbaginifolia*) stocks had been lost and therefore were no longer available for study. However, this must not necessarily be the case. The alloplasmic systems that perfectly fit the original description of both cms *tabacum* (*glutinosa*) and cms *tabacum* (*plumbaginifolia*), although presumably not always under their proper names (see the following section) in all probability have been studied and are still being maintained by at least some research establishments.

5.4.2 Controversies Regarding the Provenance and Identity of Some Plasmatypes in N. tabacum

Cms plumbaginifolia* vs. *cms tabacum (plumbaginifolia) The floral morphology of *N. tabacum* cms *plumbaginifolia* was allocated by Chaplin (1964) to type IV: slightly shortened corolla, pistil protruding above the corolla, stamens with shortened filaments and normal-appearing anthers void of pollen. In another report (Chen et al., 1977), the alloplasmic Burley 21 cms *plumbaginifolia* was likewise described as male sterile with 'defective anthers'. Anthers, defective or otherwise, implied the presence of stamens, be they normal or underdeveloped. Similar floral modifications were reported in a cms accession maintained in Russia and referred to as originating from *N. plumbaginifolia*: petaloid filaments tipped with well-developed anthers void of pollen (Grebenkin et al., 1976). Likewise, Frankel et al. (1979) describe flowers of cms plumbaginifolia as having shortened (reduced?) anthers void of pollen. Asymmetric hybrid plants combining the nucleus of *N. tabacum* with the cytoplasm of *N. plumbaginifolia* obtained by Bates et al. (1987) had shortened corollas, protruding pistils and shortened stamens, which roughly corresponded to the original description by Chaplin (1964). In contrast to the above descriptions of cms *plumbaginifolia*, Burk and Durbin (1978) reported that their alloplasmic stocks of *N. tabacum* with the cytoplasm of *N. plumbaginifolia* did not express flower anomalies associated with male sterility. Those alloplasmics could also be propagated by selfing (Moav et al., 1968; Burk and Durbin, 1978).

The first doubts concerning the validity of cms *plumbaginifolia* were raised by Bonnet and Glimelius (1983). The authors challenged the authenticity of the cms *plumbaginifolia* stock in their possession and deemed it to be an accession of cms *undulata*. However, in support of their correction they apparently misconstrued some of the data contained in the reports by Burk, 1960; Chen et al., 1977; Chaplin, 1964). Firstly, Burk (1960) described two cms lineages that involved *N. plumbaginifolia*: cms *plumbaginifolia* and cms *tabacum* (*plumbaginifolia*), differing both in origin and in floral morphology. Secondly, Chen (1977) never mentioned *N. undulata*, but he did *N. glutinosa* among the sources of his RuBisCo large subunit markers. Thirdly, Chaplin (1964) described all four: cms *undulata*, cms *tabacum* (*plumbaginifolia*), cms *tabacum* (*glutinosa*) and cms *plumbaginifolia*. Of these, the first three represented the same floral phenotype, the fourth was markedly different. So, based on these three reports plus their own description Bonnet and Glimelius (1983) were right in stating that the cms accession in their study was not cms *plumbaginifolia*. Admittedly, it could have been cms *undulata* or even cms *tabacum* (*glutinosa*). Most likely, however, it was cms *tabacum* (*plumbaginifolia*). The confusion concerning cms *plumbaginifolia* and cms *tabacum* (*plumbaginifolia*) is further discussed in the next paragraph.

Floral modifications in cms *plumbaginifolia* as described by Chaplin (1964), Bates et al. (1987) and in other reports cited in the first paragraph of this section differ substantially from those reported for the alloplasmic stocks designated cms *plumbaginifolia* acquired from a foreign source and maintained in this laboratory (Berbeć, 2001; Czubacka et al., 2016) At the same time, they suit perfectly the flower morphology of *tabacum* (*plumbaginifolia*). Chaplin (1964) classified cms *tabacum* (*plumbaginifolia*) as morphological type V, where he placed it together with cms *undulata* and cms *tabacum* (*glutinosa*) as having a considerably shortened corolla, a normal and widely protruding pistil and anthers transformed into large petals tipped with stigmatoid structures of different sizes (Burk, 1960; Chaplin, 1964). Two more recent reports from Russia (Naumenko, 2012; Baranova et al., 2015) both treat cms *plumbaginifolia* and cms *N. tabacum* cms *tabacum* (*plumbaginifolia*) as distinct plasmatypes. It appears from these studies that the two cms sources differed in many aspects of agronomic performance. Regretfully, neither report provided morphological descriptions of these accessions, i.e., cms *tabacum* (*plumbaginifolia*) or cms *plumbaginifolia*.

Tentatively, it may be concluded that the reconstitution of *the N. tabacum* genome in the plasmon of *N. plumbaginifolia* produces moderate alterations to flower organs and postmeiotic disturbances of microsporogenesis, resulting in either no pollen, aborted pollen, or a mixture of viable and aborted pollen. On the other hand, lineages that take their origin from the original cms *tabacum* (*plumbaginifolia*) are characterized by premeiotic, staminal type of male sterility and the absence of sporogenic tissue.

Floral modifications associated with cms *tabacum* (*plumbaginifolia*) can also be found in nearly identical form in several other alloplasmics of *N. tabacum*: cms *tabacum* (*glutinosa*), cms *eastii* (Tsikov & Tsikova, 1981; Berbeć & Berbeć, 1992) and cms *benthamiana* (Nikova et al., 1991; Nikova & Vladova, 2002).

***Cms glutinosa* vs *cms tabacum (glutinosa)*.** Descriptions and observations of cytoplasmic accessions that go under the names cms *glutinosa* and cms *tabacum* (*glutinosa*) provoke doubts of a different nature. The floral morphology of the alloplasmic system designated cms *glutinosa* acquired from a foreign source and maintained in this laboratory (Berbeć, 2001; Czubacka et al., 2016) corresponds faithfully to the description of cms *tabacum* (*glutinosa*) by Chaplin, 1964), who classified it, along with the previously discussed cms *tabacum* (*plumbaginifolia*), as type V. The flower modifications reported by Frankel et al. (1979) in what they called cms *glutinosa* also corresponded to flower malformations produced by the *tabacum* (*glutinosa*) system as originally described by Burk (1960) and by Chaplin (1964). Another reference to cms *glutinosa* is found in the report by Berbeć (1974), but the paper referred to (Tsikova, 1967) probably also deals with cms *tabacum* (*glutinosa*), a clue having been contained in the Bulgarian paper's title. In the early reviews of cms in *Nicotiana*, Gerstel (1980) does not list cms *glutinosa* and neither do Tsikova and Tsikova (1981), although both papers contain descriptions of cms *tabacum* (*glutinosa*). In the instance of the alleged cms *glutinosa*, alloplasmic lines of *N. tabacum* with the cytoplasm of *N. glutinosa* are known from two accounts (Burk & Durbin, 1978; Gerstel, 1980). In both, they are reported to have retained normal flower morphology and male fertility. Also in earlier studies (Clausen, 1928, Holmes, 1938), the authors did not report any male sterility issues or flower malformations while working with populations derived from the synthetic *N. digluta* repeatedly crossed with *N. tabacum* as the recurrent pollen parent. Likewise, it appears from a study by Russian interspecific breeders (Ternovsky and Nosova 1970, 1971) that the introgression of TMV resistance from *N. glutinosa* to *N. tabacum* proceeded undisturbed and without interference from male sterile floral anomalies regardless of whether the matrilineal inheritance in the transfer involved the cytoplasm from *N. tabacum* or from *N. glutinosa.* Of the early studies of cms in *N. tabacum*, the only one that indirectly links *N. glutinosa* to male sterility in *N. tabacum* is the report on cytoplasmic markers in *N. tabacum* by Chen et al. (1977) cited in the previous paragraph in connection with cms *plumbaginifolia.* All this notwithstanding, a male sterile cytoplasmic lineage designated 'cms *glutinosa'* was studied for its agronomic performance by Naumenko (2012) alongside another, named *tabacum* × *glutinosa* – a likely synonym of cms *tabacum* (*glutinosa*). Cms *glutinosa* and cms *glutinosa* (*tabacum*) are also listed as separate cms accessions in a Russian tobacco germplasm collection by Baranova et al. (2015). Neither report provides details on the provenance or morphology of either of these two cms systems. The picture is further complicated by the fact that the two alloplasmic accessions in the Puławy collection, cms *plumbaginifolia* and cms *glutinosa,* were found to be tentoxin sensitive (Berbeć, 2001), which agrees well with their declared cytoplasmic provenance.

Based on the evidence accumulated to date it can be assumed that the true alloplasmics of *N. glutinosa* x *N. tabacum* do not show changes in their floral morphology or suppresion of male fertility and cms *glutinosa* is probably a nonexistent plasmatype. On the other hand, the reports that include references to cms *glutinosa*, of which that by Zheng et al. (2018) is the most recent, cannot be dismissed offhand as mere cases of mistaken identity. In summary, a tentative conclusion is that the male sterile accessions that go under the name of 'cms glutinosa' are probably derivatives of the original cms tabacum *(glutinosa)* developed by Burk (1960) and described by Chaplin (1964).

Cms *tabacum* (*plumbaginifolia*) and cms *tabacum* (*glutinosa*) vs. cms *undulata*. Cms *tabacum* (*plumbaginifolia*) and cms *tabacum* (*glutinosa*) develop the same floral modifications that are characteristic of cms *undulata*. Given that the identities of both cms *tabacum* (*glutinosa*) and *N. tabacum* cms (*plumbaginifolia*) are correct and they are cytoplasmic descendants of the cytoplasmic male sterile forms reported by Burk (1960) their tentoxin responses and pedigrees can be reconciled on the assumption that both were products of chloroplast DNA substitution and mitochondrial recombination resulting from a rare transfer of cytoplasmic elements through pollen (Svab & Maliga, 2007; see also the subsequent comment on cms 'tabacum-mutant' vs. cms *glauca*). In the study by Naumenko (2012) on certain agronomic indices of different alloplasmic isolines of cv. 'Virginia Puławska', the plants of cms *tabacum* (*glutinosa*) were much shorther and their number of leaves slightly lower than those of cms *tabacum* (*plumbaginifolia*) or cms *undulata*. In the same study, the leaves of cms *tabacum* (*plumbaginifolia*) were smaller than those of cms (*glutinosa*) or cms *undulata*. Leaf nicotine and sugar contents were lower in cms *tabacum* (*plumbaginifolia*) than in cms *tabacum* (*glutinosa*) Chaplin and Ford (1965) observed small yield reductions in both cms *tabacum* (*plumbaginifolia*) and *tabacum* (*glutinosa*) vs. their fertile isogenic accession. In the study by Berbeć and Laskowska (2005), the cured leaf yield of both cms *tabacum* (*plumbaginifolia*) and cms *tabacum* (*glutinosa*)[2] was negatively affected compared to that of the fertile isoline, but the negative impact of the cms *tabacum* (*plumbaginifolia*) system was greater of the two and also included other components of agronomic performance such as internode length, plant height, leaf size and gross money returns (see Table 5.5). The varied responses of these three cms systems may point to their different origins and thus indirectly support the authenticity of *tabacum* (*plumbaginifolia*) and *tabacum* (*glutinosa*). The identity of cms *undulata* cannot be questioned since it consistently showed the same characteristic floral alterations in several independent studies and documented transfers, as well as over a number of different nuclear genotypes (Chaplin, 1964; Nikova & Tsikov, 1976; Burns & Gerstel, 1981; Bonnet & Glimelius, 1983; Aviv et al., 1984; Aviv & Galun, 1986; Svab & Maliga, 2007; Gurdon et al., 2016). However, until definitively disproved, the hypothesis that *tabacum* (*plumbaginifolia*) and *tabacum* (*glutinosa*) are just cytoplasmic lineages of cms *undulata* still remains a viable option.

[2]In the cited paper (Berbeć & Laskowska, 2005), these alloplasmics go under the designations cms *plumbaginifolia* and cms *glutinosa*. For the reasons expounded in the preceding paragraphs in all likelihood they represent the descendants of cms *tabacum* (*plumbaginifolia*) and cms *tabacum* (*glutinosa*), respectively.

Table 5.5 Potential of *Nicotiana* species as sources of usable cms in cultivated tobacco *N. tabacum*

Alloplasmic	Potential for tobacco improvement
cms *suaveolens*	See Table 5.6
cms *glauca*	See Table 5.6
cms *undulata*	Indicated as suitable for breeding and seed production (Chaplin & Ford, 1965; Tsikov & Tsikova, 1974; Tsikov & Nikova, 1980; Sastri et al., 1982); positive effect on leaf area (Naumenko, 2012) and seed yield (Baranova et al., 2015); no negative impact on plant growth and yield indices (Maktari, 1991; Berbeć & Laskowska, 2005)[1]; negative impact on curability of FC cultivars (Sastri et al., 1982), decreased reduced sugars and increased nicotine (Sastri et al., 1982); decreased protein content (Baranova et al., 2015); increased tolerance of PVY infection in a PVY-tolerant variety (Czubacka et al., 2019)[1]; increased susceptibility to frog eye (Berbeć & Laskowska, 2005)[1]; negative effect on smoke quality (Yamada et al., 1984), reduced seed yield (Kubo, 1985)
cms *benthamiana*	No significant impact on agronomically important traits vs. its autoplasmic isogenic counterpart (Nikova et al., 2004)
cms *repanda*	Reported as agronomically usable and with potential of being deployed in commercial tobacco production (Sun et al., 1999b)
cms *amplexicaulis*,	Generally indicated as having a potential for tobacco breeding and seed production (Nikova et al., 2004); no deleterious impact on days to flower or leaf size (Czubacka et al. 2016)
cms *longiflora*, cms *paniculata*, cms *velutina*, cms *maritima*	Generally indicated as having a potential for tobacco breeding and seed production (Nikova et al., 2004)
cms *tabacum* (*plumbaginifolia*)	Small yield reduction vs. autoplasmic isogenic counterpart reported by Chaplin & Ford, 1965); negative impact on leaf yield (Berbeć & Laskowska, 2005)[2]; indicated as suitable for tobacco breeding by Tsikov, 1982; Naumenko, 2012); delayed flowering and decreased mid-position leaf area (Czubacka et al., 2016)[2]; increased susceptibility to blue mold (Berbeć, 2001)[2]; increased nicotine content over male fertile counterparts (Baranova et al., 2015)
cms *tabacum* (*glutinosa*)	Preferable cms system for oriental varieties (Tsikov & Tsikova, 1974; Tsikov et al., 1974; small yield reduction vs. autoplasmic isogenic counterpart (Chaplin & Ford, 1965); reduced female fertility, substantially reduced internode length, plant height, leaf size, cured leaf yield and money returns vs. autoplasmic isogenic counterpart (Berbeć & Laskowska, 2005)[2]; decreased plant height and increased leaf area (Naumenko, 2012); increased substantial increase in carbohydrate content over male fertile counterparts (Baranova et al., 2015)

(continued)

Table 5.5 (continued)

Alloplasmic	Potential for tobacco improvement
cms *megalosiphon*	Depressed plant vigor, delayed growth rate (Chaplin & Ford, 1965; Dudek, 1971; Roman & Nalepa, 1970; Tsikov & Tsikova, 1974); slower growth and lower plant height (Yamada et al., 1984); delayed flowering (Kubo, 1985), and lowered leaf number (Chaplin & Ford, 1965), decreased leaf size and leaf number (Dudek, 1971; Biskup et al., 1972), conflicting reports on yield and crop quality: no negative effect (Chaplin & Ford, 1965), both yield and crop quality negatively affected (Roman & Nalepa, 1970; Berbeć & Laskowska, 2005); depressed seed yield (Chaplin, 1964; Maktari, 1991); depressed smoking quality (Kubo, 1985); increased tolerance of PVY infection in a PVY-tolerant variety (Czubacka et al., 2019); positive effect on seed yield (Baranova et al., 2015)
cms *rustica*	Unusable due to the occurrence of lineages with vestigial male fertility (Hart, 1965; Kubo, 1981, 1985); no significant impact on agronomically important traits vs. the autoplasmic isogenic counterpart (Nikova et al., 2004); potential for hybrid seed production also indicated by Pittarelli and Sisson (1989).
cms *knightiana*, cms *raimondii*	Unusable due to the occurrence of lineages with vestigial male fertility (Berbeć & Laskowska, 2005); significant growth retardation and leaf yield reduction vs. the isogenic autoplasmic counterpart (Berbeć & Laskowska, 2005; Czubacka et al., 2016)
cms *quadrivalvis* (see also cms 'bigelovii' listed in Table 5.5)	Indicated as usable by Tsikov and Nikova (1980) and Maktari (1991); fewer days to flower, increase in leaf number, carbohydrate content over male fertile counterparts and (Naumenko, 2012; Baranova et al., 2015); substantial reduction in plant height and leaf area (Naumenko, 2012); unusable due to significant negative impact on cured leaf yield and quality (Chaplin & Ford, 1965; Aycock & Erdogan, 1979)
cms *plumbaginifolia*	Indicated as suitable for seed production by Maktari (1991) and Naumenko (2012), unusable due to seed yield reduction (Chaplin, 1964; Aycock & Erdogan, 1979); reduced number of leaves and leaf size, negative impact on leaf yield (Chaplin & Ford, 1965); lowered smoking quality ((Yamada et al., 1984); increased nicotine content (Baranova et al., 2015); self-fertile lineage implied in the report by Burk and Durbin (1978)
cms *africana,* cms *excelsior*	Unusable due to heavy depression of growth, development and leaf yield indices vs. the autoplasmic isogenic counterpart (Nikova et al., 2004)
cms *debneyi*	Increased mid-position leaf area vs. autoplasmic counterpart (Czubacka et al., 2016); showed some promise for use in oriental tobaccos (Tsikov & Tsikova, 1974); higher plant height, leaf number and leaf area vs. a fertile isoline

(continued)

Table 5.5 (continued)

Alloplasmic	Potential for tobacco improvement
	(Naumenko, 2012); in most other studies unusable because of delayed flowering and delayed leaf maturity (Roman & Nalepa, 1970; Dudek, 1971; Kubo, 1985; Maktari, 1991); depressed vegetative vigor, reduced seed yield (Roman & Nalepa, 1970; Dudek, 1971; Kubo, 1985; Maktari, 1991); negative impact on cured leaf yield, quality and money returns to grower (Roman & Nalepa, 1970; Berbeć & Laskowska, 2005), depressed smoking quality (Kubo, 1985); some lineages are efficient seed producers (Tsikov, 1982)
cms *eastii*	Depressed cured leaf yield and money returns vs. the autoplasmic isogenic counterpart (Bezova & Skula, 1980); lowered plant height and leaf size (Naumenko, 2012); increased susceptibility to blue mold (Berbeć, 2001); depressed tolerance of PVY infection in a PVY-tolerant variety (Czubacka et al., 2019); increased nicotine and decreased protein contents (Baranova et al., 2015)
cms *exigua*	Slightly reduced plant height and leaf area (Naumenko, 2012); unusable due to depressed seed yield, delayed growth rate and flowering (Maktari, 1991), reduced cured leaf yield and quality vs. the autoplasmic isogenic counterpart (Berbeć & Laskowska, 2005); heavily depressed seed yield (Naumenko, 2012); considerably increased protein content (Baranova et al., 2015)
cms *goodspeedii*	Reported as suitable for commercial hybrids in oriental tobacco (Palakarcheva & Peeva, 1976; Iancheva & Palakarcheva, 1990; Palakarcheva et al., 1990); unusable due to depressed seed yield (Maktari, 1991; delayed growth rate, delayed flowering, reduced cured leaf yield and quality vs. the autoplasmic isogenic counterpart (Berbeć & Laskowska, 2005); substantial increase in carbohydrate and protein contents over male fertile counterparts (Baranova et al., 2015); increased tolerance of PVY infection in a PVY-tolerant variety (Czubacka et al., 2019)
cms *gossei*	Unusable due to depressed female fertility (Tsikov & Tsikova, 1986)
cms *occidentalis*	Unusable due to substantial reduction of most indices of agronomic performance such as seed yield, growth rate and cured leaf yield (Maktari, 1991; Berbeć & Laskowska, 2005; Naumenko, 2012); considerably increased protein content Baranova et al., 2015); depressed tolerance of PVY infection in a PVY-tolerant variety (Czubacka et al., 2019)
cms *wuttkei*	unusable due to a substantial negative impact on plant height, leaf number, leaf size and days to flower (Laskowska & Berbeć, 2007)

(continued)

Table 5.5 (continued)

Alloplasmic	Potential for tobacco improvement
cms *clevelandii*, cms *hesperis*, cms *simulans*, cms *rotundifolia*	No data on agronomic usability

After Doroszewska and Berbeć (2020) with minor corrections and additions
Cytoplasm donors roughly arranged in the order of decreasing agronomic usability
[1]Reports on the performance of cms *undulata* took as their subject of study the cms lineages that substantially depart for their flower morphology from anomalies that were usually described as associated with that type of cms genetics. Cms *undulata* was developed by Dr. D. R. Cameron (Chaplin, 1964) and was consistently reported to show the same flower anomalies (markedly shortened corolla, protruding stigma, and stamens turned into large petaloid structures) over a number of *N. tabacum* nuclear genotypes: var. 'Hicks Broadleaf' (Chaplin, 1964), 'Red Russian' (Burns & Gerstel, 1981), 'Burley 21' (Bonnet & Glimelius, 1983; Kofer et al., 1990, 1991), an unspecified genotype (Aviv & Galun, 1986; Gurdon et al., 2016). The cms source acquired under the name of cms *undulata*, transferred to the nuclear backgrounds of var. 'Zamojska 4' and,Wiślica' and studied by Berbeć, 2001; Czubacka et al., 2016, 2019), had normal corolla, normal or deformed pistil, and missing or stigmatoid stamens in which it approximated cms *suaveolens* and cms 'bigelovii' (see also Sect. 5.4.2)
[2]The cited reports originally list *N. tabacum* cms *glutinosa* or *N. tabacum* cms *plumbaginifolia* as the subject of study. In all likelihood, those lineages were misnamed at some stage and should be treated as *N. tabacum* (*glutinosa*) and *N. tabacum* (*plumbaginifolia*), respectively, due to the reasons expounded in Sect. 5.4.2

Cms 'tabacum-mutant' vs. cms *glauca*. Another case, and possibly the oldest contested case of cms origin, is an apparently spontaneous cms mutation discovered by Berbeć (1974) and listed in this volume as cms 'tabacum-mutant'. In its flower morphology, it closely resembles cms *glauca*. Cms 'tabacum-mutant' and cms *glauca* share the ct-DNA coded sensitivity to tentoxin (Berbeć, 2001), while *N. tabacum* is tentoxin-insensitive. Furthermore, the chloroplast DNA of cms 'tabacum-mutant' in two genetic backgrounds (the original cms lineage of cv. 'Nadwiślański Mały' and the 'tabacum-mutant' system transferred to cv. 'BP-210') consistently showed characteristics that made it similar to that of *N. glauca,* as shown by their large subunit polypeptide composition of RuBisCo (Chen et al., 1977) as well as by ct-DNA restriction patterns (Kung et al., 1981) in cms 'tabacum-mutant', cms *glauca*, *N. tabacum* and *N. glauca*. Admittedly, based on these considerations alone plus the recognized stability of chloroplast-based heredity (Chen et al., 1977), it was very difficult to envisage the sequence of spontaneous events that might have led to the appearance of the cms 'tabacum-mutant' system. However, it is also difficult to prove with all certainty that cms 'tabacum-mutant' arose by inadvertent confusing it with cms *glauca* as suggested by Gerstel (1980). A sequence of two unusual events accompanied the genealogy of the cms tabacum mutant. In a field-grown segregating population of an intervarietal cross, an individual was found that stood out among its sister plants for a compact, short-internoded, low-height cylindrical habit; hence, it was considered a spontaneous mutation. The plant gave rise to the dark air-cured cultivar 'Nadwiślanski Mały' (Berbeć, 1962). It is in the field planted to the immediate predecessor of Nadwiślanski Mały that the 'cms-tabacum mutant' was found. It was indistinguishable from its sister plants except for the absence of stamens (Berbeć,

1962, 1974). Hence, it was also considered a spontaneous mutation. Nadwiślański Mały, its closest sister lines and alloplasmic analogs were strikingly different from the cultivars and breeding lines maintained and studied in the laboratory at the time and could hardly be confused with any. Gerstel (1980) was right in pointing out that the work on developing cms *glauca* and cms 'tabacum-mutant' proceeded in the same laboratory and that the two projects overlapped in time (the late fifties and early sixties of the last century – Berbeć, 1972). Both male sterile cytoplasms were bred into several cultivars of *N. tabacum* (Berbeć, 1972), but cms *glauca* was never bred or attempted to be bred into Nadwiślański Mały, according to the evidence available. In addition, the two isogenomic types of the flue cured cv. 'Wiślica' vs. its tabacum-mutant analog showed some significant differences for several agronomic and economic indices in favor of cms 'tabacum-mutant' (Berbeć & Laskowska, 2005). Likewise, the two isogenomic alloplasmics of cv. 'BP-210' responded differently to different daylength conditions (Berbeć & Berbeć, 1976). Similarly, cms *glauca* and cms tabacum-mutant analogs of cv 'Zamojska 4' responded differently to infection by some PVY strains (Czubacka et al., 2019).

A clue that may help elucidate the controversial origin of the *N. tabacum* "mutant" came from a study that examined the potential routes by which plastid transgenes in cultivated genetically modified crops may escape to the environment. By crossing maternal cms *undulata* and fertile *N. tabacum* to the paternal *N. tabacum* stock transformed with chloroplast spectinomycin resistance genes, Svab and Maliga (2007) demonstrated that in 1 out of 10,000 plants, paternal chloroplast DNA was not excluded from tobacco sperm cells and in rare cases was transmitted by pollen into the next generation. Furthermore, the entire plastid DNA rather than DNA fragments was transferred, most likely within an intact chloroplast. Another piece of evidence yielded by the study was that in all those cases, paternal mitochondrial DNA was cotransmitted with chloroplast DNA. In view of these findings, the substitution of the whole cytoplasmic genome under conditions of open pollination, although admittedly very rare, is not wholly unlikely and may have indeed taken place. In the case of cms tabacum "mutant", there were ample opportunities for contamination with pollen from *N. glauca* since the species was grown both outdoors and in the greenhouse in close proximity to other breeding materials, including those from which the *N. tabacum* "mutant" had taken its origin. The weak point of the pollen transmission hypothesis is that it must also assume subsequent selective elimination of *N. glauca* chromosomes from the hybrid embryo followed by spontaneous chromosome doubling to restore both female fertility and *N. tabacum*-like morphology to the original cms "mutant" plant. Uniparental elimination of chromosomes from interspecific *Nicotiana* hybrids was reported in several studies (see Sects. 3.2.3, 3.5.2, 4.5.4), including the hybrid *N. tabacum* x *N. glauca* (Szilagyi, 1975). However, the probability that those events concurred and followed each other in an orderly sequence must be very low. On the other hand, little probable does not mean entirely impossible. If we recall the rate of 1 out of 10,000 plants containing paternal chloroplast DNA observed by Svab and Maliga (2007) and consider the reproductive power of tobacco of up to 3000 seeds per capsule and more than 50 capsules per plant (Jankowski & Poniewierski, 1969; Lewis, 2020), each single plant has a theoretical chance of yielding 15 individuals carrying paternal chloroplast DNA in their pollen. Viewed in this light, transmission of cytoplasmic elements through pollen under natural conditions is at least plausible and it may be treated as a provisional hypothesis to explain the differences in the physiological responses and agronomic performance of the two nuclear isolines – cms *glauca* and cms *tabacum* "mutant" (Berbeć & Berbeć, 1976; Berbeć & Laskowska, 2005). Yet another interesting clue came from observing stamen modifications developed by two genotypically identical cms F_1 hybrids, one involving the cytoplasm of tabacum-mutant and the other that of tabacum cms *glauca* (unpublished observations by the author of this volume). Under the same ambient conditions, both genetic systems (cms tabacum-mutant and tabacum cms *glauca*) usually show very similar morphology of reproductive organs: normal pistil, either no stamens or heavily deformed vestigial stamens. Grown side by side, the two particular cms iso-hybrids differed sharply for the extent of their stamen malformations. The cms *glauca* hybrid developed stamens with shortened or vestigial filaments capped with malformed, rudimentary structures whereas the stamens of cms tabacum-mutant iso-hybrid were almost unaffected save for shriveled anthers topping filaments of normal length. It is worth noting that Linia 8, the male parent of both hybrids was derived from the interspecific hybrid *N. exigua* x *N. tabacum* (Laskowska, 1993) that likely carried some introgression from the alien species. Therefore, it is not inconceivable that the specific nuclear-cytoplasm interactions in those two hybrid alloplasmics helped phenotypically expose inherent differences between the two plasmatypes, the differences that were not large enough to be unequivocally manifested in other nucleus-cytoplasm interactions. The pollen transmission hypothesis also may help explain the origin of *N. tabacum* (*alata*), *cms tabacum* (*plumbaginifolia*), *N. tabacum* (*glutinosa*), and *cms tabacum* (*glauca*) (see the previous comments on dubious cms cases).

Cms 'bigelovii' vs. cms *quadrivalvis* and cms *suaveolens*. Yet another doubtful case is that of cms 'bigelovii', which has been maintained in the collection of this laboratory since it was acquired from a foreign tobacco unit as an alloplasmic stock allegedly carrying the cytoplasm of *N. bigelovii*, now renamed *N. quadrivalvis*. The flower morphology of cms 'bigelovii' differs from the original descriptions of cms *quadrivalvis* provided by Burk (1960) and Chaplin (1964) but also from the flower alterations in the lineages included in the studies of Nikova and Tsikov (1976), Frankel et al. (1979), Aviv et al. (1984), Aviv and Galun (1986) and Kofer et al. (1991). In all these accounts, cms *quadrivalvis* produces a normally developed carpel and anthers with normal filaments that are tipped with feather-like structures or diminutive anthers. A brief description by Frankel et al. (1979) differs from the rest in that the petaloid malformations occur along with feathery structures. Other differences are concerned with the corolla. Normally developed corollas were reported by Chaplin (1964) and Nikova and Tsikov (1976), whereas in the reports by Burk (1960), Aviv et al. (1984), and Kofer et al. (1990), the corollas of cms *quadrivalvis* were deeply split. Cms 'bigelovii' held in this laboratory has never been observed to develop filamentous stamens with petaloid or feathery anthers or to have split corollas, regardless of the numerous nuclear backgrounds into which the system was bred. Instead, it consistently produced floral modifications very similar to those observed in cms *suaveolens* as described by Chaplin (1964), Kofer et al. (1991), Berbeć (2001) as well as in other cms *suaveolens* accessions and alloplasmic lines known to the author of this review. Original cms *quadrivalvis* was declared agronomically unusable on account of reduced leaf number, depressed cured leaf yield and increased sucker development (Chaplin & Ford, 1965; Aycock & Erdogan, 1979), which are agronomic flaws never associated with cms 'bigelovii'. It must also be noted that cms *suaveolens* and cms 'bigelovii' compared in the same nuclear background differed from each other for some agronomic traits (Berbeć & Laskowska, 2005).

Other doubtful cases of cytoplasmic male sterility in *N. tabacum* germplasm collections. The accession of cms *undulata* acquired from a foreign source and maintained in this laboratory substantially differs in its flower morphology from the lineages of cms *undulata* obtained and studied in other laboratories. Rather than the characteristic alterations that include protruding pistil, shortened corolla and large petaloid structures in place of anthers, the accession described by Berbeć (2001) and Czubacka et al. (2016) has a normally developed corolla and stamens that are transformed into small stigmatoid structures or missing altogether in which it resembled both cms *suaveolens* and cms 'bigelovii'.

There is also some confusion regarding two alloplasmic lineages of *N. tabacum* (rai-1 and rai-2) that allegedly carried the cytoplasm of *N. raimondii*. Of the two, rai-2 most likely represents true cms *raimondii* since its provenance is well documented (Berbeć, 1988), and the lineage is sensitive to tentoxin, just as is its purported cytoplasm donor (Berbeć, 2001). Unlike rai-2, rai-1 is tentoxin-insensitive (Berbeć, 2001) and, in all probability, carries the cytoplasm of *N. paniculata*, a tentoxin-insensitive species of the section Paniculatae. The floral morphology of rai-1 fits well with the description of cms *paniculata* developed and described by Nikova et al., 1991 and Nikova and Vladova (2002). The three hybrids of *N. knightiana*, *N. paniculata* and *N. raimondii* with *N. tabacum* were part of the same project (Berbeć et al., 1979). Their alloplasmic *N. tabacum* derivatives were developed in the same laboratory and displayed similar flower alterations. All these circumstances may have contributed to inadvertent mishandling of some of these seed stocks.

Experience shows that, in general terms, floral modifications related to the presence of an alien cytoplasm appear to be highly specific to a particular species or a group of species – the donors of the cytoplasm. However, within each of those specific groups, the extent of those basic modifications (shortened corolla, split corolla, vestigial or reduced stamens, malformed stamens, stigmatoid stamens, petaloid stamens) has also been found to be influenced by the nuclear genotype and ambient conditions, from almost nonexistent to very strongly expressed. Thus, floral morphology alone, although a very important clue, is by no means a decisive or sufficient criterion to verify the doubtful identity of an alloplasmic.

Cases of the mistaken identity of tobacco breeding materials and resistant lines obtained via introgression (Milla et al., 2005), including alloplasmics (Aviv et al., 1980; Bonnet & Glimelius, 1983), did occur in the past, and some of those errors were subsequently rectified. With the aid of molecular genotyping technologies, ambiguities concerning the provenance of particular cms stocks and lineages now seem to be possible to resolve, e.g., by comparing specific mt DNA sequences (Zhao et al., 2009; Zheng et al., 2018) or using single nucleotide polymorphism (SNP) markers (Priya et al., 2019). Of the questionable cms accessions maintained and used in this laboratory, cms 'bigelovii' and cms 'tabacum-mutant' possess good and often superior agronomic qualities and are deployed in the seed production and development of new hybrid varieties (Berbeć, 2017; see also Sect. 5.6.2).

5.5 Interactions of *N. tabacum* Plasmon with Nuclear Genomes of Other *Nicotianae*

Another issue, if only of academic interest, related to *Nicotiana tabacum* and cytoplasmic male sterility that has been almost completely ignored is the reverse interaction, i.e., between the plasmon of *N. tabacum* and the genome of an alien *Nicotiana* species. The author of this review was able to obtain alloplasmic forms of *N. alata, N. forgetiana* and *N. langsdorffii* with the cytoplasm of *N. tabacum*. The alloplasmics of *N. alata* were derived from a descendant of backcrossing the sesquidiploid 3x (*N. tabacum* × *N. alata*) (TTA) to *N. alata* (A). Of the six plants of that progeny that had been raised to maturity, five were phenotypically and genotypically *N. alata,* and the sixth was a hybrid of the chromosome constitution 9 II (A) + 11 I (T). Further selfing and backcrossing of the tentoxin-insensitive hybrid to tentoxin-sensitive *N. alata* restored the phenotype of the recurrent parent and retained the cytoplasm of *N. tabacum,* as determined by the insensitivity to the tentoxin test (Berbeć & Laskowska, 1997; author of this review, unpublished). The restored alloplasmics of *N. alata* showed no symptoms of male sterility or altered flower morphology, but unlike the parental accession of *N. alata,* they were self-compatible (Berbeć & Laskowska, 1997). The insensitivity to tentoxin was transferred from the alloplasmic *N. alata* to *N. langsdorffii* and to *N. forgetiana,* and the latter two alloplasmics likewise retained unaltered flower morphology and fertility (author of this review, unpublished). Similarly, the replacement of the native cytoplasm of *N. langsdorffii* by that of *N. glauca* apparently did not affect male fertility of the resulting alloplasmics of *N. langsdorffii* (Smith, 1988).

5.6 Agronomic Potential of *Nicotiana* Species as Sources of Cytoplasmic Male Sterility

5.6.1 Side Effects Associated with the Introgression of Alien Cytoplasm in Tobacco ('Cytoplasmic Drag')

Although cytoplasm-borne genetic factors account for only a very small fraction of genetic variability in tobacco, the replacement of the native cytoplasm by that from other *Nicotiana* species also may and does produce effects that extend beyond the modifications of male reproductive organs per se. This *sui generis* cytoplasmic drag has been studied in many alloplasmics of *Nicotiana tabacum*: cms *longiflora* (Nikova et al.2004a), cms *glauca* (Amankwa et al., 2014; Berbeć & Laskowska, 2005; Lawson et al., 2002), cms *paniculata* (Nikova et al., 2005; Kubo, 1985), cms *quadrivalvis* (Chaplin & Ford, 1965), cms *rustica* (Nikova et al., 2004; Kubo, 1985), cms *africana* (Nikova et al., 2004) cms *amplexicaulis* (Czubacka et al., 2016; Berbeć & Laskowska, 2005), cms *benthamiana* (Nikova et al., 2004), cms *debneyi* (Kubo, 1985), cms *exigua* (Czubacka et al., 2016; Berbeć & Laskowska, 2005), cms *goodspeedii* (Czubacka et al., 2016; Berbeć & Laskowska, 2005), cms *maritima* (Nikova et al., 2004), cms *megalosiphon* (Kubo, 1985; Chaplin & Ford, 1965), cms *occidentalis* (Naumenko, 2012; Berbeć & Laskowska, 2005) cms *suaveolens* (Amankwa et al., 2014; Berbeć & Laskowska, 2005; Lawson et al., 2002; Kubo, 1985) cms *velutina* (Nikova et al., 2004), *cms undulata* (Berbeć & Laskowska, 2005; Kubo, 1985), cms *tabacum(glutinosa)* Oczoś (1985). The effect of an alien cytoplasm on the agronomic performance of cytoplasmically sterile vs. normal male fertile breeding lines and varieties, if present, was usually, but not always, detrimental, but it varied from barely perceptible to substantial.

Berbeć and Laskowska (2005) studied 14 different alloplasmic cms isolines and their male-fertile cultivated flue-cured counterpart cv. 'Wiślica' for several indices of agronomic performance and found some significant differences that could be attributable to the effect of alien cytoplasm. Among the studied alloplasmics, cms 'bigelovii', cms 'tabacum' and cms *suaveolens* were found to be in no way inferior to their autoplasmic isolines (cv. Wiślica) in terms of selected growth parameters, i.e., plant height, days to flower, leaf number, leaf area combined with cured leaf yield, quality, and money income. Cms *undulata* and cms *glauca* were also found to be acceptable but slightly inferior to the autoplasmic variety in terms of yield indices and economic returns. Negative effects related to alien cytoplasm were particularly evident in some cms nuclear isolines: reduced leaf size in cms *raimondii*, cms *megalosiphon*, cms *occidentalis*, considerably shorter plants in cms *plumbaginifolia*, cms *tabacum (glutinosa)*, cms *eastii*, and delayed date of anthesis in cms *knightiana*, cms *raimondii*, cms *eastii*, cms *tabacum (glutinosa)* and cms *plumbaginifolia*. Save for the alloplasmics mentioned before as agronomically acceptable, depression of cured leaf yield was observed in all the others. These results are somewhat different from those reported by Lawson et al. (2002) and Amankwa et al. (2014) for cms *glauca* and cms *suaveolens*. These researchers found no important differences between the two alloplasmics in terms of agronomic suitability. The results published by Berbeć and Laskowska (2005) for cms *goodspeedii* are also in disagreement with those reported by Iancheva and Palakarcheva (1990). The latter investigators found cms *goodspeedii* to have no detrimental effects on seed production and the agronomic performance of Virginia, Burley and oriental tobacco hybrids.

Unlike the notorious example of maize, where the cytoplasmic factors linked to the cms-T system rendered the crop extremely vulnerable to infection by *Bipolar (Helminthosporium) maydis* (Bruns, 2017), the cytoplasmic effects related to disease resistance in tobacco are by far less detrimental. This notwithstanding, alien cytoplasm has been found to modify the response to pressure from some diseases (Berbeć, 2001; Berbeć & Laskowska, 2005; Czubacka et al., 2019). More details on that issue are contained in Sect. 5.6.2 and in Table 5.5, which gives information on the cms systems currently available and on what is known regarding their agronomic potential.

Variation in an alloplasmic of *N. tabacum* perpetuated by self-fertilization. A specific case of the influence that an alien cytoplasm exerts on the performance of an alloplasmic tobacco variety was studied by Berbeć (1994a, c). The alloplasmic strain of the local flue-cured cultivar 'Zamojska 4' with the cytoplasm of *N. knightiana* retained partial male fertility (see Sect. 5.2) and thus offered a unique opportunity to study the effect of cytoplasmic factors on the stability and integrity of an alloplasmic breeding line reproduced by selfing. The alloplasmic line was found to be inferior to its normal fertile counterpart for plant height, leaf width, yield and money returns, most of which were common effects of an alien cytoplasm. However, some of these parameters underwent further reduction in the successive selfed generations of the studied alloplasmic. Furthermore, a significant progeny-to-progeny variation for the characters examined in the study was found in selfed populations derived from individual alloplasmic plants. The extent of that variation was much greater than that found in the offspring of the autoplasmic Zamojska 4 or in the alloplasmic lineage reproduced using Zamojska 4 as a fertile maintainer line. It is difficult to judge how much, if any, of the observed variation was actually related to changes in the nuclear genotype or to some epigenetic effects brought on by the alien cytoplasm and whether the observed effects were specific to that particular alloplasmic combination or had a more universal character. The alloplasmic strain of Zamojska 4 had a strongly impaired ability to produce functional pollen grains. Consequently, competition among male gametophytes was greatly reduced, and minor negative mutations, including those that might have been caused by the presence of an alien cytoplasm, normally eliminated at this stage, had an increased chance of being transferred to offspring. In any case, although neither directly nor conclusively, the findings seem to support a certain tradition among Polish tobacco breeders who viewed local varieties as having a 'domesticated cytoplasm' and preferred them to foreign accessions as the maternal parents in designing their breeding projects.

5.6.2 Currently Deployed cms Systems in Commercial Hybrid Cultivars of Tobacco

Information on cms sources used in present-day tobacco hybrid cultivars is often lacking. However, it is the cytoplasm of N. suaveolens that seems to have become the cms system of choice since the first commercial hybrids started to replace traditional pure-bred cultivars (Wernsman and Rufty, 1987). It continues to be universally used because of its unquestionable merits related to yield and crop quality (Lawson et al., 2002, Amankwa et al., 2014, Zheng et al., 2018). However, alternatives to cms *suaveolens* have been intensively sought mainly because of difficulties in seed production due to depressed ovule fertility, especially in interactions with particular genotypes. In the study by Berbeć and Laskowska (2005) involving nuclear isolines of the same cv. 'Wiślica', it was not much of a problem. However, it can be a serious issue with some other genotypes to the extent of making hybrid seed production impracticable (unpublished observations by the author of this volume).

Another issue connected with cms *suaveolens* is increased susceptibility to some diseases. Under the natural field pressure from undetermined strains of PVY cv. 'Wiślica', a carrier of the 'va' locus for PVY resistance (Julio & Dorlhac de Borne, 2003), less than 1% of individuals showed PVY symptoms (Berbeć & Laskowska, 2005). Under the same conditions, 8% of cms *suaveolens* alloplasmics of 'Wiślica' but none of their cms 'bigelovii' counterparts developed PVY symptoms. In the majority of cases, the alloplasmic condition made cv. Wiślica prone to white leaf spots, mostly attributable to *Cercospora* sp. Only the plants of autoplasmic Wiślica and four alloplasmics (cms *debneyi*, cms *tabacum (glutinosa)* cms *knightiana* and cms *plumbaginifolia)* were free of those spots. In some of the alloplasmics, including cms 'bigelovii' and cms *suaveolens,* the percentage of plants with white spots was fairly high (14% in cms suaveolens and 17% in cms 'bigelovii').

Nicotiana glauca has been studied for its potential as an alternative source of cms to replace the *suaveolens* system (Lawson et al., 2002; Amankwa et al., 2014). The cms lineage named S 'K326') that deployed the cms *glauca* genetics was reported to be commercially grown on a large scale in China (Liao et al., 2017). In a more recent report, a new flue-cured hybrid cultivar with the cms system from *N. glauca* was released in Canada (Amankwa et al., 2019)

The apparently closely related cms 'tabacum–mutant' system was formerly deployed in probably the world's first commercial flue-cured tobacco hybrids (Berbeć, 1966). It continues to be used in Poland in the breeding of some flue-cured and Burley varieties and is probably also deployed in Hungarian flue-cured hybrids.

The cms 'bigelovii' system is currently the major cms source in Polish flue-cured hybrids (Berbeć (2007, 2017). cms 'bigelovii' may actually be the cms system reported to be deployed in Cuba (Lopez et al., 2008, Pino Perez, 2012). Unlike cms 'bigelovii', the true cms *quadrivalvis*, (formerly cms *bigelovii*), has a generally poor record of agronomic usability (Chaplin and Ford, 1965), (Aycock and Erdogan, 1979), (Naumenko, 2012).

The merits and flaws of the four major cms systems that have a record of deployment in commercialized cultivars are compiled in Table 5.6.

Table 5.6 The major sources of CMS deployed in the development and seed production of tobacco hybrid cultivars

Alloplasmic	Advantages and defects
cms *suaveolens*	Universally recognized as agronomically suitable and most extensively used worldwide in hybrid breeding and seed production (Zheng et al., 2018; Amankwa et al., 2014; Haji et al., 2000, 2007; Lawson et al., 2002; Yamada et al., 1984; Kubo, 1981); generally no negative impact on agronomic performance (Chaplin i Ford, 1965; Hosfield & Wernsman, 1974; Kubo, 1981, 1985; Yamada et al., 1984; Berbeć & Laskowska, 2004, 2005, Chen et al., 2012); no negative impact on smoke quality (Yamada et al., 1984); increased susceptibility to PVY (Berbeć & Laskowska, 2005; Czubacka et al., 2019) and to frog eye (Berbeć & Laskowska, 2005) vs. alloplasmic isogenic counterparts; increased leaf number and plant height, more days to flower (Hosfield & Wernsman, 1974), Berbeć and Laskowska (2004); considerably increased protein content (Baranova et al., 2015); delayed leaf maturity, depressed tolerance of PVY infection in a PVY-tolerant variety (Czubacka et al., 2019); reduced female fertility in combination with some nuclear genotypes caused by tendency to extensive malformations of entire carpels (unpublished observations by the author of this volume); heavily depressed seed yield (Naumenko, 2012)
cms 'bigelovii'	Most likely, a lineage of cms *suaveolens*; chief morphological differences: corolla slightly elongated in cms 'bigelovii' and slightly shortened in cms *suaveolens*; morphological alterations of the carpel (shortened or fasciated style with multiple stigma locules, malformed ovary) more frequent in cms *suaveolens* than in cms 'bigelovii' (Berbeć, 2001; unpublished observations by the author of this book); no depressive effect on growth, yield quality and money returns; in direct comparisons, cms 'bigelovii' system did not compromise resistance to PVY infection but was more prone to induce frog eye leaf spots vs. its autoplasmic counterpart and cms *suaveolens* (Berbeć & Laskowska, 2005); unlike in some plasmatypes known and recognized as cms *suaveolens*, in cms 'bigelovii' disturbed female fertility has not been found to be an important issue (unpublished observations of the author' of this review). At present, the most frequently used type of cms in Polish-bred hybrid cultivars, especially of the flue-cured market type.
cms 'tabacum-mutant'	Possibly, a distinct lineage of cms *glauca* or a product of uncontrolled cytoplasm transfer through pollen; no deleterious impact on cured leaf yield and quality vs. its autoplasmic male fertile counterpart; outyielded cms *glauca* (Berbeć & Laskowska, 2005); both cms *glauca* and cms 'tabacum-mutant' are very reliable seed producers; indicated as suitable for hybrid development and seed production (Roman & Nalepa, 1970; Tsikova & Nikova, 1980; Maktari, 1991; Naumenko, 2012); reduced number of leaves, hastened flowering and reduced leaf size reported by Czubacka et al. (2016); substantial increase in carbohydrate content over male fertile counterparts (Baranova et al., 2015); depressed tolerance of PVY infection in a PVY-tolerant variety (Czubacka et al., 2019); used for breeding

(continued)

Table 5.6 (continued)

Alloplasmic	Advantages and defects
	and commercial purposes in Poland since late 1960s (Berbeć, 1967, 1974) and probably deployed in some other countries (Berbeć, 2017); in the author's unpublished general experience, under Poland's conditions highly commendable for air-cured cultivars; quality-wise, in flue-cured tobacco slightly inferior to cms 'bigelovii'
cms *glauca* (see also cms 'tabacum mutant')	No negative impact on plant growth, yield and chemical composition (Maktari, 1991; Chen et al., 2012), yield of hybrid seeds (Maktari, 1991), yield and quality of cured leaves (Lawson et al., 2002; Amankwa et al., 2014); yield of hybrid seeds improved compared to that in cms *suaveolens* (Amankwa et al., 2014); slight leaf yield depression compared to the isogenic autoplasmic counterpart (Berbeć & Laskowska, 2005), increased content of reducing sugars, chlorogenic acid and rutins vs. male the fertile analog (Nikolov et al., 1984); fewer days to flower (Baranova et al., 2015); an alloplasmic variant of cv. K326, universally known and appreciated flue-cured tobacco cultivar, based on *N. glauca* cms system and designated Nta (gla.) S 'K326' is grown on commercial scale in China (Liao et al., 2017; another cms *glauca* flue-cured tobacco hybrid was released to commercial farmers in Canada (Amankwa et al., 2019)

After Doroszewska and Berbeć (2020) with minor modifications and additions

5.6.3 Concluding Notes on Nicotiana Species as cms Sources for Tobacco Improvement

Of 27 cms sources that have been obtained and studied for their agronomic merits, four have actually been deployed in commercial tobacco hybrids: cms *suaveolens* (Lawson et al., 2002; Amankwa et al., 2014; Zheng et al., 2018) and its possible cytoplasmic variant (cms 'bigelovii'), cms glauca (Liao et al., 2017; Amankwa et al., 2019; Lewis, 2020) and cms 'tabacum-mutant', a suspected cytoplasmic variant of cms *glauca*. The agronomic potential of cms *undulata* has been confirmed by at least two independent studies, but no report of its actual deployment has come to the notice of the author of this volume. Seven species were indicated as possibly usable sources of cms: *N. longiflora*, *N. paniculata*, *N. amplexicaulis*, *N. benthamiana*, *N. maritima*, and *N. velutina*. However, in the latter cases, the evaluations were performed by a single team of researchers, and no confirmation came from other sources. Seventeen cms sources were found to be agronomically unusable mostly because of negative impact on growth, yield and crop quality but in some cases also because of depressed female fertility or incomplete male sterility. For 4 reported alloplasmic combinations of *N. tabacum,* no data have been found on their agronomic performance.

It appears that only 31 alloplasmic combinations of *N. tabacum* with other species have ever been obtained, excluding that with *N. sylvestris* since the latter can be considered as autoplasmic bearing in mind the ancestry of *N. tabacum*. The vast majority of alloplasmic combinations in tobacco produce full male sterility, the only exceptions being *N. rustica, N. raimondii* and *N. knightiana,* for which lineages with vestigial pollen production have been reported (Hart, 1965; Berbeć, 1994a, b, 2001). *N. glutinosa* is a debatable case, reported and studied entries of cms *glutinosa* having been probably confused with cms *tabacum* (*glutinosa*) (see Sect. 5.4.2).

Although *N. tabacum* is known to have been successfully hybridized with 59 other *Nicotianae*, no alloplasmics of *N. tabacum* are known for 30 of these combinations. There is yet another large group of *Nicotiana* species, 40 taxa altogether including those most recently described, for which no hybrid associations with *N. tabacum* have been reported. This is a vast pool of unexplored potential alloplasmic diversity that may contain cytoplasmic male sterile combinations agronomically superior to those already known.

Cytoplasmically male sterile hybrid cultivars have been successfully replacing open-pollinated varieties of tobacco that are grown for leaves as an item of commerce. The hybrid cultivars also provided a meeting place for the two most important outcomes of interspecific breeding in tobacco: resistance to diseases and cytoplasmic male sterility. In the F_1 and, more recently, three-way hybrids, these two achievements have since complemented and reinforced each other in the improvement of tobacco as a crop.

References

Ahuja, M. R. (1962). A cytogenetic study of heritable tumors in *Nicotiana* species hybrids. *Genetics, 47*, 865–880.

Amankwa, G. A., Mishra, S., Shearer, A. D., & Van Hooren, D. L. (2014). Evaluation of two flue – Cured tobacco F_1 hybrids with different sources of male sterile cytoplasm. CORESTA Congress Quebec, Canada APPOST02.

Amankwa, G. A., Mishra, S., Shearer, A. D., Saude, C., Van Hooren, D. L., & Richmond, M. D. (2019). CTH144 flue-cured tobacco F1 hybrid. CTH144 flue-cured tobacco F_1 hybrid. *Canadian Journal of Plant Science, 99*, 966–968. https://doi.org/10.1139/cjps-2018-0328

Asahi, T., Kumashiro, T., & Kubo, T. (1988). Constitution of mitochondrial and chloroplast genomes in male sterile tobacco obtained by protoplast fusion of *N. tabacum* and *N.debneyi*. *Plant & Cell Physiology, 29*, 43–49.

Atanassov, I. (1993). Somatic hybridization in g. *Nicotiana*. Dissertation. Academy of Agriculture, Institute of Genetic Engineering, Sofia, Bulgaria.

Atanassov, I. I., Atanassova, S. A., Dragoeva, A. I., & Atanassov, A. I. (1998). A new CMS source in *Nicotiana* developed via somatic cybridization between *N. tabacum* and *N. alata*. *Theoretical and Applied Genetics, 97*, 982–985.

Aviv, D., & Galun, E. (1980). Restoration of fertility in cytoplasmic male sterile (CMS) *Nicotiana sylvestris* by fusion with X-irradiated *N. tabacum* protoplasts. *Theoretical and Applied Genetics, 58*, 121–127.

Aviv, D., & Galun, E. (1986). Restoration of male fertile *Nicotiana* by fusion of protoplasts derived from two different male sterile cybrids. *Plant Molecular Biology, 7*, 411–417.

Aviv, D., & Galun, E. (1987). Chondriome analysis in sexual progenies of *Nicotiana* cybrids. *Theoretical and Applied Genetics, 73*, 821–826.

Aviv, D., Fluhr, R., Edelman, M., & Galun, E. (1980). Progeny analysis of interspecific hybrids: *Nicotiana tabacum* (CMS) + *N.sylvestris* with respect to nuclear and chloroplast markers. *Theoretical and Applied Genetics, 56*, 145–150.

Aviv, D., Arzee-Gonen, P., Bleichmann, S., & Galun, E. (1984). Novel alloplasmic Nicotiana plants by "donor-recipient" protoplast fusion: Cybrids having *N.* tabacum or *N.* sylvestris nuclear genomes and either or both plastomes and chondriomes from alien species. *Molecular & General Genetics, 196*, 244–253.

Aycock, M. K., & Erdogan, H. R. (1979). Influence of five male sterile cytoplasm on the performance of Maryland tobacco cultivars and hybrids. *Report on Maryland Agricultural Experiment Station MP, 940*, 1–36.

Baranova, E. G., Ivanitskii, K. I., & Suchkov, V. I. (2015). Формирование Генофонда Источников цитоплазматической мужской стерильности табака (Development of the gene pool sources of cytoplasmatic male sterility of tobacco). apriori-journal.ru/seria2/2-2015 APRIORI. Серия: Естественные и технические науки. 2015. № 2. С. 9.).

Bates, G. W. (1990). Asymmetric hybridization between *Nicotiana tabacum* and *N. repanda* by donor recipient protoplast fusion: Transfer of TMV resistance. *Theoretical and Applied Genetics, 80*, 481–487.

Bates, G. W., Hasenkampf, C. A., Contolini, C. F., & Piastuch, W. C. (1987). Asymmetric hybridization in *Nicotiana* by fusion of irradiated protoplasts. *Theoretical and Applied Genetics, 74*, 718–726.

Belliard, G., Pelletier, G., Vedel, F., & Quetier, F. (1978). Morphological characteristics and chloroplast DNA distribution in different cytoplasmic parasexual hybrids of *Nicotiana tabacum*. *Molecular & General Genetics, 165*, 231–237.

Belliard, G., Vedel, F., & Pelletier, G. (1979). Mitochondrial recombination in cytoplasmic hybrids of Nicotiana tabacum by protoplast fusion. *Nature, 281*, 401–440.

Berbeć, J. (1962). Problems and major results of tobacco research conducted in Puławy as related to current tobacco production needs. Part I. Tobacco improvement issues (in Polish). *Post Nauk Roln, 5*(77), 20–42.

Berbeć, J. (1966). Badania nad krzyżówką Nicotiana glauca x N. tabacum (Investigations on the cross *Nicotiana glauca* × *N. tabacum*). (in Polish). *Inf PAN, 7*, 213–340.

Berbeć, J. (1967). Cytoplasmic male sterile variety of cultivated tobacco (*Nicotiana tabacum* L.) (in Polish). *Inf PAN, 7-195*, 299–300.

Berbeć, J. (1972). Investigations into male sterility of tobacco and its utilisation in hybrid seed production (in Russian). International Symposium Mutual Economic Assistance Council (RWPG), Warsaw, 9–13 May, 1972:1–5.

Berbeć, J. (1974). A cytoplasmic male sterile mutation form of *Nicotiana tabacum* L. *Z Pflanzenzuechtg, 73*, 204–216.

Berbeć, A. (1988). Morphology, cytogenetics and resistance to potato virus Y of the amphidiploid *Nicotiana raimondii* Macbride × *N. tabacum* L. *Genetica Polonica, 29*, 41–52.

Berbeć, A. (1994a). Effect of *Nicotiana knightiana* cytoplasm on the stability of a true-breeding tobacco variety increased by selfing. CORESTA Congress Inf Bull Harare: 61.

Berbeć, A. (1994b). Microsporogenesis in two alloplasmic isonuclear strains of tobacco *Nicotiana tabacum* L. cv. Zamojska 4 with the cytoplasms of *N. knightiana* Goodspeed and *N. raimondii* Macbride. *Genetica Polonica, 35*, 23–31.

Berbeć, A. (1994c). Variation among offspring of alloplasmic tobacco *Nicotiana tabacum* L. cv 'Zamojska 4' with the cytoplasm of *N. knightiana* Goodspeed. *Theoretical and Applied Genetics, 89*, 127–132.

Berbeć, A. (2001). Floral morphology and some other characteristics of iso-genomic alloplasmics of *Nicotiana tabacum* L. *Beiträge zur Tabakforschung International/Contributions, 19*, 309–314.

Berbeć, A. (2007). The cytoplasm of *Nicotiana bigelovii* (Torrey) Watson – A new type of cytoplasmic male sterility used in the production of commercial tobacco (*N. tabacum* L.) hybrids. Plant genetic resources and their exploitation in the plant breeding for food and agriculture. 18th EUCARPIA genetic resources section meeting, May 23–26 Piestany P2-2: 16.

Berbeć, A. (2017). Three-way crosses vs. single crosses in tobacco: First agronomic assessment. *Crop Science, 57*, 1363–1372. https://doi.org/10.2135/cropsci2016.07.0624
Berbeć, J., & Berbeć, A. (1976). Growth and development of *Nicotiana tabacum* L. form reconstituted on the cytoplasm of *Nicotiana glauca* Grah. *Gene Pool, 17*, 309–318.
Berbeć, J., & Berbeć, A. (1992). Male sterility in tobacco (*Nicotiana tabacum* L.) by one-step substitution of the cytoplasm from *Nicotiana eastii* Kostoff (in Polish). *Pamiętnik Puławski, 100*, 136–139.
Berbeć, A., & Doroszewska, T. (1992). Alloplasmic forms of cultivated tobacco (*Nicotiana tabacum* L.) with the cytoplasms of *Nicotiana amplexicaulis*, *N. knightiana* i *N. raimondii* (in Polish). *Pamiętnik Puławski, 100*, 141–150.
Berbeć, A., & Doroszewska, T. (2020). The use of *Nicotiana* species in tobacco improvement. In N. Ivanov, N. Sierro, & M. Peitsch (Eds.), *The tobacco plant genome. Compendium of plant genomes*. Springer.
Berbeć, A., & Laskowska, D. (1994). Androgenetic plants by culturing anthers of alloplasmic strains of tobacco *Nicotiana tabacum* L. cv. Zamojska 4 with the cytoplasm of *Nicotiana knightiana* Goodspeed and *N. raimondii* Mcbride. *Genetica Polonica, 35*, 33–41.
Berbeć, A., & Laskowska, D. (1997). Compatibility and tentoxin sensitivity in the off-spring of near-amphihaploid hybrid *Nicotiana tabacum* × *N. alata*. *Journal of Applied Genetics, 38*, 143–150.
Berbeć, A., & Laskowska, D. (2004). Agronomic performance of flue-cured tobacco F_1 hybrids obtained with different sources of male sterile cytoplasm. *Beiträge zur Tabakforschung International/Contributions, 21*, 235–239.
Berbeć, A., & Laskowska, D. (2005). Investigations of isogenomic alloplasmics of flue-cured tobacco *Nicotiana tabacum* cv. Wislica. *Beiträge zur Tabakforschung International/Contributions, 21*, 258–263.
Berbeć, J., Berbeć, A., & Czop-Doroszewska, T. (1979). Investigations on obtaining new varieties and breeding lines of *Nicotiana tabacum* resistant to virus leaf rib necrosis (Potato Virus Y) and virus degeneration of plants by Lycopersicum virus 3. USDA Grant no. FG-Po-341 (JB-6) Annual Report 2 (1978–1979) (unpublished).
Berbeć, A., Berbeć, J., & Doroszewska, T. (1990). Graded effects of four alien cytoplasms in *Nicotiana tabacum* L. Bul Spec CORESTA, 1990, Symposium Kallithea: 143 A34.
Bergman, P., Kofer, W., Hakansson, G., & Glimelius, K. S. (1995). A chimeric and truncated mitochondrial atpA gene is transcribed in alloplasmic cytoplasmic male-sterile tobacco with *Nicotiana bigelovii* mitochondria. *Theoretical and Applied Genetics, 91*, 603–610.
Bezova, K., & Skula, K. (1980). Fertile and sterile forms of VP9. *Bull Tabak Priem, 23*, 19–31.
Billenkamp, N. (1997). Estimation of quantitative genetic parameters in a breeding population of flue-cured tobacco. CORESTA meeting Agro-Phyto Groups, Montreux, AP16.
Biskup, J., Mazur, M., & Roman, T. (1972). Ocena wzrostu i rozwoju oraz niektorych cech użytkowych odmian i mieszańcow płodnych w porównaniu z analogicznymi formami meskojałowymi (Appraisal of growth, development and agronomic performance of some commercial cultivars and fertile hybrids vis-a-vis their male sterile analogues) (in Polish). *Biul Centr Lab Przem Tyton, 3–4*, 29–40.
Bonnet, H. T., & Glimelius, K. (1983). Somatic hybridization in *Nicotiana*: Behavior of organelles after fusion of protoplasts from male-fertile and male-sterile cultivars. *Theoretical and Applied Genetics, 65*, 213–217.
Bruns, H. A. (2017). Southern corn blight: A story worth retelling. *Agronomy Journal, 109*, 1–7.
Burk, L. G. (1960). Male-sterile flower anomalies in interspecific tobacco hybrids. *The Journal of Heredity, 51*, 27–31.
Burk, L. G. (1967). An interspecific bridge-cross – *Nicotiana repanda* through *N. sylvestris* to *N. tabacum*. *The Journal of Heredity, 58*, 215–218.
Burk, L. G., & Durbin, R. D. (1978). Reaction of *Nicotiana* species to tentoxin. *The Journal of Heredity, 69*, 117–120.

Burk, L. G., & Mann, T. J. (1970). Onset, prevention and restoration of male sterile flower anomalies in tobacco: Interaction of the *N. repanda* plasmon and *N. tabacum* genome. *The Journal of Heredity, 61*, 143–146.

Burns, J. A., & Gerstel, D. U. (1981). Role of nuclear chromosomes in anther-restoration of male sterile tobacco. *The Journal of Heredity, 72*, 413–418.

Burns, J. A., Gerstel, D. U., & Sand, S. A. (1978). Cytoplasmic male sterility in *Nicotiana*, restoration of fertility, and the nucleolus. II. *N. debneyi* cytoplasm. *Genetics, 90*, 151–159.

Cameron, D. R. (1965). Cytoplasmic effects in *Nicotiana*. Proceeding of the 11th international congress genetics 1:202–204.

Chaplin, J. F. (1959). Interspecific hybridization, gene transfer and chromosomal substitution in *Nicotiana*. PhD thesis, North Carolina State University, Raleigh.

Chaplin, J. F. (1964). Use of male sterile tobaccos in the production of hybrid seed. *Tobacco Science, 8*, 105–109.

Chaplin, J. F. (1965). Influence of different *Nicotiana* species cytoplasms on male sterile flue-cured tobacco. Proceeding of the abstracts papers addresses 62nd Ann Conv Association South Agriculture Workers, Dallas, Texas, Tob Abstract 10, 1966, p. 464 Abstract 1683.

Chaplin, J. F. (1966). Comparative performance of F_1 flue-cured tobacco hybrids and their parents. *Tobacco Sciences, 10*, 126–130.

Chaplin, J. F., & Ford, Z. T. (1965). Agronomic and chemical characteristics of male-sterile flue-cured tobacco as influenced by cytoplasms of different *Nicotiana* species. *Crop Science, 5*, 436–438.

Chen, K. S., Johal, S., & Wildman, S. G. (1977). Phenotypic markers for chloroplast DNA genes in higher plants and their use in biochemical genetics. In L. Bogorad & J. H. Weil (Eds.), *Nucleic acid synthesis in plants* (pp. 183–194). Plenum.

Chen, X., Luo, J., Li, W., Xiao, B., & Li, Y. (2012). Flow cytometry assisted selection on a CMS source (Nta(gla.)S) of flue-cured tobacco via somatic hybridisation. CORESTA Congress, Sapporo, 2012 AP 06.

Christoff, M. (1938). *On a case of male sterility in Nicotiana rustica (in Bulgarian)*. Godisznik na Sofijskaya Univ.

Clausen, R. E. (1928). Interspecific hybridization in Nicotiana. VII. The cytology of hybrids of the synthetic species, *digluta*, with its parents: *glutinosa* and *tabacum*. *University of California Publications in Botany, II*, 177–211.

Clayton, E. E. (1950). Male sterile tobacco. *The Journal of Heredity, 4*, 171–175.

Crow, J. F. (1998). 90 years ago: The beginning of hybrid maize. *Genetics, 148*, 923–928.

Czubacka, A., Depta, A., & Doroszewska, T. (2016). Agronomic performance of cytoplasmic male sterile forms of flue-cured tobacco. CORESTA Congress, Berlin, APPOST 08.

Czubacka, A., Depta, A., & Doroszewska, T. (2019). Diversification of defense response to Potato Virus Y among alloplasmic forms of tobacco (in Polish, English summary). *Polish Journal of Agronomy, 39*, 27–34.

Dexter-Boone, A., & Lewis, R. S. (2019). Heterosis in flue-cured tobacco and its utility in predicting transgressive segregation within derived populations of inbred lines. *Crop Science, 59*, 957–967.

Donaldson, P., Sproule, A., Bevis, E., Pandeya, R., Keller, W. A., & Gleddie, S. (1993). Non-random chloroplast segregation in *Nicotiana tabacum + N.rustica* somatic hybrids selected by dual nuclear-encoded resistance. *Theoretical and Applied Genetics, 86*, 465–473.

Donaldson, P. A., Bevis, E., Pandeya, R. S., & Gleddie, S. C. (1994). Random chloroplast segregation and frequent mt DNA rearrangements in fertile somatic hybrids between *Nicotiana tabacum* L. and *N. glutinosa* L. *Theoretical and Applied Genetics, 87*, 900–908.

Donaldson, P. A., Bevis, E., Pandeya, R., & Gleddie, S. C. (1995). Rare symmetric and asymmetric *Nicotiana tabacum* (+) *N. megalosiphon* somatic hybrids recovered by selection for nuclear-encoded resistance genes and in the absence of genome inactivation. *Theoretical and Applied Genetics, 91*, 747–755.

Dudek, M. (1971). Wplyw sterylnej cytoplazmy typu *debneyi* i *megalosiphon* na niektóre cechy mieszańcow w krzyżówkach wypierajacych kilku odmian tytoniu (Effect of the *debneyi* i *megalosiphon* sterile cytoplasms on some traits of the hybrids in backcrosses to several tobacco varieties) (in Polish). Biul Centr Lab Przem Tyton (1–2): 29–44.
Durbin, R. D., & Uchytil, T. F. (1977a). A survey of plant insensitivity to tentoxin. *Phytopathology, 67*, 602–603.
Durbin, R. D., & Uchytil, T. F. (1977b). Cytoplasmic inheritance of chloroplast coupling factor 1 subunits. *Biochemical Genetics, 15*, 1143–1146.
East, E. M. (1932). Studies on self-sterility. The behaviour of crosses between self-sterile and self-fertile plants. *Genetics, 17*, 175–202.
Farbos, I., Mouras, A., Bereterbide, A., & Glimelius, K. (2001). Defective cell proliferation in the floral meristem of alloplasmic plants of *Nicotiana tabacum* leads to abnormal foral organ development and male sterility. *The Plant Journal, 26*, 131–142.
Fitter, J. T., Thomas, M. R., Niu, C., & Rose, R. J. (2005). Investigation of Nicotiana tabacum (+) N. suaveolens cybrids with carpelloid stamens. *Journal of Plant Physiology, 162*, 225–235.
Flick, C. E., & Evans, D. A. (1982). Evaluation of cytoplasmic segregation in somatic hybrids of *Nicotiana*: Tentoxin sensitivity. *The Journal of Heredity, 73*, 264–266.
Frankel, R., & Galun, E. (1977). *Pollination mechanisms, reproduction and plant breeding* (Monograph in Genetics 2) (pp. 210–214). Springer.
Frankel, R., Scoweroft, W. R., & Whitfield, P. R. (1979). Chloroplast DNA variation in isonuclear male sterile lines of *Nicotiana*. *Molecular & General Genetics, 169*, 129–135.
Gerstel, D. U. (1980). Cytoplasmic male sterility in *Nicotiana* (A review). North Carolina Agricultural Research Service Technical Bulletins 263.
Gerstel, D. U., & Burns, J. A. (1974). Meiosis and staminal sterility in hybrids between *Nicotiana tomentosiformis* and *N. otophora*. *Tobacco Sciences, 18*, 157–159.
Gerstel, D. U., & Burns, J. A. (1983). Anther restoration in male sterile tobacco by nucleolus organizing chromosomes a restorer from Nicotiana bigelovii. *Tobacco Science, 27*, 151–153.
Gerstel, D. U., Burns, J. A., & Burk, L. G. (1978). Cytoplasmic male sterility in *Nicotiana*, restoration of fertility and the nucleolus. *Genetics, 89*, 157–169.
Gerstel, D. U., Burns, J. A., & Sand, S. A. (1980). Cytogenetics of flower modification in cytoplasmic male sterile tobacco. *Genetics, 96*, 223–235.
Grebenkin, A. P. (1968). Study of the reaction of certain wild Nicotiana species during the crossing with unisexual forms of tobacco. *Genetika, 4*, 5–10.
Grebenkin, A. P., Ternovsky, M. F., & Zakharenko, A. M. (1976). Results of crosses between tobacco varieties from the world collection and forms with CMS from *N. megalosiphon* and *N. plumbaginifolia*. *Soviet Genetics, 9*, 418–420. (CORESTA Inf Bul 1976 2: 6970, Abstr 6171).
Gurdon, C., Svab, Z., Feng, Y., Kumar, D., & Maliga, P. (2016). Cell-to-cell movement of mitochondria in plants. *PNAS, 113*, 3395–3400.
Haji, H. M., Mishra, S., & DeVos, M. (2000). CT681 flue-cured tobacco. *Canadian Journal of Plant Science, 80*, 167–168.
Haji, H. M., Mishra, S., & Devos, M. (2006). Host plant resistance management strategies for control of black root rot resistance. CORESTA Agro-Phyto Groups Meet, Paris. PPost 08.
Haji, H. M., Mishra, S., & DeVos, M. (2007). CTH2 flue-cured tobacco F_1 hybrid. *Canadian Journal of Plant Science, 87*, 383–384.
Hakansson, G., & Glimelius, K. (1991). Extensive nuclear influence on mitochondrial transcription and genome structure in male-fertile and male-sterile alloplasmic *Nicotiana* materials. *Molecular & General Genetics, 229*, 380–388.
Hanson, M. R., & Conde, M. F. (1985). Functioning and variation of cytoplasmic genomes: Lessons from cytoplasmic nuclear interactions affecting male sterility in plants. *International Review of Cytology, 94*, 213–267.
Hart, G. E. (1965). Studies on extrachromosomal male sterility in *Nicotiana*. Dissertation abstracts 26: 3595 (tobacco abstracts 10: 2108) Ph D thesis Univ California, Berkley.

Holmes, F. O. (1938). Inheritance of resistance to tobacco-mosaic disease in tobacco. *Phytopathology, 28*, 553–561.

Horlow, C., Goujaud, J., Lepingle, A., Missonier, C., & Boutgin, J. P. (1990). Transmission of paternal chloroplasts in tobacco. (*Nicotiana tabacum*). *Plant Cell Reports, 9*, 249–252.

Hosfield, G. L., & Wernsman, E. A. (1974). Effect of alien cytoplasm and fertility restoring factor on growth, agronomic characters and chemical constituents in a male sterile variety of flue-cured tobacco. *Crop Science, 14*, 575–577.

Iancheva, A., & Palakarcheva, M. (1990). Cytoplasmic male sterile hybrids with *Nicotiana goodspeedii* cytoplasm. *Genetics and Breeding, 4*, 316–321.

Ilcheva, V., San, L. H., Dimitrov, B., & Zagorska, N. (2000). Morphological and cytological characteristics of somatic hybrids of *Nicotiana tabacum* L. + *N. megalosiphon* Heurk. et Muell. *In Vitro Cellular & Developmental Biology. Plant, 36*, 69–73.

Ilcheva, V., San, L. H., Zagorska, N., & Dimitrov, B. (2001). Production of male sterile interspecific somatic hybrids between transgenic *N. tabacum* (bar) and *N. rotundifolia* (npt ii) and their identification by AFLP analysis. *In Vitro Cellular and Developmental Biology-Plant, 37*, 496–502.

Izard, C., & Hitier, H. (1955). Observations sur un hybride complexe susceptible de produire des plantes a sterilite male. *Ann Du Tabac Seita, Sect, 2*(2), 1–13.

Jankowski, F., & Poniewierski, F. (1969). Tobacco seed production. In J. Skiendzielewski (Ed.), *Tobacco: Production, breeding, processing* (Warszawa 1969) (pp. 419–433) (in Polish).

Jones, J. L., & Henderson, R. G. (1978). Performance of Virginia dark-fired tobacco hybrids. *Tobacco Sciences, 22*, 109–111.

Julio, E., & Dorlhac de Borne, F. (2003). Identification of AFLP fragments linked to pathogen resistance genes and conversion to SCAR markers for rapid screening in tobacco breeding. Plant & Animal Genomes XI Conference, San Diego.

Kasza, A., & Kandra, G. A. (1990). Protoplast fusion transferred cytoplasmic male sterility and its utilization in tobacco breeding. CORESTA Inf Bul CORESTA Symposium Hellas: 140 (A31).

Kaul, M. H. L. (1988). *Male sterility in higher plants* (Monographs on Theoretical and Applied Genetics) (Vol. 10). Springer.

Kobus, I. (1978). Preliminary investigations of crop improvement value of male sterile *Nicotiana tabacum* mutants obtained by exposure to gamma rays (in Polish). *Biul Inst Hod Nas, 134*, 249–259.

Kofer, W., Glimelius, K., & Bonnet, H. T. (1990). Modifications of floral development in tobacco induced by fusion of protoplasts of different male-sterile cultivars. *Theoretical and Applied Genetics, 79*, 97–102.

Kofer, W., Glimelius, K., & Bonnett, H. T. (1991). Restoration of normal stamen development and pollen formation by fusion of different cytoplasmic male-sterile cultivars of *Nicotiana tabacum*. *Theoretical and Applied Genetics, 81*, 390–396.

Kofer, W., Glimelius, K., & Bonnet, H. T. (1992). Fusion of male sterile tobacco causes modification of mtDNA leading to changes in floral morphology and restoration of fertility in cybrid plants. *Physiologia Plantarum, 85*, 334–338.

Kubo, T. (1981). Agronomic characteristics of male sterile flue-cured tobacco hybrids with the cytoplasm from *Nicotiana suaveolens*. *Bulletin of the Iwata Tobacco Experiment Station, 13*, 61–71.

Kubo, T. (1985). Studies on hybrid breeding by means of cytoplasmic male sterility in tobacco. *Bulletin of the Iwata Tobacco Experiment Station, 17*, 69–138.

Kubo, T., Kumashiro, T., & Saito, Y. (1988). Cytoplasmic male sterile lines of a tobacco variety, Tsukuba 1, developed by asymmetric protoplast fusion. *Japanese Journal of Breeding, 38*, 153–164.

Kumashiro, T., & Kubo, T. (1986). Cytoplasm transfer of *N. debneyi* to *N. tabacum* by protoplast fusion. *Japanese Journal of Breeding, 36*, 39–48.

Kumashiro, T., Asahi, T., & Komari, T. (1988). A new source of cytoplasmic male sterile tobacco obtained by fusion between *N. tabacum* and X-irradiated *N. africana* protoplasts. *Plant Sciences, 55*, 247–254. (CORESTA Inf Bul 1989:79).

Kumashiro, T., Asahi, T., & Nakakido, F. (1989). Transfer of cytoplasmic factors by asymmetric fusion from a cross-incompatible species, *N. repanda* to *N. tabacum* and characterization of cytoplasmic genomes. *Plant Science, 61*, 137–144.

Kung, S. (1977). Expression of chloroplast genomes in higher plants. *Annual Review of Plant Physiology, 28*, 401–437.

Kung, S. D., Zhu, Y. S., Chen, K., Shen, G. F., & Sisson, V. A. (1981). *Nicotiana* chloroplast genome. II Chloroplast DNA alteration. *Molecular Genetics and Genomics, 183*, 20–24.

Laskowska, D. (1993). *Katalog odmian uprawnych tytoniu (Nicotiana tabacum L.).* (Catalogue of tobacco cultivars (*Nicotiana tabacum* L.). Instytut Uprawy Nawożenia I Gleboznawstwa (in Polish)

Laskowska, D., & Berbeć, A. (2007). The new alloplasmic *Nicotiana tabacum* L. line of with *Nicotiana wuttkei* Clarkson et Symon cytoplasm (in Polish). *Biul IHAR, 244*, 289–296.

Lawson, D. M., Schaeffer, S.,Wernsman, E., & Nielsen, M. T. (2002). A comparison of *Nicotiana suaveolens* and *Nicotiana glauca* cytoplasmic male sterility systems in flue-cured tobacco. CORESTA Congress Agro-Phyto Groups, New Orleans, AP11.

Legg, P. D. (1991). Genetic variability in broadleaf dark tobacco. *Tobacco Sciences, 35*, 32–34.

Lewis, R. S. (2020). *Nicotiana tabacum* L.: Tobacco. In J. Novak & W. D. Blüthner (Eds.), *Medicinal, aromatic and stimulant plants* (Handbook of plant breeding) (Vol. 12). Springer. https://doi.org/10.1007/978-3-030-38792-1_9

Li, F.-X., Yang, A.-G., Cui, M.-M., Gong, D.-P., Wang, W.-F., & Sun, Y.-H. (2011). Mitochondrial microsatellite variability of tobacco CMS lines with four different cytoplasms. *Acta Agronomica Sinica, 37*, 2285–2292. https://doi.org/10.3724/SP.J.1006.2011.02285

Liao, J., Dai, J., Yang, S., Zhou, X., Ren, L., Chen, Z., He, H., & Chen, S. (2017). Interspecific cross-hybrids of *Nicotiana tabacum* L. cv. (gla.) S 'K326' with *Nicotiana alata*. *Plant Breeding, 136*, 427–435. https://doi.org/10.1111/pbr.12474

Lopez, M., Espino, E., & Garcia, H. (2008). Informe de nuevas variedades 'Capero-1': primer hı́brido androestéril comercial de tabaco negro cubano (*Nicotiana tabacum* L.). *Cultivos Tropicales, 29*, 51.

Maktari, A. T. A. (1991). Use of cytoplasmic male sterility in the production of hybrid seed in tobacco (in Russian). Doctoral thesis. Kubanskij Gosudarstwiennyj Agrarnyj Institut.

Matibiri, E. A., & Mantell, S. H. (1994). Cybridization in *Nicotiana tabacum* L. using double inactivation of parential protoplasts and post-fusion selection based on nuclear-encoded and chloroplast-encoded marker genes. *Theoretical and Applied Genetics, 88*, 1017–1022.

Matzinger, D. F., & Wernsman, E. A. (1967). Genetic diversity and heterosis in *Nicotiana* I. Interspecific crosses. *Theoretical and Applied Genetics, 37*(Der Zuchter), 188–191.

Matzinger, D. F., & Wernsman, E. A. (1968). Genetic diversity and heterosis in *Nicotiana*. Oriental x flue-cured variety crosses. *Tobacco Sciences, 40*, 177–180.

Medgyesy, P., Golling, R., & Nagy, F. (1985). A light sensitive recipient for the effective transfer of chloroplast and mitochondrial traits by protoplast fusion in *Nicotiana*. *Theoretical and Applied Genetics, 70*, 590–594.

Medgyesy, P., Pay, A., & Marton, L. (1986). Transmission of paternal chloroplasts in *Nicotiana*. *Molecular & General Genetics, 204*, 195–198.

Menczel, L., Nagy, F., Kiss, Z., & Maliga, P. (1981). Streptomycin resistant and sensitive somatic hybrids of *N.tabacum* + *N.knightiana*. Correlation of resistance to *N.tabacum* plastids. *Theoretical and Applied Genetics, 59*, 191–195.

Menczel, L., Polsby, L. S., Steinback, K. E., & Maliga, P. (1986). Fusion-mediated transfer of triazine-resistant chloroplasts: Characterization of *Nicotiana tabacum* cybrid plants. *Molecular & General Genetics, 205*, 201–205.

Milla, S. R., Levin, J. S., Lewis, R. S., & Rufty, R. C. (2005). RAPD and SCAR markers linked to an introgressed gene conditioning resistance to *Peronospora tabacina* D.B. Adam. in Tobacco. *Crop Science, 45*, 2346–2354.

Moav, J., Moav, R., & Zohary, D. (1968). Spontaneous morphological alterations of chromosomes in Nicotiana hybrids. *Genetics, 59*, 57–63.

Nagy, F., Torok, I., & Maliga, P. (1981). Extensive rearrangements in the mitochondrial DNA in somatic hybrids of *Nicotiana tabacum* and *Nicotiana knightiana. Molecular & General Genetics, 183*, 437–439.

Nagy, F., Lazar, G., Menczel, L., & Maliga, P. (1983). A heteroplasmic state induced by protoplast fusion is a necessary condition for detecting rearrangements in *Nicotiana* mitochondrial DNA. *Theoretical and Applied Genetics, 66*, 203–207.

Naumenko, S. A. (2012). *Особенности селекции сортов табака сортотипов Вирджиния и Берлей в Россuu (Particularities of the development of flue-cured and Burley tobacco cultivars in Russia)*. DSc Thesis,. Russian Timiryazev State Agrarian University.

Nikolov, S., Stoyanova, M., & Mihailova, A. (1984). The effect of alien cytoplasm (*Nicotiana glauca*) on some chemical indices of importance for the quality of CMS tobacco. *Genetika i Selektsiya, 17*, 362–368.

Nikova, V. M. (1979). Influence of different sources of cytoplasmic male sterility in tobacco on the development of the pistil. *Comptes Rendus de l'Académie Bulgare des Sciences, 32*, 209–212.

Nikova, V. (1984). Cytoplasmic male sterility in *Nicotiana* with cytoplasm from *Nicotiana benthamiana* Domin. *Comptes Rendus de l'Académie des Sciences Bulgarica, 37*, 1253–1256.

Nikova, V. M. (1986). *Nicotiana excelsior* Black as a source of cytoplasmic male sterility in *Nicotiana tabacum* L. *Comptes Rendus de l'Académie Bulgare des Sciences, 39*, 105–107.

Nikova, V. M., & Shabanov, D. S. (1988). *Nicotiana amplexicaulis* Burbidge as a source of cytoplasm male sterility in tobacco (*Nicotiana tabacum* L.). *Comptes Rendus de l'Académie Bulgare des Sciences, 41*(3), 83–86.

Nikova, V. M., & Shabanov, D. (1992). Development of male sterile tobacco forms by crossing *Nicotiana knightiana* × *N. tabacum*. CORESTA Inf Bul special congress Jerez de La Frontera:132.

Nikova, V., & Tsikov, D. (1976). Sravnitelji izsledovanija wrchu morfogenesisa na cweta na razlichni iztochnici na citoplazmena mzska sterilnost pri tutuni (Comparative investigations on flower morphogenesis in various tobacco sources of cytoplasmic male sterility) (in Bulgarian). *Genetics and Plant Breeding (Sofia), 9*, 362–375.

Nikova, V., & Vladova, R. (2002). Wild *Nicotiana* species as a source of cytoplasmic male sterility in *Nicotiana tabacum. Beiträge zur Tabakforschung International/Contributions, 20*, 301–311.

Nikova, V. M., Menczel, L., & Shabanov, D. (1990) Transfer of cytoplasmic male sterility to *Nicotiana tabacum* L. by protoplast fusion. Symp CORESTA, Thessaloniki 1990.

Nikova, V. M., Zagorska, N. A., & Pundeva, R. S. (1991). Development of four tobacco male sterile sources using *in vitro* techniques. *Plant Cell, Tissue and Organ Culture, 27*, 289–295.

Nikova, V., Vladova, R., Pundeva, R., & Shabanov, D. (1997). Cytoplasmic male sterility in *Nicotiana tabacum* L. obtained through interspecific hybridization. *Euphytica, 94*, 375–378.

Nikova, V., Pundeva, R., & Petkova, A. (1999). *Nicotiana tabacum* L. as a source of cytoplasmic male sterility in interspecific cross with *N. alata* link & Otto. *Euphytica, 107*, 9–12.

Nikova, V., Pundeva, R., Vladova, R., & Petkova, A. (2001). A new tobacco cytoplasmic male sterile source from the hybrid combination *Nicotiana longiflora* Gav. and *N. tabacum* L. using in vitro techniques. *Israel Journal of Plant Sciences, 49*, 9–13.

Nikova, V., Iantcheva, A. M., Pandeva, R., & Petkova, A. (2004). Investigation of alloplasmic tobacco lines by certain economically indices. CORESTA congress, Kyoto, Agro-Phyto Groups, PPOST9.

Nikova, V., Nikolov, E., Iancheva, A., Pandeva, R., & Petkova, A. (2005). Hybrid seed production in tobacco by using male sterile plants. CORESTA meet, Agro-Phyto Groups, Santa Cruz do Sul, APOST24.

Oczoś, A. (1985). Influence of cytoplasmic male sterility from *Nicotiana eastii* and *N. glutinosa* on growth, development and on certain quality indices of flue-cured tobacco *N. tabacum* L. *Pamiętnik Puławski, 84*, 219–236. (in Polish).

Palakarcheva, M. T. (1968). Some results from the interspecific hybridisation *N. goodspeedii* × *N. tabacum* (in Russian). *Academy of Agricultural Sciences Bulgarica, 1*, 315–321.

Palakarcheva, M., & Peeva, R. (1976). Male sterile tobacco lines Djebel developed on the basis of *N. Goodspeedii* cytoplasm (in Bulgarian). *Bulgarski tiutiun, 21*, 25–28.

Palakarcheva, M. T., Edreva, A. M., & Cholakova, N. I. (1980). Study of the backcrossing effect of *Nicotiana tabacum* in hybrids obtained with the amphidiploids *N. tabacum* L. × *N. debneyi* D. (2n=96) and *N. goodspeedii* × *N. tabacum* L. (2n=88). Bul spec CORESTA Congress Manila:102.

Palakarcheva, M., Yantcheva, A., & Delon, R. (1990). Study of cytoplasmic sterile male hybrids (CMS)-cytoplasm derived from *Nicotiana goodspeedii*. Bull Spec CORESTA Symposium, Kallithea, 141, A32, ISSN.0525-6240.

Pino Perez L. A. (2012). Esterilidad masculina citoplasmástica en tabaco. *Cuba Tabaco, 13*, 68–71.

Pittarelli, G. W., & Sisson, V. A. (1989). Registration of cytoplasmic male sterile tobacco germplasm Bel MS-2. *Crop Science, 29*, 836–837.

Priya, B. N. V., Venkata Reddy, T., Chidanand, U., & Saiprasad, G. V. S. (2019). Development of KASP marker for cytoplasmic male sterility in *Nicotiana tabacum* and utilization in trait introgression. *Journal of Genetics, 98*, 79.

Pscheidt, A., Lemos, R. C., Souza, J. C., Oliveira, V. B., Souza, A. M., & Pádua, J. M. V. (2021). Feasibility of using tobacco hybrids of the dark tobacco type. *Genetics and Molecular Research, 20*, 1–11.

Raineri, D., Jordan, P., & Kumar, A. (1992). Restoration of fertility in cytoplasmic male-sterile *Nicotiana tabacum* (Cytoplasm *N. bigelovii*) by protoplast fusion with X-irradiated protoplasts of *N. tabacum*, SR-1. *Journal of Experimental Botany, 43*, 195–203.

Ramavarma, K. T., Appa Rao, K., & Narayanan, A. I. (1978). Male sterile flower amomalis in an inter specific *Nicotiana* hybrid. *Tobacco Research, 4*, 29–33.

Reed, S. M., & Burns, J. A. (1986). Cross-restoration between *Nicotiana* cytoplasmic male-sterile and restored lines. *The Journal of Heredity, 77*, 159–163.

Roman, T., & Nalepa, S. (1970). Cytological research of male sterile forms of tobacco. 5th International Tobacco scientific Congress CORESTA Inf Bul Abstract A107:14.

Sand, S. A., & Christoff, G. T. (1972). Interaction of cytoplasmic and chromosomal systems in *Nicotiana*. *Genetics, 71*(II, Supplement), 54.

Sand, S. A., & Christoff, G. T. (1973). Cytoplasmic-chromosomal interactions and altered differentiation in tobacco. *The Journal of Heredity, 64*, 24–30.

Sastri, A. B., Venkata, C. R., Prasada, P. V. R., Krishnamurthy, A. S., Hanumantha, A. R., Satyanarayana, M., & Ramamohan, R. (1982). Effect of cytoplasmic male sterility from *Nicotiana undulata* on yield, yield components, chemical and physical quality characteristics of flue cured tobacco genotypes. *Tobacco Research, 8*, 173–177.

Schweppenhauser, M. A., & Mann, T. J. (1968). Restoration of staminal fertility in *Nicotiana* by introgression. *Canadian Journal of Genetics and Cytology, 10*, 401–411.

Shabanov, D., Chapkunova, Z., & Sotirov, S. I. (1986). Lines produced from crosses between *Nicotiana rustica* and *N. tabacum*. *Rasten Nauk, 23*, 44–48. (CORESTA Inf Bul (3/4):94).

Shava, J. G., Richardson-Kageler, S., Dari, S., Magama, F., & Rukuni, D. (2020). Breeding for yield, quality and associated traits in the Zimbabwean flue-cured tobacco. *Agricultural Reviews, 41*, 79–84.

Smith, H. H. (1968). Recent cytogenetic studies in the genus *Nicotiana*. *Advances in Genetics, 14*, 1–43.

Smith, H. H. (1988). The inheritance of genetic tumors in *Nicotiana* hybrids. *The Journal of Heredity, 79*, 277–283.

Spangenberg, G., Perez, R. V., Oliveira, M. M., Osusky, J., Nagel, J., Pais, M. S., & Potrykus, I. (1992). CMS system in *Nicotiana*: Flower development, patterns of mitochondrial DNA and mitochondrial gene expression. *Sexual Plant Reproduction, 5*, 13–26.

Stavely, J. R., Pittarelli, G. W., & Burk, L. G. (1973). *Nicotiana repanda* as a potential source for disease resistance in *N. Tabacum. The Journal of Heredity, 64*, 265–271.

Stoyanova, M. (1972). Investigations of the hybrids between the species *Nicotiana tabacum* and *N. glauca* (in Bulgarian). B: Otdal Hybridiz Rast C BAN:127–136.

Sun, Y., Ding, C., Zhang, L., Bu, G., & Gong, M. (1999a). The creation and application of new tobacco cytoplasmic male sterile (CMS) line 86-6. *Acta Tabacaria Sinica, 5*, 20–24.

Sun, Y., Ding, C., Zhang, L., Bu, G., & Gong, M. (1999b). Rapid creation of a new cytoplasmic male sterile line of flue-cured tobacco by interspecific cybridization. CORESTA Agro/Phyto meet, Suzhou, AP51.

Sun, Y.-H., Xue, Q.-Z., Ding, C.-M., Zhang, X.-Y., Wang, W.-F., & Ali, S. (2005). Somatic cybridization between *Nicotiana tabacum* and *N. repanda* based on a single inactivation procedure of nuclear donor parental protoplasts. *Plant Science, 168*, 303–308.

Svab, Z., & Maliga, P. (2007). Exceptional transmission of plastids and mitochondria from the transplastomic pollen parent and its impact on transgene containment. *Proceedings of the National Academy of Sciences, 104*, 7003–7008.

Svistava, H. K., & Gupta, H. S. (1991). Mitochondrial genome with special reference to male sterility. In K. R. Khanna (Ed.), *Biochemical aspects of crop improvement*. CRC Press/Ann Harbor.

Szilagyi, L. (1975). Elimination of chromosomes in an alloploid hybrid of *Nicotiana tabacum* × *N. glauca. Acta Botanica Academiae Scientiarum Hungaricae, 21*, 433–441.

Ternovsky, M. F., & Nosova, P. P. (1970). Investigation of immunity to tobacco mosaic virus (TMV) in hybrids on *N. tabacum* and *N. glutinosa* cytoplasm. CORESTA International Tobacco Sciences Congress:16 (A111).

Ternovsky, M. F., & Nosova, P. P. (1971). The inheritance of immunity to tobacco mosaic virus in interspecific hybrids with the cytoplasm of tobacco or of a wild species. *Sb Nauch-Issl Rabot WITIM, 156*, 65–69. (Plant Breed Abstr (1971) 45:5661).

Ternovsky, M. F., Moiseyeva, M. E., & Grebenkin, A. (1973). Experimentally produced new type of male sterility in interspecies *Nicotiana* hybrids. *Soviet Genetics, 9*, 693–702.

Thyssen, G., Svab, Z., & Maliga, P. (2012). Cell-to-cell movement of plastids in plants. *Proceedings of the National Academy of Sciences of the United States of America, 109*, 2439–2443.

Tsikov, D. (1982). Tobacco *Nicotiana* spp cytoplasmic male sterility hybrid breeding and hybrid seed production. *Genetika i Selektsiya, 15*, 319–324.

Tsikov, D., & Nikova, V. (1980). Male sterility in tobacco (*Nicotiana tabacum*). IV. Seed productivity. *Genetika i Selektsiya, 13*, 198–205.

Tsikov, D., & Tsikova, E. (1974). On the breeding prospects of some cytoplasms in male sterile tobaccos. *Genetika i Selektsiya, 7*, 77–81. (Tob Abstracts 19:1111–1112:2427).

Tsikov, D., & Tsikova, E. (1981). Genetics and cytology of cytoplasmic male sterility in *Nicotiana* (review). Bulg Akad Sci, Sofia (in Bulgarian with English summary).

Tsikov, D., & Tsikova, E. (1986). Male sterility in tobacco. VI. Cytoplasm of *N. gossei* (in Bulgarian). *Genetika i Selektsiya 19*, 121–131 [CORESTA Inf.Bul.1986(2):51–52].

Tsikov, D., Popivanov, I., & Kaneva, S. (1974). Tobacco hybrids of oriental type with cytoplasmic male sterility. *Genetika i Selektsiya, 7*, 425–423. (Tob Abstr 19:64274).

Tsikov, D., Tsikova, E., & Nikova, V. (1977). Male sterility in tobacco. III. On anther feminization (in Bulgarian). *Genetika i Selektsiya, 10*, 129–140.

Tsikova, E. D. (1967). Cytological study on cytoplasmic male sterility in tobacco. I. A form from crossing *N. tabacum* x *N. glutinosa* (in Bulgarian). Genet Issled Sb:19–30.

van ter Hakansson, G. M. F., Bonnett, H. T., & Glimelius, K. (1988). Variant mitochondrial protein and DNA patterns associated with cytoplasmic male-sterile lines of *Nicotiana. Theoretical and Applied Genetics, 76*, 431–437.

Verrier, J. L., Ano, G., Tancogne, J., Vidal, B., & Delon, R. (2000). The tobacco breeding program of Institut du Tabac (SEITA, ALTADIS): Methods, results and future prospects. Bull Spec CORESTA, Lisbon Congress:61, AP11.

Wang, R., Cai, X., Hu, S., Li, Y., Fan, Y., Tan, S., Liu, Q., & Zhou, W. (2020). Comparative analysis of the mitochondrial menomes of *Nicotiana tabacum*: Hints toward the key factors closely related to the cytoplasmic male sterility mechanism. *Frontiers in Genetics, 11*, 257. https://doi.org/10.3389/fgene.2020.00257

Wernsman, E. A., & Matzinger, D. F. (1966). A breeding procedure for the utilization of heterosis in tobacco-related species hybrids. *Crop Science, 6*, 298–300.

Wernsman, E. A., & Rufty, R. C. (1987). Tobacco. In W. W. Fehr (Ed.), *Principles of cultivar development. Vol. 2. Crop species* (pp. 669–698). Macmillian.

Yamada, T., Nakatogawa, H., & Kubo, T. (1984). Agronomic characteristics of some male sterile lines of Japanese domestic tobacco varieties. *Bulletin of the Iwata Tobacco Experiment Station, 16*, 47–57.

Zagorska, N. A., Tsikova, F., & Tsikov, D. (1978). Haploids and dihaploids of cytoplasmic male sterile tobacco obtained by the method of tissue culture. Bul Spec CORESTA Symp Sophia:81 (P04).

Zelcer, A., Aviv, D., & Galun, E. (1978). Interspecific transfer of male sterility by fusion between protoplasts of normal *N.sylvestris* and X-ray irradiated protoplasts of male sterile *N.tabacum. Zeitschrift für Pflanzenphysiologie, 90*, 397–407.

Zhao, T., Zhu, T., Liu, Q., Zhang, M., & Jiang, H. (2009). SNP in tobacco mitochondrial gene atp6 and its correlation with CMS. *Acta Agronomica Sinica, 35*, 1655–1661.

Zheng, Y., Liu, Z., Sun, Y., Liu, G., Yang, A., & Li, F. (2018). Characterization of genes specific to sua-CMS in *Nicotiana tabacum. Plant Cell Reports, 37*, 1245–1255. https://doi.org/10.1007/s00299-018-2309-2

Chapter 6
Closing Notes on Interspecific Hybridization in Tobacco Improvement

Nicotiana tabacum, one of the world's most important cultivated crops, is phylogenetically a relatively young form. With the possible exceptions of *N. arentsii* and *N. rustica*, the other direct ancestors of the present-day *Nicotiana* species had completed their evolutionary process much earlier than did the immediate ancestor of cultivated tobacco. Accompanied and driven by a multitude of evolutionary events, various diploid, allopolyploid and aneuploid lineages evolved along their separate or intertwined paths to ultimately take shape of the *Nicotiana* species as we know them now. No matter how geologically distant those events are now, all of them were taking place long before the first human beings arrived in and colonized the New World. The indigenous people inhabiting the Americas and the Australian continent became familiarized with and appreciated the usability of several of the *Nicotiana* species known to them (Tatemichi, 1990). In the pre-Columbian America, several species, primarily *N. tabacum*, *N. rustica*, *N. quadrivalvis* and most likely also *N. attenuata*, were grown by the natives, and it is not wholly unlikely that attempts were even made to improve them by some forms of selection (Setchell, 1921). It is very unlikely, however, that in those prehistoric days, any conscious human-aided hybridization, let alone crossing of different species, took place.

The geographical discoveries of America and Australia opened up a new era in the history of *Nicotiana*. Both wild-growing forms of tobacco and those subjected to some forms of cultivation that were encountered by the first Europeans had long been separated from each other by various mechanisms that obstructed them from crossing with one another. It took a few hundred years before the first attempts were made to bypass or overcome those obstructions and another century before sufficient motives were found to make a systematic study of *Nicotiana* hybrids worthwhile. Fortunately, for the early *Nicotiana* students, the barriers that obstructed the hybridization process were not absolutely tight. In contrast, they were loose enough to allow those researchers to synthetize quite a considerable number of *Nicotiana* hybrids without resorting to any artificial means, the latter having been invented much later.

A. Berbeć, *A Century of Interspecific Hybridization and Introgression in Tobacco*,
https://doi.org/10.1007/978-3-031-54964-9_6

In the meantime, cultivated tobacco has continued to grow in importance as one of the world's major cash crops. Therefore, it is not surprising that the wild relatives of tobacco started to attract the interest of tobacco researchers. Moreover, both the tobacco industry craving more profits and governments heavily dependent on the taxation of tobacco products for their budget revenues did not hesitate to channel an abundant flow of money into fundamental and applied studies of tobacco, including genetics and breeding. It is the half-century split by the Second World War that was the most prosperous period for tobacco science before the unfavorable turn of that prosperity occurred due to public health concerns. However, before that happened, the years from the 1930s to late 1970s were the heyday of interspecific cytogenetics and gene introgression in *Nicotiana*. In those years most *Nicotiana* hybrids were produced and studied, the most significant books on *Nicotiana* hybrids were written (Kostoff, 1943; Goodspeed, 1954), and the fundamental rules of interspecific gene transfer and introgression were disclosed. It is also in those years that some of the most momentous gene transfers in the history of interspecific introgression in *Nicotiana* were accomplished: N gene from *N. glutinosa* conferring resistance to tobacco mosaic, *Phl* and *Php* genes from *N. longiflora* and *N. plumbaginifolia*, respectively, for resistance to black shank, resistance to wildfire also from *N. longiflora*, resistance to blue mold and black root rot from *N. debneyi*, to name the best studied ones. Shifts to more virulent races made some of these resistances obsolete, as was the case with the resistance to black shank from *N. plumbaginifolia* and *N. longiflora* or, in some cases, resistance to blue mold from *N. debneyi*. The resistances to TMV from *N. glutinosa* and to black root rot from *N. debneyi* have remained unchallenged for decades. The N factor from *N. glutinosa* probably owes its long life to its linked defects. It has been deployed on a relatively small scale and has not exerted much selective pressure on the pathogen, although there is a strain (TMV-O) capable of breaking N gene-mediated resistance (Padget & Beachy, 1993). What all these interspecific transfers have in common is that most of them were accomplished by very simple, not to say crude, means before the advent of sophisticated biotechnological and molecular methods.

This is not to say that introgressive breeding in tobacco stopped altogether after that early fruitful period. Significant advances were made toward the end of the twentieth century and have continued to be made in this century. Resistance factors from *N. rustica* that combined effectiveness against the whole range of important races of wildfire, angular leaf spot and black shank without a penalty of undesirable linkages were bred into tobacco cultivars (Woodend & Mudzengerere, 1992; Drake et al., 2015). Resources of PVY-resistant germplasm were enriched and complemented by the tolerance response from *N. africana*, covering the whole range of PVY isolates, including those against which the *va* alleles had become ineffective (Lewis, 2007; Doroszewska, 2010; Korbecka-Glinka et al., 2018). Research is in progress to make resistance to TSWV from *N. alata usable* by minimizing the deleterious linkage (Laskowska & Berbeć, 2010; Trojak-Goluch et al., 2011, 2016a, b; Korbecka-Glinka et al., 2021). Simple chemical treatments were devised that could stop the development of type II lethality in tobacco hybrids (Shiragaki et al., 2020). At the same time, novel gene identification techniques based on transposable elements and genetic manipulation, such as gene editing, were recently successfully deployed for the same purpose (Ma et al., 2020).

We may not even be fully aware of how and to what extent *Nicotiana* species have contributed to the genotypes of present-day tobacco cultivated worldwide. Wild *Nicotianae*, especially those that were reputed to be of potential utility in tobacco breeding, were frequently part of in-house tobacco germplasm collections across the world. Both hints dispersed in the literature and personal accounts suggest that many tobacco breeders tried their luck with wild species. Not all of them were expert geneticists and academicians contributing to scientific journals. Not familiar with the peculiarities and intricacies of interspecific genetics, they might have handled the products of their interspecific matings just like they would the offspring of regular intervarietal crosses. This is probable and feasible on two accounts:

1. relative ease with which – under some circumstances – some hybrids of *N. tabacum* with an alien species can be obtained and their progeny repeatedly backcrossed until the phenotype of the cultivated species is successfully recovered without resorting to fertility-restoring treatments or laboratory-based manipulations, the phenomenon discussed in Sect. 3.6.3;
2. the offspring of some interspecific hybrids with *N. tabacum* have been reported to include *N. tabacum*-like fertile or partly fertile phenotypes and these apparently gynogenic or androgenetic products, which nonetheless may have carried introgression from the alien species, were likewise easily crossable with *N. tabacum*, the details having been discussed in Sect. 3.5.2;

Aided by those phenomena which are admittedly rare or exceptional, yet not wholly unlikely, quite a few breeding lines and local cultivars may have been developed, and the alien germplasm dispersed in their genetic make-up. The impact of what may be called fortuitous introgression that has passed unnoticed and unlabeled as such is difficult to evaluate, however. The molecular tools to do this are already there. They have been used with success to trace events from the remote past of wild *Nicotianae* (see Chap. 2), and they have been helpful in clarifying some introgression events in the pedigrees of present-day tobacco cultivars (Johnson et al., 2002; Milla et al., 2005). However, much work has yet to be done to fully assess the impact of *Nicotiana* species on tobacco improvement. At present, a poorly advanced genetic mapping of *Nicotiana* species (Lewis, 2011) makes such an assessment difficult.

However strange it may seem, both state-of-the-art biotechnology and genetic engineering have apparently made a moderate contribution to the genetic makeup of present-day tobacco cultivars, including introgression from wild *Nicotiana* species. It is noteworthy that two genes (the N gene from *N. glutinosa* and the nad1 gene from *N. alata*) are the only resistance-related genes from any *Nicotiana* known to be cloned (Whitham et al., 1996; Lewis, 2011; Pereg, 2013). The transfer of tobacco breeding from the public sector to private enterprises with their confidentiality policies may in part account for that dearth of information on advances in *Nicotiana* breeding technologies. There are various other reasons for which the interest in *Nicotiana* species and interspecific introgression has been and still is on the decline, maybe even on an irrevocable decline. Some of those reasons are fairly obvious, and all of them have been expounded by Lewis (2011) in his review. They are still valid and need not be repeated here.

Prophesizing about the future of tobacco and about the outlook for further *Nicotiana* research seems to be a somewhat futile exercise. One of God's most precious blessings to mankind is that He has chosen to keep the future hidden from us. Tobacco and its noncultivated *Nicotiana* relatives have always been at the forefront of plant and crop science, and hopefully, they will stay that way.

References

Doroszewska, T. (2010). Transfer of tolerance to different Potato virus Y (PVY) isolates from *Nicotiana africana* Merxm. to *Nicotiana tabacum* L. *Plant Breeding, 129*, 76–81.

Drake, K. E., Moore, J. M., Bertrand, P., Fortnum, B., Peterson, P., & Lewis, R. S. (2015). Black shank rResistance and agronomic performance of flue-cured tobacco lines and hybrids carrying the introgressed *Nicotiana rustica* region, Wz. *Crop Sci, 55*, 79–86.

Goodspeed, T. H. (1954). *The genus Nicotiana*. Chronica Botanica Co.

Johnson, E. S., Wolff, M. F., Wernsman, E. A., Atchley, W. E., & Shrew, H. D. (2002). Origin of the black shank resistance gene, Ph, in tobacco cultivar Coker 371-Gold. *Plant Disease, 86*, 1080–1084.

Korbecka-Glinka, G., Trojak-Goluch, A., Doroszewska, T., & Goepfert, S. (2018). The influence of *Nicotiana alata*-derived introgression on plant malformations of tobacco breeding lines resistant to tomato spotted wilt virus. CORESTA Congr, Agro/Phyto Groups, Kunming, AP 49.

Korbecka-Glinka, G., Trojak-Goluch, A., Bakaher, N., & Goepfert, S. (2021). Agronomic performance of Polalta-derived breeding lines resistant to tomato spotted wilt virus. CORESTA Agro/Phyto Meet, Online, AP 11.

Kostoff, D. (1943). *Cytogenetics of the genus Nicotiana*. State Printing House.

Laskowska, D., & Berbeć, A. (2010). TSWV resistance in DH lines of tobacco (*Nicotiana tabacum* L.) obtained from a hybrid between Polalta and Wiślica. *Plant Breeding, 129*, 731–733.

Lewis, R. S. (2007). Evaluation of *Nicotiana tabacum* genotypes possessing *Nicotiana africana*-derived genetic tolerance to Potato Virus Y. *Crop Science, 47*, 1975–1984.

Lewis, R. S. (2011). Nicotiana. In C. Kole (Ed.), *Wild crop relatives: Genomic and breeding resources, plantation and ornamental crops*. Springer. https://doi.org/10.1007/978-3-642-21201-7_10

Ma, J., Hancock, W. G., Nifong, J. M., Kernodle, S. P., & Lewis, R. S. (2020). Identification and editing of a hybrid lethality gene expands the range of interspecific hybridization potential in *Nicotiana. Theoretical and Applied Genetics, 133*, 2915–2925. https://doi.org/10.1007/s00122-020-03641-w

Milla, S. R., Levin, J. S., Lewis, R. S., & Rufty, R. C. (2005). RAPD and SCAR markers linked to an introgressed gene conditioning resistance to Peronospora tabacina D.B. Adam in Tobacco. *Crop Sciences, 45*, 2346–2354.

Padget, H. S., & Beachy, R. N. (1993). Analysis of o tobacco mosaic virus strain capable of breaking N-gene mediated resistance. *Plant Cell, 5*, 557–586.

Pereg, L. L. (2013). Black root rot of cotton in Australia: The host, the pathogen and disease management. *Crop and Pasture Science, 64*, 1112–1126. https://doi.org/10.1071/CP13231

Setchell, W. A. (1921). Aboriginal tobaccos. *American Anthropologist, 23*, 397–412.

Shiragaki, K., Nakamura, R., Nomura, S., He, H., Yamada, T., Marubashi, W., Oda, M., & Tezuka, T. (2020). Phenylalanine ammonia-lyase and phenolic compounds are related to hybrid lethality in the cross *Nicotiana suaveolens* × *N. tabacum*. *Plant Biotechnology (Tokyo), 37*, 327–333. https://doi.org/10.5511/plantbiotechnology

Tatemichi, Y. (1990). *Illustrated book of the genus Nicotiana*. Plant Breeding and Genetics Research Laboratory, Japan Tobacco Inc.

Trojak-Goluch, A., Laskowska, D., Agacka, M., Czarnecka, D., Kawka, M., & Czubacka, A. (2011). Effectiveness of combining resistance to *Thielaviopsis basicola* and tomato spotted wilt virus in haploid tobacco genotypes. *Breeding Science, 61*, 389–393.

Trojak-Goluch, A., Korbecka-Glinka, G., Goepfert, S., Doroszewska, T., & Berbec, A. (2016a). Tobacco breeding for TSWV resistance using RTSW-al factor derived from cultivar Polalta. CORESTA Congr Agro/Phyto Groups, Berlin, APPOST 24.

Trojak-Goluch, A., Laskowska, D., & Kursa, K. (2016b). Morphological and chemical characteristics of doubled haploids of flue-cured tobacco combining resistance to *Thielaviopsis basicola* and TSWV. *Breeding Science, 66*, 293–299.

Whitham, S., McCormick, S., & Baker, B. (1996). The N gene of tobacco confers resistance to tobacco mosaic virus in transgenic tomato. *Proceedings of the National Academy of Sciences of the United States of America, 93*, 8776–8781.

Woodend, J. J., & Mudzengerere, E. (1992). Inheritance of resistance to wildfire and angular leaf spot derived from *Nicotiana rustica* var. Brasilea. *Euphytica, 64*, 149–156.

Chapter 7
Special Supplement – List of Reported Interspecific Hybrids Within the Genus Nicotiana

Several investigators compiled comprehensive lists of *Nicotiana* hybrids. The earliest modern inventory was that by East (1928) who listed 65 interspecific combinations. It was followed by the list published by Kostoff (1943) who reported on 181 hybrids known to him. The last complete lists of interspecific Nicotiana hybrids with references were published by Goodspeed (1945, 1954). In the latter publication, the references to particular hybrids were scattered among the entries of the book's subject index. Together with several new combinations added a few years later (Goodspeed and Thompson, 1959) the number of documented interspecific hybrids stood at 243. The numbers of *Nicotiana* hybrids that were communicated later (Smith, 1968, 300 hybrids) and Apparao and Ramavarma (1974, 340 hybrids) were not supported by references to particular hybrid combinations. For the lack of more accurate data those estimations, although long outdated, have been non-the-less frequently quoted up to the present time. As more than 70 years have elapsed since Goodspeed's last inventory, the author of this volume thought it advisable to prepare a long-overdue update. Although the list below obviously exceeds the topical delimitations of this volume, annexing this updated inventory to a regular monograph on a closely related subject seems to be a good practical solution since publishing self-contained lists of this kind is not an accepted custom. The list also serves as an extension to the last column of Table 7.1.

In all probability, the list is not complete. It drew mostly on reports that were accessible through the web. The author faced the same problem as he did while working on other parts of this review: difficulty in searching materials published in remote parts of the globe or in exotic languages or/and scripts. Conversely, many of the listed hybrids have been repeatedly reported in many publications. Regardless of reporting frequency, only the first reports, a maximum of seven, were cited for each hybrid combination.

In earlier reports, hybrids that failed to reach the blooming stage did not qualify as successful hybrids and usually were listed separately. The list in this volume was extended by adding two, possibly debatable, categories:

A. Berbeć, *A Century of Interspecific Hybridization and Introgression in Tobacco*,
https://doi.org/10.1007/978-3-031-54964-9_7

Table 7.1 Interspecific hybrids within the genus *Nicotiana* reported by the year 2023

No.	Hybrid combination		Species 1 used as: F (female parent); M (male parent) (l) denotes lethal hybrid	Reported by:
	Species 1	Species 2		
1.	*N. alata*	*N. bonariensis*	F, M	Avery (1938), Kostoff (1940), Goodspeed (1945)
2.	*N. alata*	*N. langsdorffii*	F, M	Focke (1881), East and Hayes (1912), East (1916)
3.	*N. alata*	*N. longiflora*	F, M	East and Hayes (1912), Avery (1938), Goodspeed (1945), Näf (1958)
4.	*N. alata*	*N. mutabilis*	F, M	Stehmann et al. (2002)
5.	*N. alata*	*N. plumbaginifolia*	F, M	Goodspeed (1945), Takenaka (1950), Engvild (1950)
6.	*N. alata*	*N. sanderae*	F, M	D. Kostoff after Bradley and Goodspeed (1943)
7.	*N. alata*	*N. forgetiana*	F, M	East and Hayes (1912), Sorrentino et al. (2006)
8.	*N. alata*	*N. rastroensis*[1]	M	Descorbeth (2004), Descorbeth and McClure (2005)
9.	*N. alata*	*N. tabacum*	M	East and Hayes (1912), Kostoff (1930), Kostoff (1943), Ternovsky (1962)
10.	*N. alata*	*N. glauca*	M	Kostoff (1943), Goodspeed (1945), Takenaka (1953a)
11.	*N. alata*	*N. noctiflora*	M	Kostoff (1940), Goodspeed (1945), Näf (1958)
12.	*N. alata*	*N. knightiana*	M	Takenaka (1965), Takanashi and Marubashi (2017)
13.	*N. alata*	*N. paniculata*	F, M	Kostoff (1943), Goodspeed (1945), Näf (1958), Takenaka (1961)
14.	*N. alata*	*N. quadrivalvis*	M	Kostoff (1943), Goodspeed (1945)
15.	*N. alata*	*N. rustica*	F, M	Whitaker (1934), Zhukov (1939), Goodspeed (1945), Kehr (1951), Kehr and Smith (1954), Wolf (1965)
16.	*N. alata*	*N. amplexicaulis*	M	Gopinath et al. (1965, 1970)
17.	*N. alata*	*N. debneyi*	F	Zenkteler et al. (1981), Bartkowiak (1987)
18.	*N. alata*	*N. gossei*	F, M	Kostoff (1940), Goodspeed (1945), Takenaka (1962b), Gopinath et al. (1965, 1970)

(continued)

Table 7.1 (continued)

No.	Hybrid combination		Species 1 used as: F (female parent); M (male parent) (l) denotes lethal hybrid	Reported by:
	Species 1	Species 2		
19.	*N. alata*	*N. maritima*	M	Kostoff (1943), Goodspeed (1945)
20.	*N. alata*	*N. suaveolens*	M	Kostoff (1943), Goodspeed (1945)
21.	*N. alata*	*N. velutina*	M	Bradley and Goodspeed (1943)
22.	*N. alata*	*N. eastii*	M	Kostoff (1940), Goodspeed (1945)
23.	*N. alata*	*N. tomentosiformis*	M	Goodspeed, 1945
24.	*N. alata*	*N. sylvestris*	M, asexual	Berbeć et al. (1976), Fluhr, 1983
25.	*N. alata*	*N. undulata*	M	Takanashi and Marubashi (2017)
26.	*N. bonariensis*	*N. langsdorffii*	F, M	Muntzing, 1935, Avery (1938), Kostoff (1943), Goodspeed (1945), Näf (1958)
27.	*N. bonariensis*	*N. longiflora*	F, M	Avery (1938), Goodspeed (1945, 1954)
28.	*N. bonariensis*	*N. mutabilis*	M	Stehmann et al. 2002
29.	*N. bonariensis*	*N. plumbaginifolia*	F	Kostoff (1943), Goodspeed (1945), Näf (1958)
30.	*N. bonariensis*	*N. rastroensis*[1]	F, M	Lee et al. (2008)
31.	*N. bonariensis*	*N. forgetiana*	F	Butenko & Luneva, 1966
32.	*N. bonariensis*	*N. sanderae*	F, M	Kostoff (1943)
33.	*N. bonariensis*	*N. tabacum*	M?	Stavely 1979, Busconi et al. (2010)
34.	*N. bonariensis*	*N. glauca*	F	Williams and Pandey 1975[5]
35.	*N. bonariensis*	*N. quadrivalvis*	M (l)	Näf (1958)
36.	*N. bonariensis*	*N. rustica*	M	Näf (1958)
37.	*N. bonariensis*	*N. suaveolens*	M	Näf (1958)
38.	*N. forgetiana*	*N. langsdorffii*	F	East and Hayes (1912), White (1914)
39.	*N. forgetiana*	*N. mutabilis*	M	Stehmann et al. (2002)
40.	*N. forgetiana*	*N. plumbaginifolia*	M	Näf (1958), Takenaka (1965)
41.	*N. forgetiana*	*N. sanderae*	F, M	Sorrentino et al. (2006)
42.	*N. forgetiana*	*N. rastroensis*[1]	F, M	Lee et al. (2008)
43.	*N. forgetiana*	*N. tabacum*	M	Näf (1958), Ahuja (1962)

(continued)

Table 7.1 (continued)

No.	Hybrid combination		Species 1 used as: F (female parent); M (male parent) (l) denotes lethal hybrid	Reported by:
	Species 1	Species 2		
44.	*N. forgetiana*	*N. glauca*	F, M	Williams and Pandey (1975)
45.	*N. forgetiana*	*N. quadrivalvis*	M (l)	Näf (1958)
46.	*N. forgetiana*	*N. rustica*	M	Näf (1958)
47.	*N. langsdorffii*	*N. longiflora*	M	Goodspeed, 1945, Takenaka et al. (1956), Avery, 1962
48.	*N. langsdorffii*	*N. mutabilis*	M	Stehmann et al. (2002)
49.	*N. langsdorffii*	*N. plumbaginifolia*	F, M	Kostoff (1943), Goodspeed, 1945, 1954
50.	*N. langsdorffii*	*N. sanderae*	F, M	East, 1935 Smith (1937), Kostoff, 1943, Goodspeed (1945, 1954)
51.	*N. langsdorffii*	*N. rastroensis*[1]	F, M	Lee et al. (2008)
52.	*N. langsdorffii*	*N. tabacum*	M	East & Hayes, 1912, East (1935), Hu, 1956, Takenaka (1962b), Burk, 1972
53.	*N. langsdorffii*	*N. glauca*	F, M	premendelian, East (1928, 1935), Whitaker (1934), Goodspeed (1954)
54.	*N. langsdorffii*	*N. noctiflora*	M	Kostoff (1943), Goodspeed (1945, 1954)
55.	*N. langsdorffii*	*knightiana*	M	Takanashi and Marubashi (2017)
56.	*N. langsdorffii*	*N. paniculata*	M	premendelian, East and Hayes (1912), Whitaker (1934), Kostoff (1939b), Kehr and Smith, 1954, Takenaka (1962a), Takenaka and Yoneda (1964)
57.	*N. langsdorffii*	*N. miersii*	M	Kostoff (1940, 1943), Goodspeed (1945, 1954), Kehr (1951), Kehr and Smith (1954), Wolf (1965)
58.	*N. langsdorffii*	*N. quadrivalvis*	F	East and Hayes (1912), Kostoff, 1943, Näf (1958), Tezuka (2012)
59.	*N. langsdorffii*	*N. repanda*	M	Schweppenhauser et al. (1963)
60.	*N. langsdorffii*	*N. suaveolens*	M	premendelian, East (1928), Kostoff (1939b, 1943), Smith (1958), Smith and Abashian (1963), Wolf (1965)

(continued)

Table 7.1 (continued)

No.	Hybrid combination		Species 1 used as: F (female parent); M (male parent) (l) denotes lethal hybrid	
	Species 1	Species 2		Reported by:
61.	*N. langsdorffii*	*N. rustica*	M	East and Hayes (1912), Kostoff (1930), Whitaker (1934), Kehr (1951), Kehr and Smith (1954), Wolf (1965)
62.	*N. langsdorffii*	*N. otophora*	?	Berbeć et al. (1976)
63.	*N. langsdorffii*	*N. glutinosa*	M	East (1935)
64.	*N. langsdorffii*	*N. undulata*	M	Goodspeed (1945), Takanashi and Marubashi (2017)
65.	*N. langsdorffii*	*N. obtusifolia*	F (l)	Christoff (1928)
66.	*N. longiflora*	*N. sanderae*	F, M	Malloch and Malloch (1924), Brieger (1929), Kostoff (1943), Takenaka (1955)
67.	*N. longiflora*	*N. plumbaginifolia*	F, M	Avery (1938), Kostoff (1943), Goodspeed (1945), Murthy et al. (1998)
68.	*N. longiflora*	*N. tabacum*	M, F	Malloch (1924), Kostoff (1943), Takenaka (1962d), Sievert (1972)[3], Sarala et al. (2023)
69.	*N. longiflora*	*N. glauca*	F, M	East (1935), Kostoff (1943), Kehr (1951), Takenaka (1952a, 1953a, 1965), Murthy et al. (1998)
70.	*N. longiflora*	*N. paniculata*	F	Christoff (1928), Takenaka (1962a)
71.	*N. longiflora*	*N. quadrivalvis*	M (l)	Tezuka (2012)
72.	*N. longiflora*	*N. repanda*	M	Schweppenhauser et al. (1963), Davis et al. (1988)
73.	*N. longiflora*	*N. rustica*	M	McCray (1933)
74.	*N. longiflora*	*N. amplexicaulis*	not indicated	Gopinath et al. (1970)
75.	*N. longiflora*	*N. debneyi*	M	Näf (1958), Ahuja (1968)
76.	*N. longiflora*	*N. excelsior*	M	Takenaka (1965), Murthy et al. (1998)
77.	*N. longiflora*	*N. gossei*	M	Goodspeed (1945), Takenaka et al. (1956), Takenaka (1962b), Takenaka and Yoneda (1964)
78.	*N. longiflora*	*N. ingulba*	M	Kostoff (1943)
79.	*N. longiflora*	*N. megalosiphon*	M	Goodspeed (1945, 1954)

(continued)

Table 7.1 (continued)

No.	Hybrid combination		Species 1 used as: F (female parent); M (male parent) (l) denotes lethal hybrid	
	Species 1	Species 2		Reported by:
80.	*N. longiflora*	*N. suaveolens*	M	East (1935), Kostoff (1943), Goodspeed (1945, 1954), Takenaka (1952b), Takenaka and Yoneda (1964)
81.	*N. longiflora*	*N. eastii*	M	Kostoff (1940), Goodspeed (1945)
82.	*N. longiflora*	*N. sylvestris*	F	Savelli (1926)
83.	*N. longiflora*	*N. glutinosa*	F	Murthy et al. (1998)
84.	*N. longiflora*	*N. undulata*	M	Goodspeed (1945)
85.	*N. mutabilis*	*N. rastroensis*[1]	M	Lee et al. (2008)
86.	*N. plumbaginifolia*	*N. sanderae*	F, M	Goodspeed (1945), Kehr and Smith (1952), Takenaka (1953c, 1955)
87.	*N. plumbaginifolia*	*N. rastroensis*[1]	F, M	Lee et al. (2008)
88.	*N. plumbaginifolia*	*N. tabacum*	F, M	Pal and Nath (1936), Kostoff (1943), Kincaid (1949), Moav and Cameron (1960), Näf (1958), Burk (1960)
89.	*N. plumbaginifolia*	*N. glauca*	F, M	McCray (1932), Kostoff (1943), Ramanujam and Joshi (1942), Kehr and Smith (1952), Takenaka (1951a, 1953a)
90.	*N. plumbaginifolia*	*N. paniculata*	M	Christoff (1928) (l), Lilienfeld (1953), Näf (1958), Takenaka (1962a),
91.	*N. plumbaginifolia*	*N. quadrivalvis*	F	Christoff (1928) (l), Tezuka (2012) (l)
92.	*N. plumbaginifolia*	*N. rustica*	M	Kostoff (1943), Takenaka and Yoneda (1964)
93.	*N. plumbaginifolia*	*N. debneyi*	M	Kehr (1951), Kehr and Smith (1952, 1954), Smith and Abashian (1963)
94.	*N. plumbaginifolia*	*N. excelsior*	M	Sarala et al. (2008), Murthy and Rao (2008)
95.	*N. plumbaginifolia*	*N. gossei*	M	Takenaka et al. (1956), Takenaka (1962b), Murthy and Rama Prasad (2000)
96.	*N. plumbaginifolia*	*N. maritima*	F, M	Goodspeed (1945)
97.	*N. plumbaginifolia*	*N. megalosiphon*	M	Goodspeed (1945)

(continued)

Table 7.1 (continued)

No.	Hybrid combination		Species 1 used as: F (female parent); M (male parent) (l) denotes lethal hybrid	Reported by:
	Species 1	Species 2		
98.	*N. plumbaginifolia*	*N. suaveolens*	M	Goodspeed (1945), Kehr and Smith (1952), Takenaka (1952b, 1962b), Näf (1958)
99.	*N. plumbaginifolia*	*N. umbratica*	F, M	Gangadevi et al. (1987)
100.	*N. plumbaginifolia*	*N. eastii*	M	Goodspeed (1945)
101.	*N. plumbaginifolia*	*N. sylvestris*	asexual	Lin and Chen (1990), Cseplo et al.(1983)
102.	*N. plumbaginifolia*	*N. otophora*	asexual	Lin and Chen (1990)
103.	*N. plumbaginifolia*	*N. obtusifolia*	M	Murthy et al. (1998)
104.	*N. sanderae*	*N. tabacum*	F, M	Christoff (1928), Kostoff (1930), East (1935), Kehr and Smith (1954), Ternovsky (1962)
105.	*N. sanderae*	*N. glauca*	M	Kostoff (1930), Brieger and Forster (1942), Goodspeed (1945)
106.	*N. sanderae*	*N. noctiflora*	M	Kostoff (1935, 1940), Goodspeed (1945)
107.	*N. sanderae*	*N. paniculata*	F, M	Brieger (1929), East (1935), Goodspeed (1954), Kehr (1951), Kehr and Smith (1954)
108.	*N. sanderae*	*N. repanda*	M	Takenaka (1953c, 1955)
109.	*N. sanderae*	*N. miersii*	M	Näf (1958)
110.	*N. sanderae*	*N. rustica*	M	Kostoff (1930), Whitaker (1934), Goodspeed (1945), Kehr (1951), Kehr and Smith (1954)
111.	*N. sanderae*	*N. amplexicaulis*	M	Gopinath et al. (1965)
112.	*N. sanderae*	*N. debneyi*	asexual	Patel et al. (2011)
113.	*N. sanderae*	*N. gossei*	M	Goodspeed (1945)
114.	*N. sanderae*	*N. maritima*	M	Goodspeed (1945, 1954)
115.	*N. sanderae*	*N. suaveolens*	M	Goodspeed (1945), Takenaka (1955)
116.	*N. sanderae*	*N. eastii*	M	Kostoff (1939b, 1940)
117.	*N. sanderae*	*N. sylvestris*	M	Kostoff (1943)
118.	*N. sanderae*	*N. quadrivalvis*	M	Whitaker (1934), Kostoff (1943), Kehr (1951), Kehr and Smith (1954), Wolf (1965)

(continued)

Table 7.1 (continued)

No.	Hybrid combination		Species 1 used as: F (female parent); M (male parent) (l) denotes lethal hybrid	Reported by:
	Species 1	Species 2		
119.	*N. tabacum*	*N. glauca*	F, M	premendelian, East (1935)
120.	*N. tabacum*	*N. noctiflora*	F	Palakarcheva (1975), Dorossiev et al. (1978)
121.	*N. tabacum*	*N. petunioides*	F	Gisquet et al. (1940)
122.	*N. tabacum*	*N. benavidesii*	F	Goodspeed (1945)
123.	*N. tabacum*	*N. cordifolia*	M	Durbin and Uchytil (1977), Burk and Durbin (1978)
124.	*N. tabacum*	*N. knightiana*	F, M	Goodspeed (1945), Tanaka (1961), Takenaka (1962d), Berbeć et al. (1982)
125.	*N. tabacum*	*N. paniculata*	F, M	premendelian (East (1928)), Kostoff (1943), Goodspeed (1954), Takenaka (1962d)
126.	*N. tabacum*	*N. raimondii*	F, M	Kostoff (1943), Goodspeed (1954), Burk et al. (1982), Berbeć (1988)
127.	*N. tabacum*	*N. solanifolia*	F	Goodspeed (1945), Takenaka (1956a), b
128.	*N. tabacum*	*N. acuminata*	F	Gentcheff (Kostoff (1943)), Palakarcheva (1981), Iwai et al. (1986)
129.	*N. tabacum*	*N. pauciflora*	F	Gentscheff (1931), Kostoff (1943), Goodspeed (1945)
130.	*N. tabacum*	*N. clevelandii*	M	Kehr and Smith (1952), Kaul (1988)
131.	*N. tabacum*	*N. quadrivalvis*	F, M	premendelian (East (1928)), East and Hayes (1912), Kostoff (1943)
132.	*N. tabacum*	*N. nesophila*	M	Reed and Collins (1978, 1980), Huesing et al. (1989)
133.	*N. tabacum*	*N. nudicaulis*	F, M	Gentscheff (1931), Kostoff (1943)
134.	*N. tabacum*	*N. repanda*	F, M	Foster (1943), Kincaid (1949), Pittarelli and Stavely (1975), Nagao (1982[6])
135.	*N. tabacum*	*N. stocktonii*	M,	Wong (1975), Reed and Collins (1978, 1980)
136.	*N. tabacum*	*N. rustica*	F, M	premendelian, East and Hayes (1912), Eghis (1927), Savelli (1927), East (1935)

(continued)

Table 7.1 (continued)

No.	Hybrid combination		Species 1 used as: F (female parent); M (male parent) (l) denotes lethal hybrid	
	Species 1	Species 2		Reported by:
137.	*N. tabacum*	*N. africana*	F, M	Gerstel et al. (1979), Nikova and Zagorska (1990)
138.	*N. tabacum*	*N. amplexicaulis*	M, F	Wark (1970), Berbeć and Doroszewska (1981), Larkina (1980), Sarala et al. (2023)
139.	*N. tabacum*	*N. benthamiana*	F, M	Subhashini et al. (1986), DeVerna et al. (1987), Nikova et al. (1991), Zaitlin and Mundell ((2006)), Mihaylova-Kroumova et al. (2020), Sarala et al. (2023)
140.	*N. tabacum*	*N. cavicola*	M	Nikova et al. (2006)
141.	*N. tabacum*	*N. debneyi*	M	Kostoff (1943), Goodspeed (1954), Takenaka (1956a)
142.	*N. tabacum*	*N. excelsior*	M	Wark (1970), Nikova (1986), Murthy et al. (2014)
143.	*N. tabacum*	*N. exigua*	M	Wark (1970), Kobus (1971)
144.	*N. tabacum*	*N. fragrans*	M	Durbin and Uchytil (1977)
145.	*N. tabacum*	*N. goodspeedii*	M	Butenko et al. (1970), Wark (1970)
146.	*N. tabacum*	*N. gossei*	M, F	Valleau (1952), Takenaka (1962d), Sievert (1972)[4], Adachi and Inoue (1995), Murthy et al. (2014), Sarala et al. (2023)
147.	*N. tabacum*	*N. hesperis*	M	Kubo (1985)
148.	*N. tabacum*	*N. ingulba*	M	Butenko et al. (1970), Nikova et al. (1998)
149.	*N. tabacum*	*N. maritima*	M	Wark (1970), Palakarcheva (1975)
150.	*N. tabacum*	*N. megalosiphon*	M	Clayton (1950), Takenaka (1962d), Hranov (1970), Sarala et al. (2023)
151.	*N. tabacum*	*N. occidentalis*	M	Butenko et al. (1970), Ternovsky et al. (1972), Wong (1975)
152.	*N. tabacum*	*N. rosulata*	F	Ternovsky et al. (1976), Ternovsky and Larkina (1978), Larkina (2015, 2017)

(continued)

Table 7.1 (continued)

No.	Hybrid combination		Species 1 used as: F (female parent); M (male parent) (l) denotes lethal hybrid	Reported by:
	Species 1	Species 2		
153.	*N. tabacum*	*N. rotundifolia*	asexual	Ilcheva et al. (2001)
154.	*N. tabacum*	*N. simulans*	M	Kubo (1985), Tatemichi (1990), Ma et al. (2020)
155.	*N. tabacum*	*N. suaveolens*	F, M	premendelian (East (1928)), Kostoff (1943)
156.	*N. tabacum*	*N. umbratica*	M	Murthy et al. (2014), Sarala et al. (2023)
157.	*N. tabacum*	*N. velutina*	M	Wark (1970), Ma et al. (2020)
158.	*N. tabacum*	*N. wuttkei*	M	Laskowska and Berbeć (2012)
159.	*N. tabacum*	*N. eastii*	M	Chaplin & Mann, 1961
160.	*N. tabacum*	*N. sylvestris*	F, M	East and Hayes (1912), Bellair (1913), Malinowski (1916), Sachs-Skalińska (1917)
161.	*N. tabacum*	*N. kawakamii*		Ohashi (1985)
162.	*N. tabacum*	*N. otophora*	F	Goodspeed (1945), Ar-Rushdi (1955), Takenaka (1962)
163.	*N. tabacum*	*N. setchellii*	F, M	Greenleaf (1941), Goodspeed (1945, 1954)
164.	*N. tabacum*	*N. tomentosa*	F, M	Goodspeed and Clausen (1928), McCray (1932), Kostoff (1943)
165.	*N. tabacum*	*N. tomentosiformis*	F, M	Brieger (1928), Breisser (1934), Lehmann (1936), Kostoff (1943)
166.	*N. tabacum*	*N. obtusifolia*	F, M	Takenaka (1951b, 1956a), Chung et al. (1988), Choi and Lee (1991), Liu and Marubashi (2014)
167.	*N. tabacum*	*N. palmeri*	F	Goodspeed (1945), Berbeć et al. (1982)
168.	*N. tabacum*	*N. arentsii*	asexual	DeVerna (1984)
169.	*N. tabacum*	*N. glutinosa*	F, M	premendelian (East (1928)), first report by Koelreuter (Mayr (1986)), East (1935),
170.	*N. tabacum*	*N. undulata*	M	Kehr and Smith (1952), Takenaka (after Goodspeed and Thompson (1959))

(continued)

Table 7.1 (continued)

No.	Hybrid combination		Species 1 used as: F (female parent); M (male parent) (l) denotes lethal hybrid	
	Species 1	Species 2		Reported by:
171.	*N. glauca*	*N. noctiflora*	F, M	Goodspeed (1934), Kostoff (1943), Goodspeed (1945)
172.	*N. glauca*	*N. benavidesii*	F	Goodspeed (1945)
173.	*N. glauca*	*N. paniculata*	F, M	East (1935), Lehmann (1936), Kostoff (1943), Poddubnaya-Arnoldi and Ludkina (1945), Kehr and Smith (1954)
174.	*N. glauca*	*N. raimondii*	M	Goodspeed (1954)
175.	*N. glauca*	*N. solanifolia*	M	Kostoff (1940), Goodspeed (1945)
176.	*N. glauca*	*N. attenuata*	M	Gentscheff (1931)
177.	*N. glauca*	*N. miersii*	M	Kostoff (1943)
178.	*N. glauca*	*N. clevelandii*	F	Kehr and Smith (1952)
179.	*N. glauca*	*N. quadrivalvis*	M	McCray (1932), East (1935) Kostoff (1943), Goodspeed (1945), Kehr and Smith (1952), Meiselman et al. (1961)
180.	*N. glauca*	*N. nudicaulis*	M	Goodspeed (1934, 1945)
181.	*N. glauca*	*N. rustica*	M	East (1935), Bolsunov (1937), Modilevsky (1937), Kostoff (1939c), Goodspeed (1945), Ilin (1948)
182.	*N. glauca*	*N. debneyi*	F, M	Kehr (1951), Lautz (1957), Smith and Abashian (1963)
183.	*N. glauca*	*N. goodspeedii*	F	Horton (1981)
184.	*N. glauca*	*N. gossei*	M, F	Akada and Hirai (1986) (somatic), Murthy et al. (1998), Murthy and Subbarao (2004), Sarala et al. (2008)
185.	*N. glauca*	*N. megalosiphon*	F	Goodspeed (1945)
186.	*N. glauca*	*N. simulans*	F	Horton (1981)
187.	*N. glauca*	*N. suaveolens*	F, M	East (1935), Goodspeed (1945), Näf (1958), Smith & Abashian, 1963 (amphidiploid synthetic species *N.* x *flindersiensis*)

(continued)

Table 7.1 (continued)

No.	Hybrid combination		Species 1 used as: F (female parent); M (male parent) (l) denotes lethal hybrid	Reported by:
	Species 1	Species 2		
188.	*N. glauca*	*N. umbratica*	M	Gangadevi et al. ((1987))
189.	*N. glauca*	*N. velutina*	F	Jeanes (1999)
190.	*N. glauca*	*N. eastii*	M	Kostoff (1943)
191.	*N. glauca*	*N. sylvestris*	F	Ternovsky (1962), Takenaka (1965)
192.	*N. glauca*	*N. otophora*	F	Takenaka (1953b)
193.	*N. glauca*	*N. tomentosa*	F, M	Kostoff (1932), East (1935), Lehmann (1936), Smith (1942)
194.	*N. glauca*	*N. tomentosiformis*	F, M	Kostoff (1932), Lehmann (1936), Goodspeed (1945)
195.	*N. glauca*	*N. obtusifolia*	M	Murthy et al. (1998)
196.	*N. glauca*	*N. glutinosa*	F, M	Kostoff (1930), East (1935), Goodspeed (1945)
197.	*N. glauca*	*N. thyrsiflora*	M	Goodspeed (1945, 1954)
198.	*N. noctiflora*	*N. petunioides*	F	Goodspeed and Thompson (1959)
199.	*N. noctiflora*	*N. paniculata*	M	Goodspeed (1934), Kostoff (1943), Goodspeed (1945)
200.	*N. noctiflora*	*N. rustica*	M	Kostoff (1943), Näf (1958), Takenaka and Yoneda (1964), Kehr and Smith (1954), Wolf (1965), Busconi et al. (2010)
201.	*N. petunioides*	*N. corymbosa*	M	Goodspeed and Thompson (1959)
202.	*N. petunioides*	*N. miersii*	M	Goodspeed (1945)
203.	*N. benavidesii*	*N. paniculata*	M	Goodspeed (1945, 1954)
204.	*N. benavidesii*	*N. raimondii*	F	Goodspeed (1945)
205.	*N. benavidesii*	*N. solanifolia*	F	Goodspeed (1945)
206.	*N. benavidesii*	*N. rustica*	M	Goodspeed (1945)
207.	*N. benavidesii*	*N. glutinosa*	F	Goodspeed (1945)
208.	*N. benavidesii*	*N. undulata*	M	Goodspeed (1945)
209.	*N. cordifolia*	*N. knightiana*	M	Goodspeed (1945, 1954)
210.	*N. cordifolia*	*N. raimondii*	F, M	Goodspeed (1945, 1954)
211.	*N. cordifolia*	*N. solanifolia*	F, M	Goodspeed (1954)
212.	*N. cordifolia*	*N. undulata*	M	Goodspeed (1945)
213.	*N. knightiana*	*N. raimondii*	M	Durbin and Uchytil (1977)
214.	*N. knightiana*	*N. solanifolia*	M	Durbin and Uchytil (1977)

(continued)

Table 7.1 (continued)

No.	Hybrid combination		Species 1 used as: F (female parent); M (male parent) (l) denotes lethal hybrid	Reported by:
	Species 1	Species 2		
215.	*N. knightiana*	*N. rustica*	F	Takenaka (1958, 1959, 1962c)
216.	*N. knightiana*	*N. umbratica*	F	Gangadevi et al. (1985)
217.	*N. knightiana*	*N. sylvestris*	asexual	Menczel et al. (1978)
218.	*N. paniculata*	*N. raimondii*	F, M	Kostoff (1940), Goodspeed (1945)
219.	*N. paniculata*	*N. solanifolia*	F, M	Goodspeed (1945), Durbin and Uchytil (1977)
220.	*N. paniculata*	*N. miersii*	F, M	Kostoff (1943), Goodspeed (1945)
221.	*N. paniculata*	*N. pauciflora*	F	Goodspeed (1934)
222.	*N. paniculata*	*N. quadrivalvis*	F	premendelian (East (1928)), East and Hayes (1912)
223.	*N. paniculata*	*N. nudicaulis*	F (I)	Yamada et al. (1999) (inviable type III)
224.	*N. paniculata*	*N. rustica*	F, M	premendelian (East (1928)), East and Hayes (1912)
225.	*N. paniculata*	*N. gossei*	F (I)	Yamada et al. (1999)
226.	*N. paniculata*	*N. suaveolens*	F	premendelian (East (1928))
227.	*N. paniculata*	*N. umbratica*	F	Subhashini and Gopinath (1974)
228.	*N. paniculata*	*N. velutina*	M	Bradley and Goodspeed (1943)
229.	*N. paniculata*	*N. otophora*	F	Takenaka (1953b)
230.	*N. paniculata*	*N. tomentosa*	F	Lehmann (1936)
231.	*N. paniculata*	*N. tomentosiformis*	F, M (I)	Lehmann (1936)
232.	*N. paniculata*	*N. glutinosa*	F, M	premendelian (East (1928))
233.	*N. paniculata*	*N. undulata*	F, M	Fatalizade (1939), Kostoff (1943), Ternovsky (1962)
234.	*N. raimondii*	*N. solanifolia*	F, M	Kostoff (1940, 1943), Goodspeed (1945)
235.	*N. raimondii*	*N. undulata*	M	Goodspeed (1945)
236.	*N. solanifolia*	*N. quadrivalvis*	M	Goodspeed (1934), Kostoff (1943), Goodspeed (1945)
237.	*N. solanifolia*	*N. rustica*	M	Goodspeed (1934, 1945)
238.	*N. solanifolia*	*N. suaveolens*	M	East (1928), Kostoff (1943)
239.	*N. solanifolia*	*N. otophora*		Berbeć et al. (1976)
240.	*N. solanifolia*	*N. undulata*	M	Ternovsky (1962)
241.	*N. acuminata*	*N. corymbosa*	M	Goodspeed and Thompson (1959)

(continued)

Table 7.1 (continued)

No.	Hybrid combination		Species 1 used as: F (female parent); M (male parent) (l) denotes lethal hybrid	
	Species 1	Species 2		Reported by:
242.	*N. acuminata*	*N. pauciflora*	F, M	Kostoff (1943), Goodspeed and Thompson (1959), Takenaka (1965)
243.	*N. acuminata*	*N. undulata*	M	Gray (1978)
244.	*N. attenuata*	*N. miersii*	F, M	Kostoff (1943), Goodspeed and Thompson (1959), Pearse et al. (2006) Krügel (2010)
245.	*N. attenuata*	*N. clevelandii*	M	Goodspeed (1945)
246.	*N. attenuata*	*N. quadrivalvis*	M	Goodspeed (1945)
247.	*N. attenuata*	*N. rustica*	M	premendelian (East (1928)), Patel (1960). Goodspeed (1945), Tezuka (2012)(lethal)
248.	*N. attenuata*	*N. maritima*	M	Goodspeed (1945)
249.	*N. attenuata*	*N. obtusifolia*	M	Anssour et al. (2009), Navarro-Quezada et al. (2013), Navarro-Quezada (2014)
250.	*N. attenuata*	*N. undulata*	M	Gentscheff (1931), Kostoff (1943), Goodspeed (1954)
251.	*N. corymbosa*	*N. linearis*	F	Goodspeed and Thompson (1959)
252.	*N. miersii*	*N. clevelandii*	F	Goodspeed (1945)
253.	*N. miersii*	*N. quadrivalvis*	F	Kostoff (1940, 1943), Goodspeed (1945)
254.	*N. miersii*	*N. rustica*	M	East (1928), Goodspeed (1945), Patel (1960)
255.	*N. miersii*	*N. megalosiphon*	F	Kostoff (1940), Goodspeed (1945)
256.	*N. miersii*	*N. suaveolens*	F	Goodspeed (1934), Kostoff (1940), Goodspeed (1945)
257.	*N. pauciflora*	*N. rustica*	M	Goodspeed (1945)
258.	*N. clevelandii*	*N. quadrivalvis*	F, M	Masuta et al. (1993), Kiraly and Schoeltz (1995)
259.	*N. clevelandii*	*N. nudicaulis*	F	Goodspeed (1945)
260.	*N. clevelandii*	*N. debneyi*	M	Kehr and Smith (1952), Lautz (1957), Smith and Abashian (1963), Smith et al. (1970)
261.	*N. clevelandii*	*N. exigua*	F	Kostoff (1943), Goodspeed (1945)

(continued)

Table 7.1 (continued)

No.	Hybrid combination		Species 1 used as: F (female parent); M (male parent) (l) denotes lethal hybrid	Reported by:
	Species 1	Species 2		
262.	*N. clevelandii*	*N. umbratica*	F	Gangadevi et al. (1987)
263.	*N. clevelandii*	*N. glutinosa*	F, M	Kehr and Smith (1952), Lautz (1957), Smith and Abashian (1963), Smith et al. (1970)
264.	*N. quadrivalvis*	*N. nudicaulis*	F, M	Goodspeed (1945)
265.	*N. quadrivalvis*	*N. rustica*	F, M	premendelian (East (1928)), Maryanovich (1939), Fardy and Hitier (1945)
266.	*N. quadrivalvis*	*N. amplexicaulis*	F	Gopinath et al. (1965, 1970)
267.	*N. quadrivalvis*	*N. benthamiana*	F	Kostoff (1943), Goodspeed (1945), Apparao and Ramavarma (1974)
268.	*N. quadrivalvis*	*N. debneyi*	F	Kehr and Smith (1952)
269.	*N. quadrivalvis*	*N. excelsior*	M	Zaitlin and Mundell (2006), Mihaylova-Kroumova et al. (2020)
270.	*N. quadrivalvis*	*N. megalosiphon*	F	Kostoff (1943), Goodspeed (1945), Takenaka (1959)
271.	*N. quadrivalvis*	*N. suaveolens*	F, M	premendelian (East (1928)), Goodspeed and Clausen (1927), East (1935), Fardy and Hitier (1945), Näf (1958)
272.	*N. quadrivalvis*	*N. eastii*	M	Goodspeed (1945)
273.	*N. quadrivalvis*	*N. sylvestris*	F	Goodspeed (1934), Kostoff (1943), Goodspeed (1945)
274.	*N. quadrivalvis*	*N. tomentosa*	F	East (1935), Kostoff (1943), Goodspeed (1945)
275.	*N. quadrivalvis*	*N. tomentosiformis*	F	Kostoff (1943), Goodspeed (1945)
276.	*N. quadrivalvis*	*N. glutinosa*	F, M	premendelian (East (1928)), East (1935)
277.	*N. quadrivalvis*	*N. undulata*	asexual	Razdan (2003)
278.	*N. nesophila*	*N. repanda*	F	Goodspeed (1945)
279.	*N. nesophila*	*N. stocktonii*	F	Goodspeed (1945)
280.	*N. nesophila*	*N. umbratica*	M	Raju et al. (2008), Murthy et al. (2014)
281.	*N. nesophila*	*N. sylvestris*	F	Unnikrishnan et al. (1981)

(continued)

Table 7.1 (continued)

No.	Hybrid combination		Species 1 used as: F (female parent); M (male parent) (l) denotes lethal hybrid	Reported by:
	Species 1	Species 2		
282.	*N. nudicaulis*	*N. repanda*	F	Kostoff (1943)
283.	*N. nudicaulis*	*N. rustica*	F, M	Kostoff (1943), Goodspeed (1954)
284.	*N. nudicaulis*	*N. debneyi*	F, M	Goodspeed (1934, 1945)
285.	*N. nudicaulis*	*N. exigua*	F, M	Kostoff (1943), Goodspeed (1945)
286.	*N. nudicaulis*	*N. suaveolens*	F, M	Kostoff (1943), Goodspeed (1945)
287.	*N. nudicaulis*	*N. sylvestris*	F	McCray (1932), Kostoff (1943)
288.	*N. nudicaulis*	*N. tomentosa*	F	Kostoff (1940, 1943), Goodspeed (1945, 1954)
289.	*N. nudicaulis*	*N. tomentosiformis*	F	Brieger (1929), Goodspeed (1945, 1954)
290.	*N. nudicaulis*	*N. obtusifolia*	F, M	Kostoff (1943)
291.	*N. nudicaulis*	*N. palmeri*	F	Kostoff (1943), Goodspeed (1945)
292.	*N. nudicaulis*	*N. glutinosa*	F	Christoff (1928), Kostoff (1943), Goodspeed (1945)
293.	*N. repanda*	*N. stocktonii*	F, M	Kostoff (1943)
294.	*N. repanda*	*N. africana*	F	Murthy and Rama Prasad (2000), Murthy and Subhashini (2000)
295.	*N. repanda*	*N. benthamiana*	M	Subhashini et al. (1998), Raju et al. (2008, 2009), Murthy et al. (2014)
296.	*N. repanda*	*N. debneyi*	M (l)	Yamada et al. (1999)
297.	*N. repanda*	*N. exigua*	M	Goodspeed (1945)
298.	*N. repanda*	*N. umbratica*	F	Gangadevi et al. (1987), Murthy and Rama Prasad (2000)
299.	*N. repanda*	*N. sylvestris*	F	Kostoff (1940), Goodspeed (1945), Burk and Dropkin (1961)
300.	*N. repanda*	*N. tomentosiformis*	F (l)	Kobori and Marubashi (2004)
301.	*N. repanda*	*N. palmeri*	F	Goodspeed (1945), Takenaka (1958, 1959), Takenaka, 1962c, Schweppenhauser et al. (1963)

(continued)

Table 7.1 (continued)

No.	Hybrid combination		Species 1 used as: F (female parent); M (male parent) (l) denotes lethal hybrid	Reported by:
	Species 1	Species 2		
302.	*N. repanda*	*N. glutinosa*	M	Kehr and Smith (1952)
303.	*N. stocktonii*	*N. sylvestris*	F (l)	Muraida and Marubashi (2015)
304.	*N. stocktonii*	*N. tomentosiformis*	F (l)	Muraida and Marubashi (2015)
305.	*N. rustica*	*N. benthamiana*	F	Sarala et al. (2023)
306.	*N. rustica*	*N. exigua*	F	Bolsunov (1970)
307.	*N. rustica*	*N. gossei*	F	Takenaka (1965)
308.	*N. rustica*	*N. megalosiphon*	F	Takenaka (1965)
309.	*N. rustica*	*N. suaveolens*	F	premendelian (East (1928)), Christoff (1938), Goodspeed (1945)
310.	*N. rustica*	*N. sylvestris*	asexual	Gleddie et al. (1983)
311.	*N. rustica*	*N. kawakamii*	F	Japan Tobacco Inc. (2022)
312.	*N. rustica*	*N. tomentosa*	F	Japan Tobacco Inc. (2022)
313.	*N. rustica*	*N. tomentosiformis*	F	Kostoff (1929)
314.	*N. rustica*	*N. palmeri*	F	Kostoff (1930), Goodspeed (1945)
315.	*N. rustica*	*N. glutinosa*	F	Lehmann (1936), Calitz and Milne (1962), Bannikova (1965)
316.	*N. rustica*	*N. undulata*	F	Kostoff (1940, 1943); Goodspeed (1945)
317.	*N. rustica*	*N. wigandioides*	F	Goodspeed (1934, 1945)
318.	*N. africana*	*N. benthamiana*	M	Iizuka et al. (2012)
319.	*N. africana*	*N. fragrans*	M	Gerstel et al. (1979)
320.	*N. africana*	*N. gossei*	M	Murthy and Subhashini (2000)
321.	*N. amplexicaulis*	*N. excelsior*	F	Gopinath et al. (1970), Gangadevi et al. (1987)
322.	*N. amplexicaulis*	*N. gossei*	not indicated	Gopinath et al. (1970)
323.	*N. amplexicaulis*	*N. megalosiphon*	F	Gopinath et al. (1965, 1970)
324.	*N. amplexicaulis*	*N. occidentalis*	M	Gopinath et al. (1965, 1970)
325.	*N. amplexicaulis*	*N. suaveolens*	F	Gopinath et al. (1965, 1970)
326.	*N. amplexicaulis*	*N. umbratica*	F	Gangadevi et al. (1987)
327.	*N. amplexicaulis*	*N. velutina*	F	Gopinath et al. (1965, 1970)

(continued)

Table 7.1 (continued)

No.	Hybrid combination		Species 1 used as: F (female parent); M (male parent) (l) denotes lethal hybrid	Reported by:
	Species 1	Species 2		
328.	*N. amplexicaulis*	*N. glutinosa*	F	Gopinath et al. (1965, 1970)
329.	*N. benthamiana*	*N. cavivola*	F	Williams (1975)
330.	*N. benthamiana*	*N. debneyi*	F	Bradley and Goodspeed (1943), Goodspeed (1945), Apparao and Ramavarma (1974)
331.	*N. benthamiana*	*N. excelsior*	F, M[2]	Fitzmaurice (2002), Tezuka et al. (2021b)
332.	*N. benthamiana*	*N. fragrans*	F	Williams (1975)
333.	*N. benthamiana*	*N. gossei*	M	Goodspeed (1945), Takenaka (1965), Durbin and Uchytil (1977)
334.	*N. benthamiana*	*N. ingulba*	M	Apparao and Ramavarma (1974), Ramavarma and Apparao (1976)
335.	*N. benthamiana*	*N. megalosiphon*	F	Goodspeed (1945), Apparao and Ramavarma (1974)
336.	*N. benthamiana*	*N. suaveolens*	M	Wheeler (1945), Goodspeed (1945), Apparao and Ramavarma (1974)
337.	*N. benthamiana*	*N. umbratica*	M	Gangadevi et al. (1987)
338.	*N. benthamiana*	*N. velutina*	F	Horton (1981)
339.	*N. benthamiana*	*N. glutinosa*	F	Apparao and Ramavarma (1974)
340.	*N. burbidgeae*	*N. velutina*	F	Symon (1984)
341.	*N. cavicola*	*N. debneyi*	M	Williams (1975)
342.	*N. cavicola*	*N. umbratica*	F, M	Williams (1975), Gangadevi et al. (1987)
343.	*N. debneyi (forsteri)*	*N. exigua*	F, M	Kostoff (1943)
344.	*N. debneyi (forsteri)*	*N. fragrans*	F	Iizuka et al. (2012), Tezuka et al. (2021a)
345.	*N. debneyi (forsteri)*	*N. goodspeedii*	M	Goodspeed (1945)
346.	*N. debneyi (forsteri)*	*N. maritima*	F, M	Kostoff (1943), Goodspeed (1945)
347.	*N. debneyi (forsteri)*	*N. megalosiphon*	M	Goodspeed (1945, 1954)

(continued)

Table 7.1 (continued)

No.	Hybrid combination		Species 1 used as: F (female parent); M (male parent) (l) denotes lethal hybrid	
	Species 1	Species 2		Reported by:
348.	*N. debneyi (forsteri)*	*N. occidentalis*	F	Gangadevi et al. (1987)
349.	*N. debneyi (forsteri)*	*N. simulans*	M	Gangadevi et al. (1987)
350.	*N. debneyi (forsteri)*	*N. suaveolens*	F	Kostoff (1943), Williams (1975)
351.	*N. debneyi (forsteri)*	*N. umbratica*	F	Williams (1975), Gangadevi and Satyanarayana (1982)
352.	*N. debneyi (forsteri)*	*N. velutina*	F, M	Gillham et al. (1977)
353.	*N. debneyi (forsteri)*	*N. sylvestris*	F, M (l)	Yamada et al.(1999), Tezuka et al. (2007)
354.	*N. debneyi (forsteri)*	*N. otophora*	F	Takenaka (1959), Takenaka (1962c)
355.	*N. debneyi (forsteri)*	*N. tomentosiformis*	F	Takenaka (1965), Tezuka et al. (2007)
356.	*N. debneyi (forsteri)*	*N. obtusifolia*	F, M	Kostoff (1940, 1943), Goodspeed (1945)
357.	*N. debneyi (forsteri)*	*N. palmeri*	F, M	Kostoff (1943), Goodspeed (1945)
358.	*N. debneyi (forsteri)*	*N. thyrsiflora*		Ohashi et al. (1992)
359.	*N. debneyi (forsteri)*	*N. glutinosa*	F, M	Dusseau et al. (1944), Fardy and Hitier (1945), Kehr and Smith (1952)
360.	*N. excelsior*	*N. gossei*	M, F	Goodspeed (1945), Murthy et al. (1998)
361.	*N. excelsior*	*N. suaveolens*	M	Näf (1958), Takenaka (1965)
362.	*N. excelsior*	*N. velutina*	F	Bradley and Goodspeed (1943)
363.	*N. excelsior*	*N. obtusifolia*	M	Murthy et al. (1998)
364.	*N. excelsior*	*N. glutinosa*	F	Murthy & Rao, 2008
365.	*N. exigua*	*N. goodspeedii*	F	Wheeler (1945), Goodspeed (1954)
366.	*N. exigua*	*N. gossei*	M	Goodspeed (1945)
367.	*N. exigua*	*N. maritima*	F, M	Kostoff (1940), Goodspeed (1945)

(continued)

Table 7.1 (continued)

No.	Hybrid combination		Species 1 used as: F (female parent); M (male parent) (l) denotes lethal hybrid	
	Species 1	Species 2		Reported by:
368.	*N. exigua*	*N. megalosiphon*	F	Wheeler (1945), Goodspeed (1945)
369.	*N. exigua*	*N. suaveolens*	F	Kostoff (1940), Goodspeed (1945)
370.	*N. exigua*	*N. umbratica*	F	Gangadevi et al. (1987)
371.	*N. exigua*	*N. velutina*	F, M	Kostoff (1943), Goodspeed (1954)
372.	*N. exigua*	*N. wuttkei*	M (l)	Laskowska and Berbeć (2003)
373.	*N. fragrans*	*N. umbratica*	M	Williams (1975)
374.	*N. fragrans*	*N. glutinosa*	M	Williams (1975)
375.	*N. goodspeedii*	*N. megalosiphon*	F	Goodspeed (1945)
376.	*N. goodspeedii*	*N. rotundifolia*	F	Goodspeed (1945)
377.	*N. goodspeedii*	*N. suaveolens*	F	Wheeler (1945), Goodspeed (1945), Takenaka (1965)
378.	*N. goodspeedii*	*N. umbratica*	F	Gangadevi et al. (1987)
379.	*N. gossei*	*N. maritima*	M	Wheeler (1945), Goodspeed (1954)
380.	*N. gossei*	*N. megalosiphon*	F	Kostoff (1943), Goodspeed (1945, 1954)
381.	*N. gossei*	*N. suaveolens*	M	Kostoff (1943), Goodspeed (1945), Takenaka (1952b)
382.	*N. gossei*	*N. occidentalis*	M	Wheeler (1945), Goodspeed (1945, 1954)
383.	*N. gossei*	*N. umbratica*	F	Gangadevi et al. (1987)
384.	*N. gossei*	*N. velutina*	M	Wheeler (1945), Goodspeed (1945)
385.	*N. gossei*	*N. eastii*	M	Kostoff (1943), Goodspeed (1945, 1954)
386.	*N. gossei*	*N. glutinosa*	M	Murthy et al. (1998)
387.	*N. hesperis*	*N. umbratica*	F	Gangadevi et al. (1987)
388.	*N. ingulba*	*N. megalosiphon*	F	Ramavarma and Apparao (1976)
389.	*N. ingulba*	*N. velutina*	F	Ramavarma and Apparao (1976)
390.	*N. ingulba*	*N. glutinosa*	F	Subhashini and Unnikrishnan (1978)
391.	*N. maritima*	*N. megalosiphon*	F	Goodspeed (1945, 1954)
392.	*N. maritima*	*N. suaveolens*	F, M	Kostoff (1939a), Goodspeed (1945)

(continued)

Table 7.1 (continued)

No.	Hybrid combination		Species 1 used as: F (female parent); M (male parent) (l) denotes lethal hybrid	Reported by:
	Species 1	Species 2		
393.	*N. maritima*	*N. velutina*	F, M	Kostoff (1943), Goodspeed (1945)
394.	*N. maritima*	*N. wuttkei*	M	Laskowska and Berbeć (2003)
395.	*N. maritima*	*N. eastii*	M	Kostoff (1939a, 1940), Goodspeed (1945, 1954)
396.	*N. maritima*	*N. undulata*	F	Goodspeed (1945)
397.	*N. megalosiphon*	*N. suaveolens*	F, M	Kostoff (1940), Goodspeed (1945), Durbin and Uchytil (1977)
398.	*N. megalosiphon*	*N. umbratica*	F	Subhashini (1975), Gangadevi et al. (1987)
399.	*N. megalosiphon*	*N. velutina*	M	Wheeler (1945), Goodspeed (1945)
400.	*N. megalosiphon*	*N. eastii*	M	Kostoff (1943), Goodspeed (1945)
401.	*N. megalosiphon*	*N. glutinosa*	F, M	Satyanarayana and Subhashini (1964), Subhashini et al. (1985)
402.	*N. occidentalis*	*N. simulans*	F, M	Gangadevi et al. (1987)
403.	*N. occidentalis*	*N. umbratica*	F	Gangadevi et al. (1987)
404.	*N. occidentalis*	*N. sylvestris*	M(l), F	Tezuka and Marubashi (2012), Kawaguchi et al. (2021)
405.	*N. occidentalis*	*N. tomentosiformis*	F	Tezuka and Marubashi (2012), Kawaguchi et al. (2021)
406.	*N. occidentalis*	*N. undulata*	M	Takenaka (1965)
407.	*N. rosulata*	*N. umbratica*	F	Gangadevi et al. (1987)
408.	*N. rotundifolia*	*N. umbratica*	F	Gangadevi et al. (1987)
409.	*N. rotundifolia*	*N. velutina*	M	Bradley & Goodspeed, 1943, Goodspeed (1945)
410.	*N. simulans*	*N. umbratica*	F	Gangadevi et al. (1987)
411.	*N. suaveolens*	*N. velutina*	M	Goodspeed (1945, 1954), Chase et al. (2021)
412.	*N. suaveolens*	*N. wuttkei*	M (l)	Laskowska and Berbeć (2003)
413.	*N. suaveolens*	*N. eastii*	M	Kostoff (1939a), Goodspeed (1945, 1954)

(continued)

Table 7.1 (continued)

No.	Hybrid combination		Species 1 used as: F (female parent); M (male parent) (l) denotes lethal hybrid	Reported by:
	Species 1	Species 2		
414.	*N. suaveolens*	*N. sylvestris*	asexual, F (l)	Fluhr (1983), Inoue et al. (1996)
415.	*N. suaveolens*	*N. tomentosiformis*	F	Inoue et al. (1996)
416.	*N. suaveolens*	*N. glutinosa*	F, M	premendelian (East (1928)), Dusseau et al. (1943), Goodspeed (1945), Gupta and Gupta (1973)
417.	*N. umbratica*	*N. velutina*	M	Gangadevi et al. (1987)
418.	*N. umbratica*	*N. kawakamii*	F	Japan Tobacco Inc. (2021)
419.	*N. umbratica*	*N. glutinosa*	F	Subhashini (1974)
420.	*N. velutina*	*N. wuttkei*	M	Laskowska and Berbeć (2003)
421.	*N. velutina*	*N. eastii*	F, M	Kostoff (1943)
422.	*N. sylvestris*	*N. kawakamii*	F	Ogura (1980)
423.	*N. sylvestris*	*N. otophora*	F, M	Goodspeed (1945, 1954), Takenaka (1954, 1956a)
424.	*N. sylvestris*	*N. setchellii*	F	Greenleaf (1942), Goodspeed (1945)
425.	*N. sylvestris*	*N. tomentosa*	F, M	Goodspeed (1934, 1945), East (1935), Takenaka (1956a)
426.	*N. sylvestris*	*N. tomentosiformis*	F, M	Kostoff (1930), Goodspeed (1945)
427.	*N. sylvestris*	*N. obtusifolia*	F	Liu and Marubashi (2014)
428.	*N. sylvestris*	*N. glutinosa*	F, M	Goodspeed (1934, 1945), East (1935), Fardy and Hitier (1945)
429.	*N. sylvestris*	*N. undulata*	cybrid	Aviv et al. (1984)
430.	*N. kawakamii*	*N. otophora*	F	Ogura (1980)
431.	*N. kawakamii*	*N. tomentosa*	F	Ogura (1980)
432.	*N. kawakamii*	*N. tomentosiformis*	F	Ogura (1980)
433.	*N. otophora*	*N. setchellii*	F, M	Goodspeed (1945), Bawolska et al. (1978)
434.	*N. otophora*	*N. tomentosa*	F	Goodspeed (1945)
435.	*N. otophora*	*N. tomentosiformis*	M	Gerstel and Burns (1974)
436.	*N. otophora*	*N. obtusifolia*	M (l)	Yamada et al. (1999)
437.	*N. otophora*	*N. glutinosa*	M	Goodspeed (1945)
438.	*N. setchellii*	*N. tomentosa*	M	Goodspeed (1945)
439.	*N. setchellii*	*N. tomentosiformis*	M	Clausen (1932), Goodspeed (1945)

(continued)

Table 7.1 (continued)

No.	Hybrid combination		Species 1 used as: F (female parent); M (male parent) (l) denotes lethal hybrid	
	Species 1	Species 2		Reported by:
440.	*N. setchelli*	*N. obtusifolia*	M	Goodspeed (1945)
441.	*N. setchellii*	*N. palmeri*	M	Goodspeed (1945)
442.	*N. tomentosa*	*N. tomentosiformis*	F	Goodspeed (1945, 1954)
443.	*N. tomentosa*	*N. obtusifolia*	M	Goodspeed (1934, 1945), Kostoff (1943)
444.	*N. tomentosa*	*N. palmeri*	M	Kostoff (1943), Goodspeed (1954)
445.	*N. tomentosa*	*N. glutinosa*	M	Goodspeed (1934), Greenleaf (1941), Goodspeed (1945)
446.	*N. tomentosiformis*	*N. obtusifolia*	M	Goodspeed and Bradley (1942), Kostoff (1943), Goodspeed (1954), Liu and Marubashi (2014)
447.	*N. tomentosiformis*	*N. palmeri*	M	McCray (1933), Kostoff (1943)
448.	*N. tomentosiformis*	*N. glutinosa*	M	Kostoff (1932), Goodspeed (1945, 1954)
449.	*N. obtusifolia*	*N. palmeri*	F, M	Goodspeed (1945), Takenaka (1959, 1962c)
450.	*N. obtusifolia*	*N. glutinosa*	M, F	Krishnamoorthy and Bhat (1957), Appa Rao and Krishnamurthy (1963), Murthy et al. (1998)
451.	*N. arentsii*	*N. undulata*	M	Goodspeed (1945, 1954)
452.	*N. arentsii*	*N. wigandioides*	F	Goodspeed (1945)
453.	*N. glutinosa*	*N. wigandioides*	F	Elvers (1934), Goodspeed (1945)
454.	*N. glutinosa*	*N. undulata*	M	Goodspeed (1954)
455.	*N. undulata*	*N. wigandioides*	F	Goodspeed (1945, 1954)

[1]hybrid seeds were obtained but plants were not attempted to be grown therefrom.
[2]amphidiploid form of that hybrid is known as Nicotiana *excelsiana*.
[3]amphidiploid 4x (*N. longiflora* x *N. tabacum*) included in the study
[4]amphidiploid 4x *N. tabacum* x *N. gossei* included in the study
[5]regular sexual hybrid mentioned once but with no details
[6]somatic sterile aneuploids

1. five reports on the intrasectional hybrids with *N. rastroensis* – the hybrids did not develop beyond the seed stage and seeds were not sown. The author assumes that in this particular case successful fertilization and viable seeds may substitute for an actual living hybrid plant since there has been no report on post-zygotic crossability barriers between 9-chromosome species of the section Alatae;
2. reporting on 12 lethal hybrids that may have not actually been grown past the seedling stage. The argument is that those hybrids mostly belong to lethality type II, the crossability barrier that has been repeatedly demonstrated to be removable by tissue culture and, more recently, by chemical treatment and genetic engineering. Those two categories put aside, the inventory still includes 434 viable hybrid combinations including a few that may have not grown to full maturity.

References

Adachi, A., & Inoue, M. (1995). Temperature-response of interspecific hybrid plants between *Nicotiana gossei* Domin and *N. tabacum* L., obtained from ovule culture. *Japanese Journal of Breeding, suppl 2*, 287.

Ahuja, M. R. (1962). A cytogenetic study of heritable tumors in *Nicotiana* species hybrids. *Genetics, 47*, 865–880.

Ahuja, M. R. (1968). An hypothesis and evidence concerning the genetic components controlling tumour formation in *Nicotiana. Molecular & General Genetics, 103*, 176–184.

Akada, S., & Hirai, A. (1986). Studies on the mode of separation of chloroplast genomes in parasexual hybrid calli. III. Random separation of two types of chloroplast genomes in a hybrid callus. *Japanese Journal of Genetics, 61*, 437–445.

Anssour, S., Krugel, T., Sharbel, T. F., Saluz, H. P., Bonaventure, G., & Baldwin, I. T. (2009). Phenotypic, genetic and genomic consequences of natural and synthetic polyploidization of *Nicotiana attenuata* and *Nicotiana obtusifolia. Annals of Botany (Lond), 103*, 1207–1217. https://doi.org/10.1093/aob/mcp058

Appa Rao, K., & Krishna Murthy, K. V. (1963). Studies on multiple polyploids in *Nicotiana. Genetica, 34*, 66–78.

Apparao, K., & Ramavarma, K. T. (1974). Cytogenetics of the amphidiploid *Nicotiana benthamiana* × *N. Glutinosa. Cytologia, 39*, 379–382.

Ar-Rushdi, A. H. (1955). Inheritance in *Nicotiana tabacum*. XXVI. Sterility genes from tomentosae species. *Journal of Genetics, 54*, 9–22.

Avery, P. (1938). Cytogenetic evidences of *Nicotiana* phylesis in the *alata*-group. *University of California Publications in Botany, 18*, 153–194.

Aviv, D., Arzee-Gonen, P., Bleichmann, S., & Galun, E. (1984). Novel alloplasmic *Nicotiana* plants by "donor-recipient" protoplast fusion: cybrids having *N. tabacum* or *N.sylvestris* nuclear genomes and either or both plastomes and chondriomes from alien species. *Molecular & General Genetics, 196*, 244–253.

Bannikova, V. P. (1965). Cytoembryology of distant hybrids. II. Disturbances in the fertilization process on crossing *Nicotiana glutinosa* L. with *N. rustica. Ukrainian Botanical Journal, 22*, 40–46.

Bartkowiak, E. (1987). Metody embriologii eksperymentalnej i hybrydyzacji somatycznej w otrzymywaniu mieszańców międzygatunkowych u roślin. *Postępy Biologii Komórki, 14*, 1–16.

Bawolska, M., Kobus, I., & Jurzysta, A. (1978). A characteristic of spontaneous hybrid *Nicotiana setchellii* × *N. otophora. Genetica Polonica, 19*, 49–59.

Bellair, G. (1913). Recroissees entre elles deux species qui sont degagees d'un hybride n'obeissent plus a la loi Mendelienne de la dominance. *Comptes rendus et rapports: IVe Conférence internationale de génétique*, 201–203.

Berbeć, A. (1988). Morphology, cytogenetics and resistance to potato virus Y of the amphidiploid *Nicotiana raimondii* Macbride × *N. tabacum* L. *Genetica Polonica, 29*, 41–52.

Berbeć, A., & Doroszewska, T. (1981). Investigations of the interspecific hybrid *Nicotiana amplexicaulis* Burbidge × *Nicotiana tabacum. Genetica Polonica, 22*, 197–207.

Berbeć, J., Berbeć, A., Glazewska, Z., Opoka, B., Kobus, I., & Czop, T. (1976). *Investigations on obtaining new varieties and breeding lines of Nicotiana tabacum resistant to virus leaf rib necrosis (potato virus Y) and virus degeneration of plants by Lycopersicum virus 3* (USDA Grant no. FG-Po-341 (JB-6) report no 2). Institute of Soils Science and Plant Cultivation, Pulawy, Poland (unpublished).

Berbeć, J., Berbeć, A., Glazewska, Z., Opoka, B., Kobus, I., & Czop, T. (1982). *Investigations on obtaining new varieties and breeding lines of Nicotiana tabacum resistant to virus leaf rib necrosis (Potato Virus Y) and virus degeneration of plants by Lycopersicum virus 3* (USDA Grant no. FG-Po-341 (JB-6) Final Report). Institute of Soils Science and Plant Cultivation. (unpublished).

Bolsunov, I. (1937). Sur les caractères des hybrides de première génération entre *N. rustica* et *N. glauca. Journal of Institution Botany Academic Science RSS Ukraine, 11*, 51–57.

Bolsunov, I. (1970). New Nicotiana amphidiploid as a valuable starting material for the control of virulent blue mould lines. In *International Tobacco Scientific Congress, Hamburg 1970 Proceedings 1971* (p. 160)

Bradley, M. V., & Goodspeed, T. H. (1943). Colchicine-induced Allo- and autopolyploidy in *Nicotiana. Proceedings of the National Academy of Sciences of the United States of America, 29*, 295–301.

Breisser, E. (1934). Zytologisch-genetische Untersuchungen an der Bastarden *Nicotiana tabacum* × *N. rusbyi* und ihren Rueckkreuzungen zu *N. rusbyi* und *N. sylvestris* I. Die Nachcommen aus reduzierten Eizelle. N. *Zeitschrift fuer indukt. Abstamm. u. Vererbungsl, 67*, 115–151.

Brieger, F. (1928). Uber die Vermehrung der Chromosomenzahl bei dem Bastard *Nicotiana tabacum* L. × Rusbyi Britt. *Zeitschrift Inductive Abstam u Vererbungslehre, 47*, 1–53.

Brieger, F. G. (1929). Vererbung bei Artbastarden unter besonderer Beruecksichtigung der Gattung *Nicotiana. Der Zuechter, 1*, 140–152.

Brieger, F. G., & Forster, R. (1942). Tumores em certos híbridos do gênero *Nicotiana. Bragantia*, 2259–2274.

Burk, L. G. (1960). Male-sterile flower anomalies in interspecific tobacco hybrids. *The Journal of Heredity, 51*, 27–31.

Burk, L. G. (1972). Viable hybrids from monosomics *N. tabacum* by *N. langsdorffii. Tobacco Science, 16*, 43–45.

Burk, L. G., & Dropkin, V. H. (1961). Response of *N. repanda, N. sylvestris* and their amphidiploid to root knot nematodes. *Plant Disease Report, 45*, 734–735.

Burk, L. G., & Durbin, R. D. (1978). Reaction of *Nicotiana* species to tentoxin. *The Journal of Heredity, 69*, 117–120.

Burk, L. G., Gooding, G. V., & Chaplin, J. F. (1982). Reaction of *Nicotiana* species and cultivars or breeding lines of *Nicotiana tabacum* to three strains of Potato Virus Y. *Tobacco Science, 26*, 85–88.

Busconi, M., Reggi, S., Lorenzoni, C., & Fogher, C. (2010). *Interspecific crosses in the genus Nicotiana*. Proceedings of the 56th Italian Society of Agricultural Genetics Annual Congress, Post 7.02.

Butenko, R. G., & Luneva, M. Z. (1966). Application of the sterile culture technique for the culturing of distant hybrids of *Nicotiana* (in Russian). *Fizjologija Rastenij, 13*, 733–736.

Butenko, R. G., Ternovsky, M. F., Semenova, I. V., & Moiseeva, M. E. (1970). *5th International Tobacco Science Congress CORESTA Inf Bul 12*. Abstr A102.

Calitz, P. C., & Milne, D. L. (1962). Reaction of *Nicotiana* species and species crosses to the root knot nematode, Meloidogyne javanica. *South African Journal of Agricultural Science, 5*, 123–126.

Chaplin, J. F., & Mann, T. J. (1961). Interspecific hybridization, gene transfer and chromosomal substitution in *Nicotiana. North Carolina Agricultural Experiment Station Technical Bulletin, 128*.

Chase, M. W., Fay, M. F., Nollet, F., & Christenhusz, M. (2021). Nicotiana notha. Solanaceae. *Curtis's Botanical Magazine, 38*. https://doi.org/10.1111/curt.12401

Choi, S. J., & Lee, S. C. (1991). Interspecific hybrid of *Nicotiana trigonophylla* X *N. tabacum* through i*n vitro* culture of fertilized ovules. *Journal of Korean Society of Tobacco Science, 13*, 52–58.

Christoff, M. (1928). Cytological studies in the genus *Nicotiana. Genetics, 13*, 233–277.

Christoff, M. (1938). *Wrkhu edin sluchai na mzhkka sterilnost pri Nicotiana rustica* (On a case of male sterility in *Nicotiana rustica*. Godisznik Sofijskiya University.

Chung, C. S., Nakjima, T., & Takeda, G. (1988). Interspecific hybridization between *Nicotiana trigonophylla* Dun. and *N. tabacum* L. through ovule culture. *Japanese Journal of Breeding, 38*, 319–326.

Clausen, R. E. (1932). Interspecific hybridization in *Nicotiana*. XIII. Further data as to the origin of *Nicotiana tabacum. Svensk Botanisk Tidskrift, 26*, 123–136.

Clayton, E. E. (1950). Male sterile tobacco. *Journal of Heredity, 4*, 171–175.

Clayton, E. E. (1954). Identifying disease resistance suited to interspecific transfer. *Journal of Heredity, 45*, 273–277.

Cseplo, A., Nagy, F., & Maliga, P. (1983). *Rescue of a cytoplasmic lincomycin resistance factor from Nicotiana sylvestris into Nicotiana plumbaginifolia by protoplast fusion* (pp. 126–127). International Protoplast Symposium Poster Proceedings, Basel.

Davis, E. L., Rich, J. R., & Gwynn, G. R. (1988). Reaction of selected *Nicotiana* spp. × *Nicotiana tabacum* crosses grown in microplots to 3 *Meloidogyne* spp. *Nematropica, 18*(2), 109–116.

Descorbeth, M. P. (2004). *Genotype and phenotype of Nicotiana alata × Nicotina rastroensis*. University of Missouri–Columbia. Office of Undergraduate Research.

Descorbeth, P., & McClure, B. (2005). Genotype and Phenotype of *Nicotiana alata* X *Nicotiana rastroensis*. In Vivo. *Cancer and the Environment 37th Annual MACUB Conference, Long Island University, 26*(2), 12.

DeVerna, J. W. (1984). *In vitro-facilitated wide hybridization in Nicotiana*. PhD dissertation, University of Kentucky.

DeVerna, J. W., Myers, J. R., & Collins, G. B. (1987). Bypassing prefertilization barriers to hybridization in *Nicotiana* using in vitro pollination and fertilizatio*n. Theoretical and Applied Genetics, 73*, 665–671.

Dorossiev, L., Palakarcheva, M., Stanoeva, M., & Petkova, M. (1978). Overcoming the sterility in F_1 of interspecific hybrids of the genus *Nicotiana* using the methods of tissue culture. *Bul Spéc CORESTA Symposium Sofia*, 80–81. (P03).

Durbin, R. D., & Uchytil, T. F. (1977). Cytoplasmic inheritance of chloroplast coupling factor 1 subunits. *Biochemical Genetics, 15*, 1143–1146.

Dusseau, A., Hitier, H., & Fardy, A. (1943). Transformation d'un hybride interspecifique sterile *Nicotiana suaveolens* Lehm (n=16) - a meiose asyndetique, en un hybride fertile rigoureusement autosyndetique par l'action de la colchicine. *Comptes Rendus de l'Académie des Sciences, 217*, 704–705.

Dusseau, A., Fardy, A., & Hitier, H. (1944). *Nicotiana* polyploides: especes tetraploides et hybrides interspecifiques amphiploides, obtenus par l'action de la colchicine. *Comptes Rendus de l'Académie des Sciences, 218*, 124–126.

East, E. M. (1916). Inheritance in crosses between *Nicotiana langsdorffii* and *Nicotiana alata*. *Genetics, 1*, 311–333.

East, E. M. (1928). The genetics of the genus *Nicotiana. Bibliography Genetics, 4*, 243–318.

East, E. M. (1935). Genetic reactions in *Nicotiana.* I. Compatibility. *Genetics, 20*, 403–413.

East, E. M., & Hayes, H. K. (1912). Heterozygosis in evolution and in plant breeding. *USDA Bulletin, 243*, 26–32.

Eghis, S. A. (1927). Experiments on interspecific hybridation in the genus *Nicotiana* (in Russian with English summary). *Bulletin of Applied Botany, Leningrad, 17*, 151–190.

Elvers, I. (1934). Interspecific hybridization in Nicotiana. XIV. The cytology of F_1 *glutinosa* x *tomentosa. University of California Publications in Botany, 17*, 341–354.

Engvild, Y. (1950). Cytogenetic studies on the genus *Nicotiana. National Institute of Genetics Annual Report, 1*, 35.

Fardy, A., & Hitier, H. (1945). Espèces tétraploides et hybrides interspécifiques amphidiploides et triples diploides ed *Nicotiana*, obtenus par l'action de la colchicine. *Mémorial du Serv d'Expl Industr des Tabacs et des Allumettes (SEITA) Ser B*, 1–117.

Fatalizade, F. A. (1939). Acenaphtene-induced polyploidy in *Nicotiana. Doklady Akademii Nauk URSS, 22*, 180–183. (after Kostoff 1943).

Fitzmaurice, W. P. (2002). *Interspecific Nicotiana hybrids and their progeny.* US Patent 6344597.

Fluhr, R. (1983). The segregation of organelles and cytoplasmic traits in higher plant somatic fusion hybrids. Protoplasts 1983, Experientia 46: Suppl, Birkhauser Verlag, Basel:69–84

Focke, W. O. (1881). *Die Pflanzen-Mischlinge*. Borntraeger.

Foster, H. H. (1943). Resistance in the genus *Nicotiana* to *Phytophtora parasitica* Dastur var. *nicotianae* Tucker. *Phytopathology, 33*, 403–404.

Gangadevi, T., & Satyanarayana, K. V. (1982). Cytogenetic study of the interspecific cross of *Nicotiana debneyi* x *N. umbratica. Theoretical and Applied Genetics, 63*, 177–181.

Gangadevi, T., Rao, P. N., Rao, B. H., et al. (1985). A study of morphology, cytology and sterility in interspecific hybrids and amphidiploids of *Nicotiana knightiana* x *N. umbratica. Theoretical and Applied Genetics, 70*, 330–332.

Gangadevi, T., Panuganti, N. R., & Satyanarayana, K. V. (1987). Morphological and cytological studies of interspecific hybrids involving *N. umbratica* Burbidge. *Cytologia, 52*, 475–486.

Gentscheff, G. (1931). Investigations on interspecific hybridisation in the genus *Nicotiana* (in Bulgarian). *Zemedlska Misl, 2*, 91–111.

Gerstel, D. U., & Burns, J. A. (1974). Meiosis and staminal sterility in hybrids between *Nicotiana tomentosiformis* and *N. otophora. Tobacco Science, 18*, 157–159.

Gerstel, D. U., Burns, J. A., & Burk, L. G. (1979). Interspecific hybridizations with an african tobacco, *Nicotiana africana* Merxm. *Journal of Heredity, 70*, 342–344.

Gillham, F. E. M., Wark, D. C., & Harrigan, E. K. S. (1977). Disease resistant flue-cured tobacco breeding lines for north Queensland. I. Resistance to blue mould, *Peronospora tabacina. Australian Journal of Experimental Agriculture and Animal Husbandry, 17*, 652–558.

Gisquet, P., Dufrenoy, J., & Dusseau, A. (1940). Hybrides interspecifiques de *Nicotiana tabacum* var. Purpurea × *N. petunoides* var. Sylvestris. *Science Genetics (Torino), 2*, 67–78.

Gleddie, S., Keller, W. A., Setterfield, G., & Wetter, L. R. (1983). Somatic hybridization between *Nicotiana rustica* and *N. sylvestris. Plant Cell, Tissue and Organ Culture, 2*, 269–283.

Goodspeed, T. H. (1934). *Nicotiana* phylesis in the light of chromosome number, morphology and behaviour. *University of California Publications in Botany, 17*, 369–398.

Goodspeed, T. H. (1945). Cytotaxonomy of *Nicotiana. Botanical Review, 11*, 533–592.

Goodspeed, T. H. (1954). *The Genus Nicotiana*. Chronica Botanica Co.

Goodspeed, T., & Bradley, M. (1942). Amphidiploidy. *Botanical Review, 8*, 272–316.

Goodspeed TH, Clausen RE (1927) Interspecific hybridization in Nicotiana V. Cytological features of two F_1 hybrids made with *Nicotiana bigelovii* as a parent. University of California, Berkeley, University of California Press 11 (6)

Goodspeed, T. H., & Clausen, R. E. (1928). Interspecific hybridization in *Nicotiana*. VIII. The *sylvestris-tomentosa-tabacum* triangle and its bearing on the origin of tobacco. *University of California Publications in Botany, 11*, 127–140.

Goodspeed, T. H., & Thompson, M. C. (1959). *Cytotaxonomy of Nicotiana II.* University of California Botanical Garden.

Gopinath, D. M., Krishnamurthy, K. V., & Krishnamurthy, A. S. (1965). Cytological studies in interspecific hybrids of *Nicotiana* involving a new Australian species, *Nicotiana amplexicaulis. Canadian Journal of Genetics and Cytology, 7*, 328–340.

Gopinath, D. M., Krishnamurthy, K. V., & Krishnamurthy, A. S. (1970). A note on the cytogenetic status of *Nicotiana amplexicaulis* Burbidge. *Current Science, 39*, 410–411.

Gray, J. C. (1978). Serological reaction of fraction I proteins from interspecific hybrids in the genus *Nicotiana. Plant Systematics and Evolution, 129*, 177–183.

Greenleaf, W. H. (1941). Sterile amphidiploids: Their possible relation to the origin of *Nicotiana tabacum. The American Naturalist, 75*, 394–399.

Greenleaf, W. H. (1942). Genic sterility in tabacum-like amphidiploids of *Nicotiana. Journal of Genetics, 43*, 69–96.

Gupta, S. B., & Gupta, P. (1973). Selective somatic elimination of *Nicotiana* glutinosa chromosomes in the F_1 hybrids of *N. suaveolens* and *N. glutinosa. Genetics, 73*, 605–612.

Horton, P. (1981). A taxonomic revision of *Nicotiana* (Solanaceae) in Australia. *Journal of the Adelaide Botanic Gardens, 3*, 1–56.

Hranov, A. (1970). Some problems on crossability between *Nicotiana tabacum* L. and *N. megalosiphon* Heurk et Muell (in Bulgarian). *Bulgaria Tobacco, 10*, 45–48.

Hu, C. (1956). Cytogenetic Studies in *Nicotiana*, XII. Reduction divisions in hybrids between *N. langsdorffii* and three other species. *Japanese Journal of Breeding, 6*, 117–121.

Huesing, J., Jones, D., DeVerna, J., Myers, J., Collins, G., Severson, R., & Sisson, V. (1989). Biochemical investigations of antiobiosis material in leaf exudate of wild *Nicotiana* species and interspecific hybrids. *Journal of Chemical Ecology, 15*, 1203–1217.

Iizuka, T., Kuboyama, T., Marubashi, W., Oda, M., & Tezuka, T. (2012). *Nicotiana debneyi* has a single dominant gene causing hybrid lethality in crosses with *N.* tabacum. *Euphytica, 186*, 321–328.

Ilcheva, V., San, L. H., Zagorska, N., & Dimitrov, B. (2001). Production of male sterile interspecific somatic hybrids between transgenic *N. tabacum* (bar) and *N. rotundifolia* (npt ii) and their identification by AFLP analysis. *In Vitro Cellular & Developmental Biology. Plant, 37*, 496–502.

Ilin, G. (1948). Alkaloids of the amphidiploid *Nicotiana rustica* x *N. glauca. Doklady Akademii Nauk SSSR, 59*, 279–280.

Inoue, E., Marubashi, W., & Niwa, M. (1996). Genomic factors controlling the lethality exhibited in the hybrid between *Nicotiana suaveolens* Lehm. and *N. tabacum* L. *Theoretical and Applied Genetics, 93*, 341–347.

Iwai, S., Kishi, C., Nakata, K., & Kawashima, N. (1986). Production of *Nicotiana tabacum* x *N. acuminata* hybrid by ovule culture. *Plant Cell Reports, 5*, 403–404. [(Tob Abstr 31:639 (1987)].

Japan Tobacco Inc. (2021). *Nicotiana* F_1 hybrid and its use. https://patents.google.com/patent/JP6917478B2/en?inventor=%E4%BF%8A+%E5%B1%B1%E5%86%85

Japan Tobacco Inc. (2022). Genus *Nicotiana* F_1 hybrid and use thereof. https://patents.google.com/patent/WO2022014699A1/en?inventor=%E4%BF%8A+%E5%B1%B1%E5%86%85

Jeanes, J. A. (1999). Solanaceae (in) Walsh NG. In T. J. Entwisle (Ed.), *Flora of Victoria vol 4, Cornaceae to Asteraceae*. Inkata Press.

Kaul, M. H. L. (1988). *Male sterility in higher plants* (p. 10). Springer.

Kawaguchi, K., Ohya, Y., Maekawa, M., Iizuka, T., Hasegawa, A., et al. (2021). Two *Nicotiana occidentalis* accessions enable gene identification for Type II hybrid lethality by the cross to *N. sylvestris. Science Reports, 11*, 17093. https://doi.org/10.1038/s41598-021-96482-6

Kehr, A. E. (1951). Genetic tumors in *Nicotiana. The American Naturalist, 99*, 73–79.

Kehr, A. E., & Smith, H. H. (1952). Multiple genome relationships in *Nicotiana. Cornell University Agricultural Experiment Station Memoir, 311*, 3–19.

Kehr, A. E., & Smith, H. H. (1954). Genetic tumors in *Nicotiana* hybrids. *Brookhaven Symposia in Biology, 6*, 55–78.

Kincaid, R. R. (1949). Three interspecific hybrids of tobacco. *Phytopathology, 39*, 284–287.

Kiraly, L., & Schoeltz, J. E. (1995). The interaction of a single, recessive gene in *Nicotiana clevelandii* and gene VI of cauliflower mosaic virus results in systemic necrosis. *Phytopathology, 85*, 1152.

Kobori, S., & Marubashi, W. (2004). Programmed death cell detected in interspecific hybrids of *Nicotiana repanda* x *N. tomentosiformis* expressing hybrid lethality. *Breeding Science, 54*, 347–350.

Kobus, I. (1971). Investigations in the polyploids of *Nicotiana tabacum* L. and their hybrids with wild tobacco species. *Genetica Polonica, 12*, 323–328.

Kostoff, D. (1929). Acquired immunity in plants. *Genetics, 14*, 37–77.

Kostoff, D. (1930). Ontogeny, genetics, and cytology of *Nicotiana* hybrids. *Genetica, 12*, 33–139.

Kostoff, D. (1932). Pollen abortion in species hybrids. *Cytologia, 5*, 337–339.

Kostoff, D. (1935). On the Increase of Mutation Frequency following Inter Specific Hybridization in *Nicotiana*. *Current Science, 3*, 302–304.

Kostoff, D. (1939a). The origin of the tetraploid *Nicotiana* from Bathurst. *Current Science, 8*, 110–112.

Kostoff, D. (1939b). Abnormal mitosis in tobacco plants forming hereditary tumours. *Nature, 144*, 599.

Kostoff, D. (1939c). Nicotine and citric acid content in the progeny of the allopolyploid hybrid *Nicotiana rustica L.* x *N. glauca* Grah. *Doklady Akademii Nauk SSSR, 23*, 121–124.

Kostoff, D. (1940). Relation degrees and phylesis of certain *Nicotiana* species determined by cytogenetic analysis. *Genetica, 22*, 215–230.

Kostoff, D. (1943). *Cytogenetics of the genus Nicotiana*. State Printing House.

Krishnamoorthy, K. V., & Bhat, N. R. (1957). An amphidiploid *Nicotiana glutinosa* L. *Nicotiana trigonophylla* Dun. *Hybrid. Current Science, 26*, 120.

Krügel, T. (2010). *Nicotiana* × *mierata* und N. × *obtusiata*: Zwei neue synthetische Hybriden. *Haussknechtia, 12*, 23–25.

Kubo, T. (1985). Studies on hybrid breeding by means of cytoplasmic male sterility in tobacco. *Bulletin Iwata Tobacco Experiment Station, 17*, 69–138.

Larkina, N. I. (1980). Overcoming incompatibility between *Nicotiana* species by means of *in vitro* pollination. *Plant Breed Abstract, 50*(8067), 690.

Larkina, N. I. (2015). *Scientific principles of interspecific hybridization using maternal Nicotiana tabacum*. Monograph. All-Russian Institute of Tobacco and Makhorka. Krasnodar ISBN 978-5-93491-681-8.

Larkina, N. I. (2017). Creation of amphidiploids – fertile interspecific *Nicotiana* hybrids. *Sciences of Europe. Biological Sciences, 2*, 95–105.

Laskowska, D., & Berbeć, A. (2003). Preliminary study of the newly discovered tobacco species *Nicotiana wuttkei* Clarkson and Symon. *Genetic Resources and Crop Evolution, 50*, 835–839.

Laskowska, D., & Berbeć, A. (2012). Production and characterization of amphihaploid hybrids between *Nicotiana wuttkei* Clarkson et Symon and *N. tabacum*. *Euphytica, 183*, 75–82.

Lautz, W. (1957). Resistance to black shank of 51 species of *Nicotiana* and of 13 interspecific hybrids. *Plant Disease Report, 41*, 95–98.

Lee, C. B., Page, L. E., McClure, B. A., & Holtsford, T. P. (2008). Post-pollination hybridization barriers in *Nicotiana* section *Alatae*. *Sexual Plant Reproduction, 21*, 183–195.

Lehmann, H. (1936). On interspecific cross sterility in the genus Nicotiana. *Plant Breed Abstract, 7*(1936), abst 1023 (in German)

Lilienfeld, F. A. (1953). Cytogenetical problems in *Nicotiana*. a. A new amphidiploid. *National Institute of Genetics Japanese Annual Report, 4*, 32–33.

Lin, R. F., & Chen, C. C. (1990). Cytological studies of interspecific somatic hybrids in Nicotiana. *Botanical Bulletin of Academia Sinica, 31*, 179–187.

Liu, H., & Marubashi, W. (2014). Species origin of genomic factors in *Nicotiana nudicaulis* Watson controlling hybrid lethality in interspecific hybrids between *N. nudicaulis* Watson and *N. tabacum* L. *PLoS One, 9*, e97004. https://doi.org/10.1371/journal.pone.0097004

Ma, J., Hancock, W. G., Nifong, J. M., Kernodle, S. P., & Lewis, R. S. (2020). Identification and editing of a hybrid lethality gene expands the range of interspecific hybridization potential in *Nicotiana*. *Theoretical and Applied Genetics, 133*, 2915–2925. https://doi.org/10.1007/s00122-020-03641-w

Malinowski, E. (1916). On the occurrence of new forms in the progeny of the hybrids *Nicotiana atropurpurea* × *Nicotiana* sylvestris (in Polish). *Comptes Rendus des Séances de la Société des Sciences et des Lettres de Varsovie, 9*, 827–864.

Malloch, W. S., & Malloch, F. W. (1924). Species crosses in *Nicotiana* with special reference to *N. longiflora* × *N. tabacum, N. longiflora* × *N. sanderae, N. tabacum* × *N. glauca*. *Genetics, 9*, 261–291.

Maryanovich, O. (1939). On the cytology and embryology of the hybrid F_1 *Nicotiana rustica* (variety Unterwalden) x *Nicotiana quadrivalvis* (in Russian). *Zhurnal Instytutu Botaniky Ukrayins'koyi Akademiyi Nauk*. USSR (21-22 (29-30)):187–195.

Masuta, C., Suzuki, M., Takanami, Y., & Koiwai, A. (1993). Yellow mosaic symptoms induced by Y satellite RNA of Cucumber Mosaic Virus is regulated by a single incompletely dominant gene in wild *Nicotiana* species. *Phytopathology, 83*, 411–413.

Mayr, E. (1986). Joseph Gottlieb Koelreuter's contribution to biology. *Osiris, 2*, 135–176. University of Chicago Press.

McCray, F. A. (1932). Compatibility of certain *Nicotiana species*. *Genetics, 17*, 621–636.

McCray, F. A. (1933). Embryo development in *Nicotiana* species hybrids. *Genetics, 18*, 95–110.

Meiselman, N., Gunckel, J. E., & Sparrow, A. H. (1961). The general morphology and growth responses of two species of *Nicotiana* and their interspecific hybrid after chronic gamma irradiation. *Radiation Botany, 1*, 69–79.

Menczel, L., Lazar, G., & Maliga, P. (1978). Isolation of somatic hybrids by cloning *Nicotiana* heterokaryons in nurse cultures. *Planta, 143*, 29–32.

Mihaylova-Kroumova, A. B., Atriouchine, I., Korenkov, V. D., & Wagner, J. G. (2020). *Patterns of inheritance of acylsugar acyl groups in selected Interspecific hybrids of genus Nicotiana* (p. 23). Kentucky Tobacco Research and Development Center Faculty Publications.

Moav, R., & Cameron, D. R. (1960). Genetic instability in *Nicotiana* hybrids. I. The expression of instability in *N. tabacum* × *N. plumbaginifolia*. *American Journal of Botany, 47*, 87–93.

Modilevsky, J. (1937). Cytogenetic investigation of the genus Nicotiana. III. Cytology and embryology of the hybrid F1 *N.* rustica L. x *N.* glauca Grah (in Russian). *Zhurnal Instytutu Botaniky Ukrayins'koyi Akademiyi Nauk USSR, 11*, 59–71.

Muntzing, A. (1935). Chromosome behaviour in some *Nicotiana hybrids*. *Hereditas, 20*, 251–271.

Muraida, N., & Marubashi, W. (2015). Characterization of hybrid seedlings from crosses of *Nicotiana stocktonii* Brandegee × *N. tabacum* L. and *N. stocktonii* × progenitors of *N. tabacum*. *Plant Biotechnology, 32*, 139–147.

Murthy, T. G. K., & Rama Prasad, G. (2000). Aphid resistance in some *Nicotiana* species and interspecific hybrids. *Tobacco Research, 26*, 52–56.

Murthy, T. G. K., & Rao, R. V. S. (2008). *Maintenance of the genus Nicotiana* (pp. 18–20). Central Tobacco Research Institute Annual Report 2007–2008, Rajamundry, India.

Murthy, T. G. K., & Subbarao, I. V. (2004). Some new interspecific hybrids in the genus *Nicotiana* – Characterization and utilization. *Tobacco Research, 30*, 33–41.

Murthy, T. G. K., & Subhashini, U. (2000). Studies on some interspecific hybrids of *Nicotiana africana* Merxm. *Tobacco Research, 26*, 12–15.

Murthy, T. G. K., Subhashini, U., & Subba Rao, I. V. (1998). New interspecific hybrids in the genus *Nicotiana* – Characterization and utilization. *Tobacco Symposium, 8*, abstr B10.

Murthy, T. G. K., Sredhaar, U., & Siva, R. K. (2014). *Incorporation of aphid resistance from N. gossei. N. repanda, N x. umbratica-nesophila and N × benthamiana–repanda* (pp. 17–20). Central Tobacco Research Institute Annual Report 2013-2014, Rajamundry, India.

Näf, U. (1958). Studies on tumour formation in *Nicotiana* hybrids I. The classification of parents into two etiologically significant groups. *Growth, 22*, 167–180.

Navarro-Quezada, A. (2014). *Quantification of homeolog expression bias upon herbivory of synthetic and natural allopolyploids from N. attenuata x N.obtusifolia*. ICE Symposium MPI for Chemical Ecology, Jena, DE.

Navarro-Quezada, A., Alejandro Perez, N., Gaquerel, E., & Baldwin, I. T. (2013). *Herbivore resistance in natural- and artificial Nicotiana attenuata x N. obtusifolia allopolyploids: A transcriptome and metabolome analysis*. Conference on Plant Genome Evolution, Elsevier.

Nikova, V. M. (1986). *Nicotiana excelsior* Black as a source of cytoplasmic male sterility in *Nicotiana tabacum* L. *Comptes Rendus de L'Academie Bulgare des Sciencesopen, 39*, 105–107.

Nikova, V. M., & Zagorska, N. A. (1990). Overcoming hybrid incompatibility between *Nicotiana africana* Merxm and *Nicotiana tabacum* and development of cytoplasmically male sterile tobacco forms. *Plant Cell, Tissue and Organ Culture, 32*, 71–75.

Nikova, V. M., Zagorska, N. A., & Pundeva, R. S. (1991). Development of four tobacco male sterile sources using *in vitro* techniques. *Plant Cell, Tissue and Organ Culture, 27*, 289–295.

Nikova, V., Vladova, R., Pundeva, R., Petkova, A., & Vladovska, A. (1998). *Study of Nicotiana ingulba Black. and its crossability with N. tabacum L* (p. 150). CORESTA Congr Bull Spec, Brighton. PPOST 12 (b).

Nikova, V., Iancheva, A. M., Vladova, R., Pandeva, R., & Petkova, A. (2006). *Genetic improvement of Nicotiana tabacum by applying remote hybridization and biotechnological methods* (p. 07). CORESTA Congr Agro-Phyto Groups, Paris, APOST.

Ogura, H. (1980). Cytogenetic studies on a recently identified *Nicotiana* species, *N. kawakamii* Ohashi. *Cytologia, 45*, 33–42.

Ohashi, Y. (1985). Breeding studies of wild species related to tobacco, with special reference to disease resistance. *Bulletin Iwata Tobacco Experiment Station, 17*, 1–62.

Ohashi, Y., Ohshima, M., & Itoh, H. (1992). Constitutive expression of stress-inducible genes, including pathogenesis-related 1 protein gene in a transgenic interspecific hybrid of *Nicotiana glutinosa* x *Nicotiana debneyi*. *Plant & Cell Physiology, 33-2*, 177–187.

Pal, B. P., & Nath, P. (1936). A note on the sterile hybrid between *N. tabacum* and *N. plumbaginifolia*. *Indian Journal of Agricultural Sciences, 6*, 828–832.

Palakarcheva, M. (1975). Interspecific hybridization in tobacco (in Bulgarian). *Bulgarian Tiutun, 11*, 30–34.

Palakarcheva, M. (1981). Genetic peculiarities of tobacco breeding for disease resistance (in Bulgarian). *Genetics Selection, 14*, 277–284.

Patel, M. S. (1960). *Indian tobacco. A monograph*. Indian Central Tobacco Committee.

Patel, D., Power, J. B., Anthony, P., Badakshi, F., Hesslop-Harison, J. S., & Davey, M. R. (2011). Somatic hybrid plants *of Nicotiana* × *sanderae* (+) *N. debneyi* with fungal resistance to *Peronospora tabacina*. *Annals of Botany, 108*, 809–819.

Pearse, I. S., Krugel, T., & Baldwin, I. T. (2006). Innovation in anti-herbivore defense systems during neopolypoloidy – The functional consequences of instantaneous speciation. *The Plant Journal, 47*, 196–210.

Pittarelli, G. W., & Stavely, J. R. (1975). Direct hybridization of *Nicotiana repanda* × *N. tabacum*. *The Journal of Heredity, 66*, 281–284.

Poddubnaya-Arnoldi, V. A., & Ludkina, M. M. (1945). Embryogenesis in hybridization of distant species and polyploidy in the genus *Nicotiana* (in Russian). *Zhurnal Instytutu Botaniky Ukrayins'koyi Akademiyi Nauk USSR, 30*, 195–216.

Raju, K. S., Sheshumadhav, M., & Murthy, T. G. K. (2008). Molecular diversity in the genus *Nicotiana* as revealed by randomly amplified polymorphic DNA. *Physiology and Molecular Biology of Plants, 14*, 377–382.

Raju, K. S., Sheshumadav, M., Chandrasekhararao, C., & Murthy, T. G. K. (2009). Molecular diversity in genus *Nicotiana* as revealed by randomly amplified polymorphic DNA. *Indian Journal of Biotechnology, 8*, 61–66.

Ramanujam, S., & Joshi, A. B. (1942). Interspecific hybridization in *Nicotiana*. A cytological study of the hybrid *N. glauca* Grah. X *N. plumbaginifolia*. *Indian Journal Of Genetics And Plant Breeding, 2*, 80–97.

Ramavarma, K. T., & Apparao, K. (1976). Studies on interspecific hybrids in genus *Nicotiana* involving the species *Nicotiana ingulba*. *Tobacco Research, 2*, 137–140.

Razdan, M. K. (2003). Chapter XII. Somatic hybridisation and cybridisation (in) Introduction to plant tissue Culture. *Science Publishers, Appendix, 12*(2), 174.

Reed, S., & Collins, G. B. (1978). Interspecific hybrids in *Nicotiana* through in vitro culture of fertilized ovules. *The Journal of Heredity, 69*, 311–315.

Reed, S., & Collins, G. B. (1980). Chromosome pairing relationship and black shank resistance in three *Nicotiana* interspecific hybrids. *The Journal of Heredity, 71*, 423–426.

Sachs-Skalińska, M. (1917). Beitrag zur Cytologie des sterilen Bastardes *Nicotiana* atropurpurea × *Nicotiana* silvestris. *Comptes Rendus des Séances de la Société des Sciences et des Lettres de Varsovie, 10*(6).

Sarala, K., Rao, R. V. S., Murthy, T. D. K., & Satyavani, J. V. R. (2008). Micropropagation of sterile and non-flowering *Nicotiana* lines. *Cytologia, 73*, 9–13.

Sarala, K., Prabhakara Rao, K., & Sheshu Madhav, M. (2023). *Interspecific hybridization. Division of crop improvement* (pp. 11–14). ICAR -Central Tobacco Research Institute Technical Bulletin.

Satyanarayana, K. V., & Subhashini, L. (1964). Interspecific hybridization and aneuploidy in the genus *Nicotiana*. *Indian Journal of Genetics and Plant Breeding, 24*, 264–271.

Savelli, R. (1926). Un ibrido di tipo eccezionale nella F1 di *Nicotiana longiflora* Cav. X *N. sylvestris* Sp. Et Comes. *Arch Bot Sist Modena, 2*, 228–240.

Savelli, R. (1927). Method of measuring the flowers of Nicotiana and biometric data on the hybrid *Nicotiana rustica* x. *N. tabacum*. *Arch Bot Sist Fitogeogr Genet, 3*, 95–106.

Schweppenhauser, M. A., Raeber, J. G., & Daulton, R. A. C. (1963). *Resistance to the root knot nematode Meloidogyne javanica in the genus Nicotiana* (pp. 222–229). Proceedings of III World Tobacco Science Congress, Salisbury, S Rhodesia.

Sievert, R. C. (1972). Resistance to anthracnose in the genus *Nicotiana*. *Tobacco Science, 16*, 32–34.

Smith, H. H. (1937). Quantitative genetic interactions. I. Genome combinations of *Nicotiana longiflora* and *N. sanderae*. *Genetics, 22*, 361–375.

Smith, H. H. (1942). Alkaloids in certain species and interspecific hybrids of *Nicotiana*. *Journal of Agricultural Research*, 347–359.

Smith, H. H. (1958). Genetic plant tumors in *Nicotiana*. *Annals of the New York Academy of Sciences, 71*, 1163–1178.

Smith, H. H. (1968). Recent cytogenetic studies in the genus *Nicotiana*. *Advances in Genetics, 14*, 1–43.

Smith, H. H., & Abashian, D. V. (1963). Chromatographic investigations on the alkaloid content of *Nicotiana* species and interspecific combinations. *American Journal of Botany, 50*, 435–447.

Smith, H. H., Hamill, D. E., Weaver, E. A., & Thompson, K. H. (1970). Multiple molecular forms of peroxidase and esterase among Nicotiana species and amphidiploids. *The Journal of Heredity, 61*, 203–212.

Sorrentino, C., Del Piano, L., Crimaldi, M., Abet, M., & Cuciniello, A. (2006). *Constitutional of ornamental hybrids of Nicotiana ssp*. Proceedings of the 50th Italian Society of Agricultural Genetics Annual Congress, Ischia, Italy.

Stavely, J. R. (1979). Disease resistance. In R. D. Durbin (Ed.), *Nicotiana. Procedures for Experimental Use* (U.S. Department of Agriculture Technical Bulletin 1586) (pp. 111–120).

Stehmann, J. R., Semir, J., & Ippolito, A. (2002). *Nicotiana mutabilis* (Solanaceae), a new species from southern Brazil. *Kew Bulletin, 57*, 639–646.

Subhashini, U. (1974). Study of the cytoplasmic effects in the interspecific cross *Nicotiana umbratica* Burbidge & *N. glutinosa* Linn. *Euphytica, 23*, 289–293.

Subhashini, U. (1975). Interspecific Hybridization in the Genus Nicotiana Involving the Species *N. umbratica* B and *N. megalosiphon* H and M Arg. *Cytologia, 40*, 409–413.

Subhashini, U., & Gopinath, D. M. (1974). Interspecific Hybridization in the Genus *Nicotiana* Involving a New Australian Species *N. umbratica*. *Cytologia, 39*, 403–409.

Subhashini, U., & Unnikrishnan, M. (1978). Cytogenetic relationship of *Nicotiana ingulba* B. with *N. glutinosa* L. *Tobacco Research, 2*, 72–74.

Subhashini, U., Venkateswarlu, T., Anjani, K., & Prasad, G. S. R. (1985). In vitro hybridization in an incompatible cross *Nicotiana glutinosa* times *Nicotiana megalosiphon*. *Theoretical and Applied Genetics, 71*, 545–549.

Subhashini, U., Venkateswarlu, T., & Anjani, K. (1986). Embryo rescue in *Nicotiana* hybrids by in vitro culture. *Plant Science, 43*, 219–222.

Subhashini, U., Murthy, T. G. K., & Siva Raju, K. (1998). Genetic transformation in *N. tabacum* using direct gene transfer from an amphidiploid (*N. benthamiana* x *N. repanda*). *Tobacco Symposium, 7*. abstr. B9.

Symon, D. E. (1984). A new species of *Nicotiana* (Solanaceae) from Dalhousie Springs. *South Australian Journal of the Adelaide Botanic Gardens, 7*, 117–121.

Takanashi, H., & Marubashi, W. (2017). Tumorigenesis inheritance from the putative progenitor species of Nicotiana rustica. *Plant Biotechnology, 34*, 79–87.

Takenaka, Y. (1950). Cytogenetic studies on the Genus *Nicotiana*. *National Institute of Genetics Annual Report, 1*, 35.

Takenaka, Y. (1951a). Cytogenetic studies on the genus *Nicotiana* II (c). Hybrid between *N. glauca* and *N. plumbaginifolia*. *National Institute of Genetics Annual Report, 2*, 28.

Takenaka, Y. (1951b). Cytogenetic studies in *Nicotiana*. On three hybrids between *N.tabacum* and other species. *La Kromosoma, 9-10*, 384–392.

Takenaka, Y. (1952a). Cytogenetic studies in the genus *Nicotiana*. IV. (a) Reduction divisions in hybrids between *N. glauca* and the *N. alata* group. *National Institute of Genetics Annual Report, 3*, 40–41.

Takenaka, Y. (1952b). Cytogenetic studies in the genus *Nicotiana*. IV. (c) Reduction divisions in hybrids between *N. suaveolens* and three other species. *National Institute of Genetics Annual Report, 3*, 42.

Takenaka, Y. (1953a). Cytogenetic studies of *Nicotiana*. III. Reduction divisions in three hybrids between *N. glauca* and the alata group. *Japanese Journal of Genetics, 28*, 155–162.

Takenaka, Y. (1953b). Cytogenetic studies in the genus *Nicotiana*. V. (b) Reduction divisions in the hybrids between the tomentosa group and three other species. *National Institute of Genetics Annual Report, 4*, 34–35.

Takenaka, Y. (1953c). Cytogenetic Studies on the genus *Nicotiana*. V. (c) Reduction divisions in hybrids between *N. sanderae* and three other species. *National Institute of Genetics Annual Report, 4*, 35–36.

Takenaka, Y. (1954). Cytogenetic studies in the genus *Nicotiana*. VI. b. Reduction divisions in several hybrids between *Nicotiana* species. *National Institute of Genetics Annual Report, 5*, 64–65.

Takenaka. (1955). Cytogenetic studies of Nicotiana. VIII. Reduction divisions in hybrids between *N. sanderae* and three other species. *Japanese Journal of Genetics, 30*, 236–241.

Takenaka, Y. (1956a). Cytogenetic studies in *Nicotiana*. XIV. Reduction divisions in five interspecific hybrids. *Japanese Journal of Genetics, 31*, 155–161.

Takenaka, Y. (1956b). Cytogenetic studies in Nicotiana. VI. On three hybrids between *N. tabacum* and other species. *Journal of Japanese Botany, 68*, 358–362.

Takenaka, Y. (1959). Cytogenetic studies of the genus *Nicotiana* XII. *National Institute of Genetics Annual Report, 10*, 77–79.

Takenaka, Y. (1961). Cytogenetic studies in the genus Nicotiana XIV. *National Institute of Genetics Annual Report, 12*, 68–69.

Takenaka, Y. (1962a). Cytogenetic studies in Nicotiana, XIX. The reduction division in four hybrids between *N. paniculata* and other species of the section Alatae. *Japanese Journal of Breeding, 12*, 278–280.

Takenaka, Y. (1962b). Cytogenetic Studies in *Nicotiana*, XV. Reduction divisions in five Interspecific hybrids between sections Alatae and Suaveolentes. *Japanese Journal of Genetics, 37*, 80–85.

Takenaka, Y. (1962c). Cytogenetic Studies in Nicotiana. XVII. Reduction Divisions in Five Interspecific Hybrids. *Japanese Journal of Genetics, 37*, 343–347.

Takenaka, Y. (1962d). Cytogenetic studies in *Nicotiana* XVI. Reduction divisions in six interspecific hybrids between *N. tabacum* and other six species. *Shokubutsugaku Zasshi, 75*(888), 237–241.

Takenaka, Y. (1965). Cytogenetic studies in the genus *Nicotiana. National Institute of Geneticst Japanese Annual Report, 16*, 75–76.

Takenaka, Y., & Yoneda, Y. (1964). Tumor formation in interspecific hybrids of *Nicotiana*, especially in relation to injury. *National Institute of Genetics Annual Report, 15*, 87.

Takenaka, Y., Amano, Y., & Tateoka, T. (1956). Cytogenetic studies in the genus *Nicotiana*. IX. *Japanese National Institute of Genetics Annual Report, 7*, 42–44.

Tanaka, M. (1961). The effect of irradiated pollen grains on species crosses of *Nicotiana. Bulletin of the Hatano Tobacco Experiment Station, 51*, 1–38.

Tatemichi, Y. (Ed.). (1990). *Illustrated book of the genus Nicotiana*. Plant Breeding and Genetics Research Laboratory, Japan Tobacco Inc.

Ternovsky, M. F. (1962). *Summary and prospects of interspecific hybridization in the genus Nicotiana (in) Wide hybridization of plants* (pp. 331–345). Proceedings of the Conference on Wide Hybridization of Plants and Animals; collection of reports.

Ternovsky, M. F., & Larkina, N. I. (1978). Hybridization of native tobacco cultivars with the wild species *Nicotiana rosulata. Soviet Genetics, 14*, 1038–1045.

Ternovsky, M. F., Butenko, R. G., & Moiseeva, M. E. (1972). The use of tissue culture to overcome the barrier of incompatibility between species and sterility of interspecific hybrids. *Soviet Genetics, 8*, 27–33.

Ternovsky, M. F., Shinkareva, I. K., & Larkina, N. I. (1976). Production of interspecific tobacco hybrids by the pollination of ovules *in vitro. Soviet Genetics, 12*, 1209–1213.

Tezuka, T. (2012). Hybrid lethality in the Genus *Nicotiana*. In J. K. Mworia (Ed.), *Botany* (pp. 191–210). Rijeka InTech.

Tezuka, T., & Marubashi, W. (2012). Genes in S and T subgenomes are responsible for hybrid lethality in interspecific hybrids between *Nicotiana tabacum* and *Nicotiana occidentalis. Online Plos One*. https://doi.org/10.1371/journal.pone.0036204

Tezuka, T., Kuboyama, T., Matsuda, T., & Marubashi, W. (2007). Possible involvement of genes on the Q chromosome of *Nicotiana tabacum* in expression of hybrid lethality and programmed cell death during interspecific hybridization to *Nicotiana debneyi. Planta, 226*, 753–764.

Tezuka, T., Kitamura, N., Imagawa, S., Hasegawa, A., et al. (2021a). Genetic Mapping of the HLA1 locus causing hybrid lethality in *Nicotiana* interspecific hybrids. *Plants, 10*, 2062. https://doi.org/10.3390/plants10102062

Tezuka, T., Kitamura, N., Yanase, M., & Morikawa, T. (2021b). Evaluation of crossability between *Nicotiana benthamiana* and *Nicotiana excelsior. Agronomy, 11*(12), 2583. https://doi.org/10.3390/agronomy11122583

Unnikrishnan, M., Subhashini, U., & Satyanarayana, K. V. (1981). Tissue culture in interspecific hybridization in *Nicotiana. Tobacco Research, 7*, 62–67.

Valleau, W. D. (1952). Breeding tobacco for disease resistance. *Economic Botany, 6*, 69–102.

Wark, D. C. (1970). Development of flue cured tobacco cultivars resistant to common strain of blue mold. *Tobacco Science, 14*, 147–150.

Wheeler, H. M. (1945). A Contribution to the Cytology of the Australian-South Pacific Species of *Nicotiana. Proceedings of the National Academy of Sciences of the United States of America, 31*, 177–185.

Whitaker, T. W. (1934). The occurrence of tumors on certain *Nicotiana* hybrids. *Journal of the Arnold Arboretum, 15*, 144–153.

White, O. E. (1914). Studies of teratological phenomena in their relation to evolution and the problems of heredity. I. A study of certain floral abnormalities in nicotiana and their bearing on theories of dominance. *American Journal of Botany, 1*, 23–36.

Williams, E. (1975). A new chromosome number in the Australian species *Nicotiana cavicola* L. (Burbidge). *New Zealand Journal of Botany, 13*, 811–812.

Williams, E., & Pandey, K. K. (1975). Meiotic chromosome pairing in interspecific hybrids of *Nicotiana*. II. South American species hybrids: The influence of genotype on pairing. *New Zealand Journal of Botany, 13*, 611.

Wolf, P. A. (1965). Hereditary abnormalities of tobacco. *Journal of the Elisha Mitchell Scientific Society, 81*(85–92), 144–172.

Wong, Y. N. (1975). Genome analysis of tobacco interspecific hybrids. *Taiwan Tob Wine Monop Bur Tob Res Inst Res Rep, 2*, 22–31.

Yamada, T., Marubashi, W., & Niwa, M. (1999). Detection of four lethality types in interspecific crosses among Nicotiana species through the use of three rescue methods for lethality. *Breeding Science, 49*, 203–210.

Zaitlin, D., & Mundell, R. (2006). *Nicotiana hybrids and plant varieties for use in production of pharmaceutics*. US Patent 20060236433 A1.

Zenkteler, M., Slusarkiewicz Jarzina, A., & Wozna, J. (1981). Cytological investigations of hybrid plants of *Nicotiana alata* and *Nicotiana debneyi* obtained by *in vitro* pollination of ovules. *Bulletin de la Soci?t? des Amis des Sciences et des Lettres de Poznan, Serie D, 21*, 79–84.

Zhukov, N. I. (1939). Hybrids of *Nicotiana rustica* L. x *N. alata* Link et Otto (in Russian). *Sb Nauch Issled Rabot WITIM, 139*, 236–244.

MIX
Papier aus verantwortungsvollen Quellen
Paper from responsible sources
FSC® C105338

If you have any concerns about our products,
you can contact us on
ProductSafety@springernature.com

In case Publisher is established outside the EU,
the EU authorized representative is:
Springer Nature Customer Service Center GmbH
Europaplatz 3, 69115 Heidelberg, Germany

Printed by Libri Plureos GmbH
in Hamburg, Germany